Highlights in Ubiquinone Research

To the memory of Professor Alfredo Rabbi

Highlights in Ubiquinone Research

Edited by

Giorgio Lenaz

Ottavio Barnabei

Alfredo Rabbi

University of Bologna, Bologna, Italy

Maurizio Battino

University of Ancona, Ancona, Italy

Proceedings of the International Symposium 'Biochemistry, Bioenergetics and Clinical Applications of Ubiquinone', held in Bologna on May 18-22, 1989, in the occasion of the celebrations for the Ninth Centenary of foundation of the Univeristy of Bologna.

M·CM·LXXXVIII

Alma Mater Studiorum
Sæcularia Nona

Taylor & Francis
London · New York · Philadelphia
1990

UK Taylor & Francis Ltd, 4 John St, London WC1N 2ET

USA Taylor & Francis Inc., 1900 Frost Road, Suite 101, Bristol, PA 19007

British Library Cataloguing in Publication Data

Biochemistry, Bioenergetics and Clinical Applications of Ubiquinone.
Highlights in ubiquinone research.
1. Organisms. Bioenergetics. Role of quinones
I. Title II. Lenaz, G. (Giorgio)
574.19121
ISBN 0-85066-848-4

Printed in Great Britain by Taylor & Francis (Printers) Ltd. Basingstoke, Hampshire

CONTENTS

2. Quinones in Bioenergetics

FOREWORD

This book contains most of the invited papers and of contributed presentations in the International Symposium 'Biochemistry, Bioenergetics and Clinical Applications of Ubiquinone', held at the University of Bologna within the celebrations of the Ninth Centenary of foundation of the 'Alma Mater Studiorum', the oldest University of the world.

The Symposium received scientific contributions concerning the most important and actively investigated areas of research on ubiquinone (or Coenzyme Q) and related quinones. Coenzyme Q was discovered in 1958 and was immediately recognized as an essential component of the respiratory chain in mitochondria. Since then this and related lipophilic quinones were found as key constituents of electron transfer chains in energy-conserving members of most living organisms.

Ubiquinone research has been implemented by a series of outstanding theories and discoveries: its role in proton translocation and energy conservation in the Q-cycle hypothesis by the Nobel Laureate Peter Mitchell; its mobile character in the random collision model of electron transfer enunicated by Charles Hackenbrock on the basis of longstanding investigations since the first predictions of the discoverers of Coenzyme Q, Fred Crane and David Green and the kinetic analysis by Kröger and Klingenberg; its presence in animal cells in other membranes besides mitochondria, with as yet unidentified roles: its possible action as an antioxidant as well as a prooxidant, an area of biochemistry receiving strong impulses by several investigators, first of all Lars Ernster, in view of the important pathological implications including the mechanism of ageing; and, last but not least, its efficacy in the therapy of human diseases, which has found in Karl Folkers its first enthusiastic advocate soon after the biological role of the quinone had been identified.

Many of the scientists that have contributed to the history of ubiquinone and are still active in the field, as well as most of those presently attempting to uncover the uncertainties still existing, were present at the Symposuim and are Authors of this book. As such, the book represents an invaluable source of up-to-date first-hand information on the properties and roles of lipophilic quinones.

Following the format of the Symposium, these Proceedings are organized into three parts. The first Section, 'Biochemistry and Biophysics of Quinones', concerns the localization and mobility of the molecules in mitochondria and other energy-conserving membranes, as well as model systems. The second one, 'Quinones in Bioenergetics', is dedicated to the role of ubiquinone as an electron transfer carrier related to the action of respiratory and photosynthetic enzymes in energy-conserving systems. Finally, the third Section, 'Biomedical and Clinical Applications of Ubiquinone', describes the more recently recognized role of quinones as antioxidants in vitro and in vivo, as well as the most interesting and promising clinical studies performed on the effect of Coenenzyme Q_{10} in cardiovascular and other diseases.

The Editors have chosen to have contributions from all active participants rather than fewer selected papers: for this reason the contributions are mostly concise and restricted to 4-6 pages, with few exceptions. The body of the different articles is such to give a complete panorama of the present research on quinones, and represents a useful source of references since each bibliography explicitly quotes pertinent previous work.

The Symposium was made possible by the sponsorship of the University of Bologna and by the combined financial efforts of the pharmaceutical companies Inverni della Beffa, Master-Pharma, Simes, Zambon and Zyma from Italy and Eisai from Japan. To all of them our most sincere thanks.

During the course of editing this book, one of the Editors, Alfredo Rabbi, Professor of Biochemistry at the University of Bologna, suddenly deceased. He was the first one who supported the idea of holding an International Symposium on Ubiquinone within the celebrations of the Ninth Centenary of the University of Bologna, and his encouragements and activities have made the Symposium possible. Professor Rabbi was a person of great scientific and humanistic culture and his death is a great loss for all of us. To him the Coeditors wish to dedicate this book.

Giorgio Lenaz

Bologna, December 4th, 1989

Pasta al Coenzyme Q_{10}-Pesto a la Bolognese (or A Means of Administering Coenzyme Q_{10} to Patients in the Form of a Delicious Meal).

Robert E. Beyer

Department of Biology, The University of Michigan, Ann Arbor, MI 48109-1048, USA.

One of the major themes of this International Symposium on Coenzyme Q has been a number of reports on the efficacy and desirability of administering coenzyme Q_{10} to patients presenting with a number of clinical signs arising from a deficiency of this compound. The receipe which follows was conceived during late afternoon of the fifth day of the Symposium and may be prepared as follows.

Ingredients:
5 ounces (150 grams) fresh basil leaves or 3 Tbs. (45 ml) crushed dry basil
1 ounce (30 grams) fresh oregano leaves or ½ Tbs. (7.5 ml) crushed dry oregano
½ cup (125 ml) pine nuts
½ cup (125 ml) grated Romano (or Parmesan) cheese
2 large or 3 small cloves garlic
5 Tbs. (75 ml) olive oil containing 50 mg coenzyme Q_{10}/ml
Salt and pepper to taste

Begin to prepare the pasta of your choice (the author prefers spinach noodles). Place all ingredients except for olive oil in a blender. Pulverise the solid ingredients and slowly add oil while blending at high speed until a smooth, homogeneous sauce is obtained. As this receipe will make enough Coenzyme Q_{10}-Pesto sauce for 6 hungry people, it may be stored refrigerated for several weeks. The garlic contains aliin of diallyl sulfide (Block, 1985; Wargovich et al., 1988), a natural antioxidant, which will serve to preserve the quality of the sauce.

ACKNOWLEDGEMENTS
Travel funds were provided by The University of Michigan Vice President for Research, the LSA Associate Dean for Research, and a grant from the Elsa U. Pardee Foundation.

REFERENCES

Block, E., 1985, The chemistry of garlic and onion. *Scientific American,* 252, 114-119.

Wargovich, M. J., Woods, C., Eng, V. W. S., Stephens, L. C., and Gray, K., 1988, Chemoprevention of N-nitrosomethylbenzymaline-induced esophageal cancer in rats by the naturally occurring thioether, diallyl sulfide. *Cancer Research,* 48, 6872-6875.

Opening lecture delivered by F.L. Crane

DEVELOPMENT OF CONCEPTS FOR THE ROLE OF UBIQUINONES IN BIOLOGICAL MEMBRANES

Frederick L. Crane

Department of Biological Sciences
Purdue University
West Lafayette, Indiana 47907 U.S.A.

The development of concepts concerning the function and requirements of coenzyme Q in biological systems will be considered under the following headings:

1. The mechanism of electron transport and proton movement in relation to quinone location in (a) fluid pools in the membrane bilayer, (b) tightly bound semiquinone radicals, and (c) wormholes where the transition from bound radical to pool quinone can occur.
2. Distribution of ubiquinones in cellular membranes in relation to site of synthesis, membrane flow, and new functions.
3. The contrast between oxygen radical generation by semiquinone and antioxidant action of quinone or hydroquinone.
4. Ubiquinone deficiency or misfunction related to external poisons, aging, or mutation of mitochondrial DNA.

Keywords: Coenzyme Q, Quinone radicals, Ubiquinone deficiency, Ubiquinone location in cells.

<u>The Mechanism of Function</u>

The advent of coenzyme Q in 1957 led to a complete change in ideas about organization of the mitochondria, and other energy coupling electron transport systems (Crane *et al.*, 1957; Mitchell, 1966)

The name ubiquinone (Fig. 1) was proposed by Morton (1961) based on the ubiquitous presence of the quinone in mammalian tissue. Later study showed lipophilic polyterpenoid quinones with modification of the quinoid ring or terpenoid tail are present in all membrane-bound energy

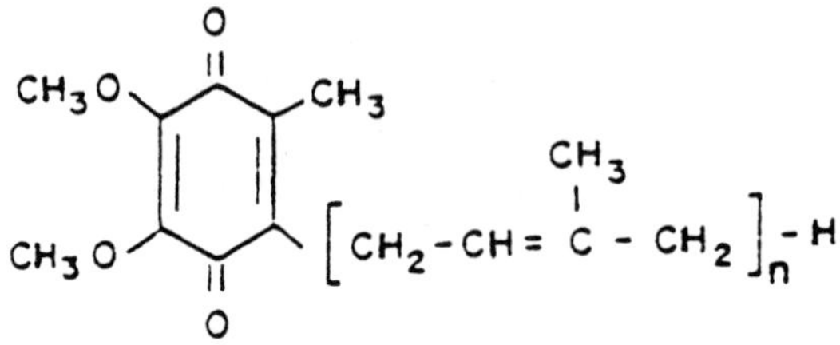

Figure 1. Ubiquinone [Coenzyme Q_n (n = 6-10)]

coupling systems, whether heterotropic or photosynthetic (Table 1) [Collins, 1985].

Table 1. The ubiquinone family of redox carriers.

Quinone	Solvent	γ_{max} (nm)
Ubiquinone	Ethanol	375,409 (inf)
Rhodoquinone	Cyclohexane	251,280,320,500 (inf)
	Ethanol	253,283,320,500 (inf)
Plastoquinone	Iso-octane	254,262
Menaquinone	Iso-octane	242,248,260,269, 326,328 (inf)
Phylloquinone	Iso-octane	242,248,260,269, 326,328 (inf)
Demethylmenaquinone	Iso-octane	242,248,254,263, 326,238 (inf)
Chlorobiumquinone	Ethanol	254,265 (inf)
Thermoplasmaquinone[a]	Iso-octane	242,248,259,269,337
Caldariellaquinone	Methanol	241,283,333
	Iso-octane	237,272,278,322
Methionaquinone[b]	Ethanol	238,262,310

[a]Thermoplasmaquinone-7(2.[5 or 8]-dimethyl-3-heptaprenyl-1,4-naphthoquinone) from *Thermoplasma acidophilum*.
[b]2-methylthio-3-VI,VII-tetrahydroheptaprenyl-1,4-naphthoquinone.
Based on Collins (1985).

Thus, the term ubiquinone is applicable to a general concept of lipophilic electron carrier or prosthetic group in all energy transducing membranes. Coenzyme Q (Crane, 1989) specifically introduced a lipophilic component into a mitochondrial electron transport chain thought to be located on the hydrophilic surface of the lipid bilayer membrane (Lehninger, 1964) [Fig. 2].

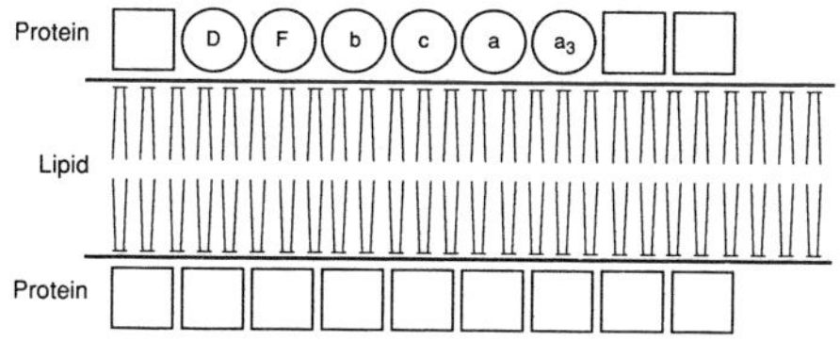

Figure 2. Electron transport organized on the surface of a unit membrane in early 1960s (Lehninger, 1964).

The lipophilic nature of coenzyme Q led Green and Wharton (1963) to propose that the quinone acted as an electron carrier pool between dehydrogenases and cytochrome *c* reductase with the enzymes also inserted into the lipid to allow interaction with the quinone (Fig. 3).

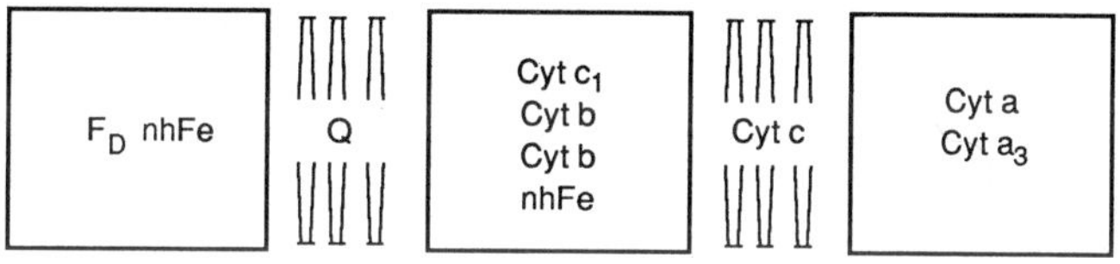

Figure 3. Coenzyme Q as electron carrier inside the membrane lipid bilayer (based on Green and Wharton, 1963).

The insertion of the protein into the lipid led to the new concept of a mobile mosaic membrane structure (Green, 1972). The quinone pool concept was clearly defined by kinetic studies of Kroger and Klingenberg and others (Klingenberg, 1986; Gutman, 1980; Reed and Regan, 1987). Phospholipid dilution of membranes to separate carrier complexes and measurement of coenzyme Q diffusion rates have also supported a Q pool (Hackenbrock *et al.*, 1986; Chazotte and Hackenbrock, 1989). On the other hand, a direct stoichiometric interaction between mobile dehydrogenases and cytochrome b-c_1 complex in the membrane using interaction with bound coenzyme Q has been proposed (Crofts and Wraight, 1983; Crofts, 1986; Heron *et al.*, 1978; Fato *et al.*, 1988) [Fig. 4].

Semiquinone formation during oxidoreduction of coenzyme Q was proposed by Baum *et al.* (1967) and demonstrated by Bäckström *et al.* (1970). The discovery of Q binding proteins which could stabilize the semiquinone strengthened the idea of sequential one electron transfer (King, 1985; Yu and Yu, 1981). The introduction of a Q cycle by Mitchell and Moyle (1985) to increase anisotropic proton movement essential for energy transduction in mitochondria further emphasized the role of quinone radicals (Fig. 4).

Two step reduction through two quinones bound at different sites in photosynthetic reactions centers in order

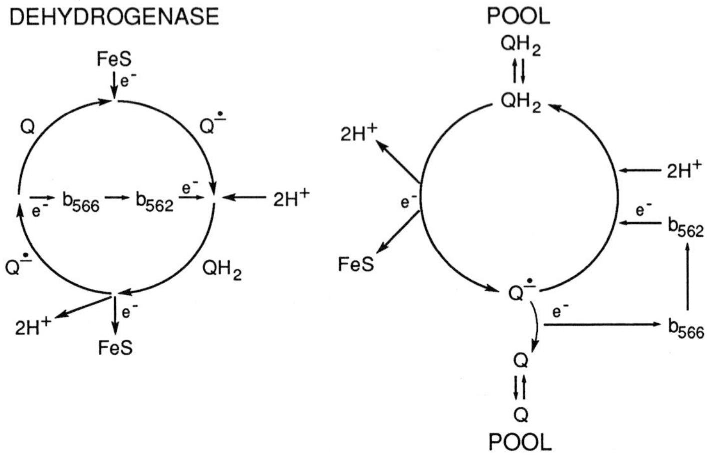

Fig. 4. Proposed interaction between dehydrogenases and bound coenzyme Q in electron transport complexes.
A. Direct interaction between dehydrogenase and bound Q.
B. Interaction of bound Q with a quinone pool.

for reduced quinone to enter the membrane pool was recognized in the period 1973 to 1977 (rev. by Crofts and Wraight, 1983). The two step reaction was observed since single flashes of light could be used for each electron transfer. Release of hydroquinone to the pool occurs only after reduction of the semiquinone at the second binding site with two light flashes. Crystallographic resolution of the reaction center from purple bacteria shows two bound quinones, Q_A and Q_B. Q_A is tightly bound and is reduced by one light flash to the semiquinone. This radical can reduce Q_B which also bound. The second flash reduces Q_B to the hydroquinone with another electron from Q_A. The protonated hydroquinone formed at B can now leave the reaction center through a pocket or wormhole in the protein to enter the Q pool. The Q_A does not interchange with the pool. Thus, the reaction center has a two electron gate for reduction of pool quinone (Michel *et al.*, 1986; Feher *et al.*, 1989) [Fig. 5].

A similar structure can be postulated for the reaction center II in chloroplasts. Tightly bound PQ_A would reduce PQ_B which would escape as a hydroquinine through a wormhole into the membrane lipid (Crofts and Wraight, 1983; Critchley, 1985).

It is worth considering if the free radical quinone binding sites in mitochondria have a function analogous to the Q_A and Q_B binding sites in chromatophores and chloroplasts. A similar two electron gate would be difficult to

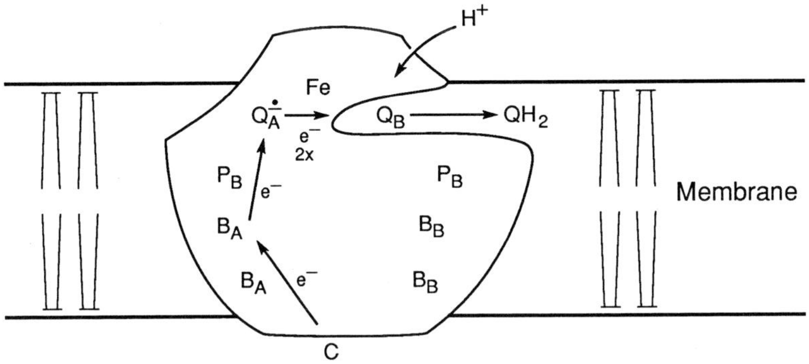

Fig. 5. Diagram to show the two electron gate in the purple bacteria photoreaction center. Electrons from bacteriochlorophyll (B_A) reduce Q_A through pheophytin (P_A). Q_A then transfers the electron to Q_B to form Q_B^-. A second electron from Q_A produces $Q_B^{=}$ which is protonated to the hydroquinone and released (based on Feher *et al.*, 1989).

detect since mitochondrial substrates provide pairs of electrons and controlled supply of single electrons would be difficult. The site where the two electrons reduce a bound quinone would be a wormhole similar to the chromatophore Q_B site which would provide access to the membrane Q pool. Likewise the b-c_1 complex could have a tightly bound semiquinone which would reduce a second bound quinone by sequential one electron transfer through cytochrome b_{566} and b_{562}. The second quinone would be bound in a wormhole so it could be released to the Q pool. This sequence would eliminate semiquinone entry into the pool (Fig. 6) and consequent oxygen radical formation.

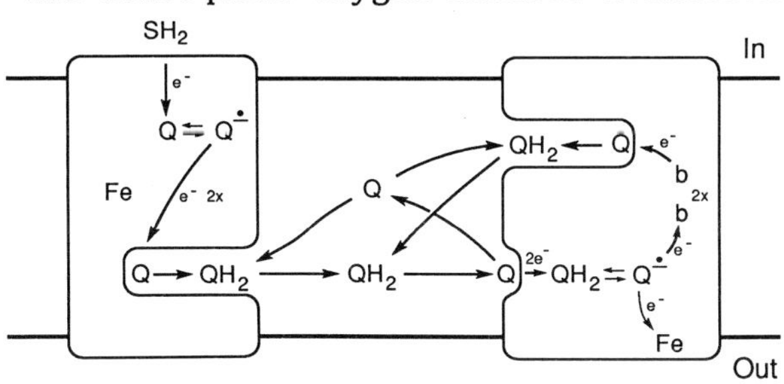

Fig. 6. Suggestion for a two electron gate applied to coenzyme Q reduction in mitochondrial electron transport complexes. In complex I or II sequential transfer of electrons through tightly bound quinone would reduce a second bound quinone which can then leave the complex. A similar two electron transfer could occur in complex III. The first quinone would correspond to Q_A and the second to Q_B in photoreaction centers.

In conclusion, there are now three states of coenzyme Q which have been found in energy coupling membranes:

1. Fluid pool quinones and hydroquinones move electrons and protons between catalytic subunits in a lipid bilayer.
2. Tightly bound quinone-semiquinone for one electron transfer with no exchange into the fluid pool.
3. Wormhole quinone-semiquinone-quinol bound at sites which would function to transfer quinol into the fluid phase and take up quinone from the fluid phase through two-stage reduction of quinone.

Binding sites for quinones have been recognized on complex I, II and III in mitochondria, and these sites have been associated with semiquinone radicals (Table 2).

Table 2. Quinone binding sites in mitochondrial complexes and inhibitors which act at these sights (King, 1985).

Subunit	Binding Site	Inhibitor
II. Succinic dehydrogenase	Q_S	TTFA
	CII-3,4	
	Q^{-}-Q^{-}	
I. NADH dehydrogenase	Q_N	rotenone
		piericidin A
III. QH_2 cytochrome c reductase	Q_i (Q_c)	Antimycin A
	Q_o (cyt b)	myxothiazole

Quinone binding sites are now recognized as the site of action of the most well-known inhibitors of mitochondrial or photosynthetic electron transport. Rotenone and piericidin compete with semiquinone at the Q_N binding site in complex I. Thenoyltrifluoroacetone competes at the Q_S site. Atrazines and DCMU compete with coenzyme Q or plastoquinone at the Q_B site in chromatophores and chloroplasts. Mutations in amino acids of the Q_B wormhole can change the response of the photocenter to these inhibitors (Fig. 7).

Recent studies show that functional failure may be associated with agents which inhibit coenzyme Q binding sites (Table 3).

Another way in which ubiquinones can act to control energy conversion is by redox control of protein phosphorylation, as in chloroplasts where the redox state of the PQ pool controls a kinase which phosphorylates a protein to regulate light harvesting efficiency (Coughlan, 1988).

230 225

his ILE GLY leu THR gly val

Q_B SER TYR

tyr PHE arg asp leu met gly

215 220

Fig. 7. Amino acids in the Q_B binding site of *Rb. capsulatus* showing sites of mutations (caps) which prevent atrazine inhibition (based on Bylina *et al*., 1989).

Table 3. Agents which affect coenzyme Q_N function.

Compound	Disease	Site of Action
MPTP[a] (MPP^+)[b]	Parkinson's	Q_N
Capsaicin	Hepatic cirrhosis	Q_N

Capsaicin: CH_3O ... H O
HO- O $-CH_2N-C-(CH_2)_4-CH-CHCH(CH_3)_2$

[a]1-methyl-4-phenyl-1,2,3,6-tetrahydropyridine
[b]1-methyl-4-phenylpyridinium
(Shimomera *et al*., 1989; Ramsay *et al*., 1989.)

Distribution of Quinones in Membranes

Coenzyme Q is not restricted to mitochondrial membranes. It is found in all endomembranes, and especially high amounts are found in Golgi and lysosomal membranes (Crane and Morré, 1977; Kalén *et al*., 1987) (Table 4).

Table 4. Coenzyme Q content of membrane fractions from rat liver (based on Kalén *et al*., 1987).

Membrane	Coenzyme Q nmoles/mg protein
Mitochondria	1.8
Inner mitochondrial membrane	2.3
Endoplasmic reticulum	0.2
Golgi apparatus	3.3
Lysosomes	2.3
Plasma membranes	0.9
Peroxisomes	0.4

Since separate biosynthetic pathways may be found for coenzyme Q formation in mitochondria and endoplasmic reticulum (Table 5), the endomembrane coenzyme Q is not a likely source of mitochondrial quinone unless carrier proteins are available for intermembrane transfer.

The relation of coenzyme Q in endomembranes to biosynthetic sites in the cell is diagramed in Fig. 8. Normal membrane flow can account for the presence of coenzyme Q from the endoplasmic reticulum in Golgi, lysosomes and plasma membrane, but not in mitochondria.

Table 5. Incorporation of radioactivity from ^3H tyrosine and ^3H mevalonate into coenzyme Q by rat liver slices.

Isolated Fraction	^3H Mevalonate	^3H Tyrosine
	dpm/μpm/μgQ_9 per 30 min.	
Endoplasmic reticulum	1303	796
Golgi	37	23
Plasma membrane	22	24
Lysosomes	24	16
Mitochondria	305	203

Based on Kalén *et al.* (1987).

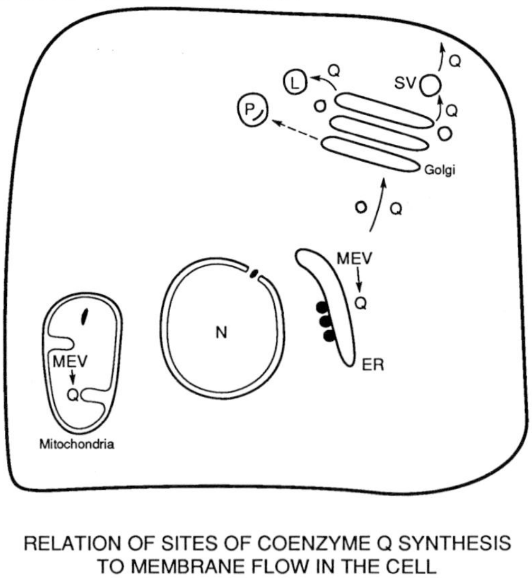

Fig. 8. Diagram of coenzyme Q distribution in the cellular membranes. Separate biosynthetic sites for mitochondria and endomembranes are indicated (Kalén *et al.*, 1987). P-peroxisome, L-lysosome, SV-secretory vesicle, N-nucleus, ER-endoplasmic reticulum.

There is evidence that coenzyme Q in endomembranes can participate in electron transport reactions on the basis of the following results:

a) Addition of NADH reduces coenzyme Q in endoplasmic reticulum (Table 6; Takada *et al.*, 1982).

Table 6. Reduction of coenzyme Q in microsomes from guinea pig.

Addition	% reduced $Q_{10}H_2$
None	47
NADH	68
NADH + Rotenone	65

b) NADH indophenol reductase activity can be restored by addition of coenzyme Q_{10} after extraction of Golgi membranes with heptane (Crane *et al.*, 1984).

c) Analogs of coenzyme Q inhibit NADH indophenol reductase activity in endoplasmic reticulum, Golgi and plasma membranes (Table 7; Crane *et al.*, 1984).

d) Radiolabeled coenzyme Q was taken up by mitochondria and plasma membrane in myocardial cells. The beating of the cells improved without increase in ATP. Since plasma membrane redox activity is inotropic in heart (Crane *et al.*, 1985), the coenzyme Q may exert its effect at the plasma membrane (Hosano *et al.*, 1981).

Table 7. Inhibition of indophenol reductase activity by coenzyme Q analog 3N-dodecylmercapto-2-hydroxy-1,4-naphthoquinone.

Membrane fraction	NADH Indophenol Reductase percent inhibition 17 μg/ml DHNQ
Mitochondria	28
Endoplasmic reticulum	87
Golgi apparatus	66
Plasma membrane	39
Milk fat membrane	0

<u>Radical Formation and Antioxidant Action</u>

Coenzyme Q in membranes can serve as a generator of oxygen radicals or as an antioxidant (Beyer *et al.*, 1987). This is illustrated in equation one where oxidation of semiquinone by oxygen will produce superoxide which can generate peroxide or other oxygen radicals. Coenzyme Q_{10} can also remove superoxide to release oxygen and semiquinone. If the membrane has a dehydrogenase which can reduce the semiquinone, the hydroquinone is produced which can also act as a radical scavenger. Thus, both coenzyme Q and reduced coenzyme Q can act as antioxidants (Landi *et al.*, 1987). The fact that coenzyme Q_{10} is deeply buried in membranes would favor its antioxidant action at deeper levels than less hydrophobic compounds (Ulrich *et al.*, 1985).

Equation I

$$CoQ_{10}^{\dot{-}} + O_2 \rightleftharpoons CoQ_{10} + O_2^{\dot{-}}$$
$$2O_2^{\dot{-}} + 2H^{+2} \xrightarrow[SOD]{} H_2O_2 + O_2$$

The generation of quinone radicals opens up the possible role of ubiquinone as a source of oxygen radicals in cells. Radical formation is known in mitochondria. The oxidation in other membranes may also generate radicals, as shown with NADH cytochrome P450 reductase from ER (Sugioka *et al.*, 1988).

Ubiquinone Deficiency or Misfunction

Evidence for therapeutic effects of coenzyme in several diseases has been slowly accumulating over several years and is documented in the series of books edited by Folkers and Yamamura entitled *Biomedical and Clinical Aspects of Coenzyme Q*, volumes I-V, published by Elsevier. Interest in coenzyme Q peaked with the publication of the book by E. G. Bliznakov and G. L. Hunt (1987) entitled *The Miracle Nutrient, Coenzyme Q_{10}*, and the subsequent headline in the Weekly World News for March 3, 1987,

MIRACLE VITAMIN STOPS AGING

which emphasized the role of coenzyme Q in energy conversion. It is clear that favorable response to administration of coenzyme Q has been recorded in some cases of congestive heart failure, peridontal disease and other conditions. A logical basis for these effects can be decrease of coenzyme Q in tissue with aging (Table 8; Beyer *et al.*, 1987).

Effects of external environment or nutrition on coenzyme Q levels in rats have also been shown (Table 9; Ramasarma, 1985).

These studies show changes in mitochondria and microsomes which can be independent of each other. Most studies of coenzyme Q content of tissue have not differentiated between coenzyme Q in mitochondria and in other membranes. Serious decrease in one membrane may be hidden by increases in other membranes.

Table 8. Changes in coenzyme Q concentration with age in rat heart.

Age (months)	Coenzyme Q (nmoles/heart)
2	180
5	300
9	480
18	580
25	200

Data from Beyer *et al.* (1987).

Table 9. External effects on coenzyme Q content of rat liver mitochondria and microsomes.

Treatment or Condition	Change in Coenzyme Q Content (%) Total	Mitochondria	Microsome
Protein deficiency	-80%	-85%	-85%
Fed coenzyme Q_9	+240	+294	+213
Vitamin A deficiency	+270	+233	+314
Thyrotoxicosis	+287	+282	+236
Cold	+252	+215	+501

Based on Ramasarma (1985).

There are many possible reasons for functional deficiencies in coenzyme Q as discussed by Lenaz (1985). Examples of changes which can affect function are listed in Table 10. Recent interest in the accumulation of muta-

Table 10. Basis for functional deficiency of coenzyme Q.

1. Low concentration (synthesis failure or degradation)
2. Bilayer viscosity change (fatty acid saturation or peroxidation)
3. Lipid hydrophobicity (peroxidation)
4. Excessive respiratory rate
5. Sequestered quinone (micelles or fibrils)
6. Modification of enzymes (mutation)
    a. Loss of affinity site
    b. Change in wormhole structure
7. Change in distribution in membranes or organelles
8. Interaction with drugs
9. Low protein diet

tions in mitochondrial DNA with aging emphasize the importance of understanding how coenzyme Q interacts with mitochondrial proteins (Linnane *et al.*, 1989).

In summary, what has coenzyme Q done for bioenergetics? The highlights are listed in Table 11 in more or less chronological order. The influence has spread from a major advance in understanding mitochondrial and photosynthetic electron transport and a logical basis for energy coupling to introducing new approaches for control of disease.

Table 11. What did coenzyme Q do to bioenergetics?

a.	Introduced mosaic membrane concept
b.	Membrane lipid fluidity emphasized
c.	Isolation of mitochondrial enzymatic subunits
d.	Provided for proton transfer in energy coupling
e.	Explain strange kinetics of cytochrome *b*
f.	Introduced radical formation
g.	Supported a Nobel Prize based on bioenergetics
h.	Opened commerical opportunity in redox research
i.	Rationalized treatment of heart failure

REFERENCES

Bäckström, D., Norling, B., Ehrenberg, A. and Ernster, L., 1970, Identification of a quinone free radical in mitochondria. Biochimica et Biophysica Acta, 197, 108-111.

Baum, H., Rieske, J.S., Silman, H.I. and Lipton, S.H., 1967, On the mechanism of electron transfer in complex III of the electron transfer chain. Proceedings of National Academy of Science, U.S., 57, 798-805.

Beyer, R.E., Nordenbrand, K. and Ernster, L., 1987, The function of coenzyme Q in free radical production and as an antioxidant: a review. Chemica Scripta, 27, 145-153.

Bliznakov, E.G. and Hunt, G.L., The Miracle Nutrient, Coenzyme Q_{10}. (Toronto: Bantam Press), p. 240.

Bylina, E.J., Jovine, R.V.M. and Youvan, D.C., 1989, A genetic system for rapidly assessing herbicides that compete for quinone binding site of photosynthetic reaction centers. Biotechnology, 7, 69-74.

Chazotte, B. and Hackenbrock, C.R., 1989, Lateral diffusion as a rate limiting step in ubiquinone mediated mitochondrial electron transport. Journal of Biological Chemistry, 264, 4978-4985.

Collins, M.D., 1985, Analysis of isoprenoid quinones. Methods in Microbiology, 18, 329-360.

Coughlan, S.J., 1988, Chloroplast thylakoid protein phosphorylation is influenced by mutation in the b_6-f complex. Biochimica et Biophysica Acta, 933, 413-422.

Crane, F.L., 1989, Remarks on the discovery of coenzyme Q. Biochimica et Biophysica Acta, 1000 (in press).

Crane, F.L. and Morré, D.J., 1977, Evidence for coenzyme Q function in Golgi membranes. In: Biomedical and Clinical Aspects of Coenzyme Q, Vol. 1, edited by K. Folkers and Y. Yamamura (Amsterdam: Elsevier), pp. 3-14.

Crane, F.L., Hatefi, Y., Lester, R.L. and Widmer, C., 1957, Isolation of a quinone from beef heart mitochondria. Biochimica et Biophysica Acta, 25, 220-221.

Crane, F.L., Sun, I.L., Barr, R. and Morré, D.J., 1984, Coenzyme Q in Golgi apparatus membrane redox activity and proton uptake. In: Biomedical and Clinical Aspects of Coenzyme Q, Vol. 4, edited by K. Folkers and Y. Yamamura (Amsterdam: Elsevier), pp. 77-86.

Crane, F.L., Sun, I.L., Clark, M.G., Grebing, C. and Löw, H., 1985, Transplasma membrane redox systems in growth and development. Biochimica et Biophysica Acta, 811, 233-264.

Critchley, C., 1985, The role of chloride in photosystem II. Biochimica et Biophysica Acta, 811, 33-46.

Crofts, A.R., 1986, Reaction center and UQH_2 cytochrome c_2 reductase act as independent enzymes in *Rps. viridis*. Journal of Bioenergetics and Biomembranes, 18, 437-445.

Crofts, A.R. and Wraight, C., 1983, The electrochemical basis of photosynthesis. Biochimica et Biophysica Acta, 726, 149-244.

Fato, R., Castellucio, C., Amardi, S., Contarini, A., Parenti-Castelli, G. and Lenaz, G., 1988, Diffusional effects in the steady state kinetics of ubiquinone cytochrome c reductase in beef heart submitochondrial particles. Biochemical and Biophysical Research Communications, 155, 1145-1152.

Feher, G., Allen, J.P., Okamura, M.Y. and Rees, D.C., 1989, Structure and function of the bacterial photosynthetic reaction centers. Nature, 339, 111-116.

Green, D.E., ed., 1972, Membrane structure and its biological applications. Annals New York Academy of Sciences, Vol. 195, p. 519.

Green, D.E. and Wharton, D., 1963, Stoichiometry of the oxidoreduction components of the electron transport chain in beef heart mitochondria. Biochemische Zeitschrift, 338, 335-346.

Gutman, M., 1980, Electron flux through the mitochondrial ubiquinone. Biochimica et Biophysica Acta, 594, 53-84.
Hackenbrock, C.R., Chazotte, B. and Gupte, S.S., 1986, A random collision model and critical assessment of diffusion and collision in mitochondrial electron transport. Journal of Bioenergetics and Biomembranes, 18, 331-368.
Heron, C., Regan, C.I. and Trumpower, B.L., 1978, The interaction between mitochondrial NADH ubiquinone oxidoreductase and ubiquinone-cytochrome *c* oxidoreductase. Biochemical Journal, 174, 791-800.
Hosano, K., Ishida, H., Okuda, M., Suzuki, N. and Nakamura, T., 1981, Protective effects of coenzyme Q against arhythmia and its intracellular distribution: A study on the cultured myocardial cell. In: Biomedical and Clinical Aspects of Coenzyme Q, Vol. 3, edited by K. Folkers and Y. Yamamura (Amsterdam: Elsevier), pp. 269-278.
Kalén, A., Norling, B., Appelkvist, E.L. and Dallner, G., 1987, Ubiquinone synthesis in the microsomal fraction of rat liver. Biochimica et Biophysica Acta, 926, 70-80.
King, T., 1985, Ubiquinone proteins. In: Coenzyme Q, edited by G. Lenaz (Chichester: Wiley), pp. 391-408.
Klingenberg, M., 1986, The role of physical parameters in the regulation of electron transport. Journal of Bioenergetics and Biomembranes, 18, 447-451.
Landi, L., Pasquali, P., Bassi, P. and Cabrini, L., 1987, Effect of oxygen free radicals on ubiquinone in aqueous solution and phospholipid vesicles. Biochimica et Biophysica Acta, 902, 200-206.
Lehninger, A.L., 1964, The Mitochondrion, (New York: W.A. Benjamin), p. 263.
Lenaz, G., 1985, A biochemical rationale for the therapeutic effects of coenzyme Q. In: Coenzyme Q, edited by G. Lenaz (Chichester:Wiley), pp. 435-440.
Linnane, A.W., Marzuki, S., Ozawa, T. and Tanaka, M., 1989, Mitochondrial DNA mutations as an important contributor to aging and degenerative diseases. The Lancet, 1989(1), 642-645.
Michel, H., Epp, O. and Deisenhofer, J., 1986, Pigment protein interactions in the photosynthetic reaction center from *Rps. viridis*. EMBO Journal, 5, 2445-2451.
Mitchell, P., 1966, Chemiosmotic Coupling in Oxidative and Photosynthetic Phosphorylation. (Bodmin: Glynn Laboratory), p. 192.

Mitchell, P. and Moyle, J., 1985, The role of ubiquinone and plastoquinone in chemiosmotic coupling between electron transfer and proton translocation. In: Coenzyme Q, edited by G. Lenaz (Chichester: Wiley), pp. 145-164.

Morton, R.A., 1961, Isolation and characterization of ubiquinone (coenzyme Q) and ubichromanol. In: Quinones in Electron Transport, edited by G.E.W. Wolstenholme and C.M. O'Connor (London: J. & A. Churchill), pp. 5-30.

Ramasarma, T., 1985, Natural occurrence and distribution of coenzyme Q. In: Coenzyme Q, edited by G. Lenaz (Chichester: Wiley), pp. 67-82.

Ramsay, R.R., Mehlhorn, R.J. and Singer, T.P., 1989, Enhancement by tetraphenylboron of the interaction of the 1-methyl-4-phenylpyridinium ion (MPP^+) with mitochondria. Biochemical and Biophysical Research Communications, 159, 983-990.

Reed, J.S. and Regan, C.I., 1987, The effect of rate limitation by cytochrome *c* on the redox state of the ubiquinone pool in NADH cytochrome *c* reductase. Biochemical Journal, 247, 657-662.

Regan, C.I. and Reed, J.S., 1986, Regulation of electron transfer by the quinone pool. Journal of Bioenegetics and Biomembranes, 18, 403-418.

Shimomura, Y., Kawada, T. and Suzuki, M., 1989, Capsaicin and its analogs inhibit the activity of NADH coenzyme Q reductase of the mitochondrial electron transport chain. Archives of Biochemistry and Biophysics, 270, 573-577.

Sugioka, K., Nakano, M., Tolsune-Kakano, H., Minakami, H., Tero-Kubota, S. and Ikegami, Y., 1988, Mechanism of $O_2^{\cdot-}$ generation in reduction and oxidation of ubiquinone in a model of mitochondrial electron transport systems. Biochimica et Biophysica Acta, 936, 377-385.

Takada, M., Ikenoya, S., Yazurika, T. and Katayama, K., 1982, Studies on reduced and oxidized coenzyme Q. Biochimica et Biophysica Acta, 679, 308-314.

Ulrich, E.L., Girvin, M.E., Cramer, W.A. and Markley, J.L., 1985, Location and mobility of ubiquinone of different chain length in artificial membrane vesicles. Biochemistry, 24, 2501-2508.

Yu, C.-A. and Yu, L., 1981, Ubiquinone binding proteins. Biochimica et Biophysica Acta, 639, 99-128.

1. Biophysics and Biochemistry of Quinones

PHYSICO-CHEMICAL PROPERTIES OF UBIQUINONE-10

Peter J. Quinn

Department of Biochemistry, King's College London, Campden Hill, London W8 7AH, U.K.

INTRODUCTION

Ubiquinones are unique components of electron transport chains in that they represent the only lipid constituent of these multicomponent systems. Although ubiquinones and their counterpart in the chloroplast membrane, plastoquinones, are regarded as mobile electron and proton carriers, the manner in which they are integrated into the energy-transducing membrane and the motional constraints within the structure are somewhat conjectural. In order to obtain basic information upon which the behaviour of ubiquinone in complex biological membranes may be predicted, a number of physico-chemical studies have been conducted to investigate the properties of ubiquinone-10.

PHASE BEHAVIOUR OF UBIQUINONE

In common with most lipids, ubiquinones exist in one of a number of physical states or phases. The phases are dependent on temperature and transition between phases is associated with either an uptake or release of heat. Characterization of the phase behaviour can therefore be determined by calorimetry which can provide the temperature at which transitions take place as well as a measure of the enthalpy of the transition. Changes in phase also imply changes in the arrangement or conformation of molecules and these change can be followed by X-ray diffraction methods.

Calorimetric studies

Melting point determinations show that ubiquinone-10

Keywords: Thermal characteristics; phase behaviour; X-ray diffraction; calorimetric studies.

is transformed from a solid to a liquid phase in the temperature region 49-50°C. Detailed differential scanning calorimetry of ubiquinones (Katsikas and Quinn, 1983) have indicated that the phase behaviour of pure ubiquinone-10 is highly complex and displays marked hysteretic effects.

This is illustrated in Fig. 1 which shows differential scanning calorimetric curves of ubiquinone-10. The first heating scan of the crystalline material shows a single endothermic transition at about 50°C which corresponds to the melting point of the compound. Subsequent cooling of the melted ubiquinone produces a single exothermic peak at about 10°C indicating a marked hysteresis in the recrystallization process. The precise temperature of crystallization depends on the rate at which the sample is cooled from the melt; preparations cooled rapidly (10°/min) crystallize at lower temperatures compared with samples cooled more slowly, indicating the existence of a

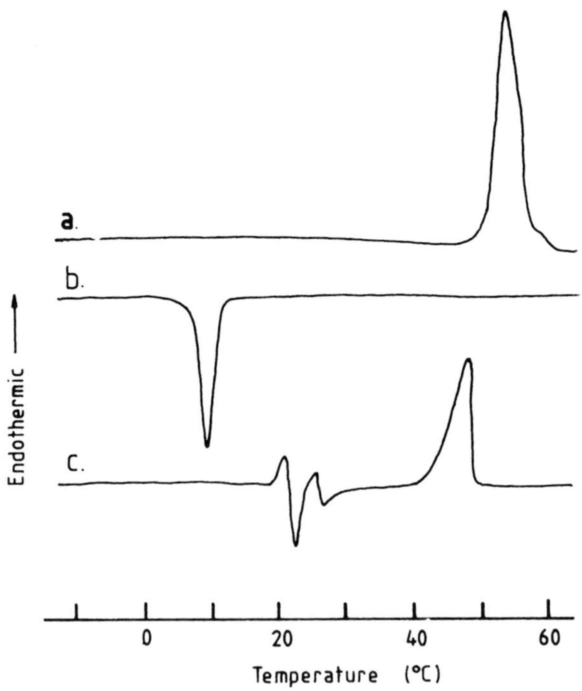

Fig. 1 - Thermograms of ubiquinone-10 (a) first heating (b) cooling (c) immediate reheating at 5°/min.

supercooling phenomenon. Preparations cooled from the melt and held isothermally at 30°C recrystallize over a period of days. If a preparation is reheated immediately after cooling from the melt to below the crystallization temperature the thermogram shows quite different features. The main endothermic transition shifts down to about 44°C. A consistent feature of the thermogram is a series of thermal events that precede the melting process. Two separate endothermic and two exothermic transitions can be distinguished in the temperature region 17-27°C.

The thermal behaviour of ubiquinone-10 is similar to homologues of ubiquinone with shorter polyisoprene chains. In general, the shorter the chain the lower is the crystallization temperature and endotherm on reheating. Similar hysteretic behaviour is observed and also the complex series of thermal events that precede the main melting endotherm.

Dynamic X-ray diffraction measurements

Calorimetric studies do not provide evidence for structural changes that take place during thermal events.

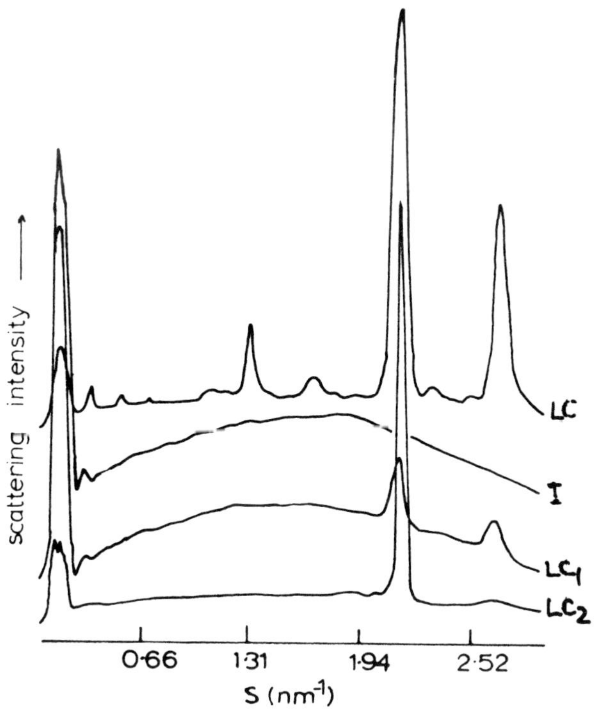

Fig. 2 - Static X-ray diffraction patterns of designated phases of ubiquinone-10

Nor can the nature of the phases themselves be characterised by this method. Phase characterization characterization of ubiquinone by wide-angle X-ray has been reported (Katsikas and Quinn, 1982). These have since been extended to diffraction in the low-angle region. Characteristic X-ray diffraction patterns recorded on a position-sensitive detector to record scattered synchrotron X-rays is shown in Fig. 2. The crystalline phase (LC) shows at least four orders of a lamellar repeat spacing of 6.27 nm. Five prominent wide-angle scattering peaks are also observed centred at 0.76, 0.69, 0.46, 0.42 and 0.37 nm. A diffraction pattern recorded at 60°C (melt or I phase) shows a relatively featureless scattering profile with a broad band centred at approximately 0.46 nm and a single broad scattering band in the small-angle region which can be indexed to a spacing 7 nm. The phase which forms immediately upon cooling the isotropic liquid phase shows a small-angle scattering band unchanged at approximately 7 nm and two wide-angle scattering bands centred at 0.46 and 0.37 nm;this is designated as LC_1 phase. Finally, the metastable crystal phase which precedes the transformation to the isotropic liquid phase is characterised by a very intense scattering band centred at 0.46 nm and is designated LC_2 phase.

The structural changes that take place during phase transitions have been investigated by real-time X-ray diffraction measurements using synchrotron radiation (Lis and Quinn, 1989). Transition between LC and I phase is shown in Fig. 3. In this experiment the sample was heated at a rate of 10°/min from 0°C and diffraction patterns accumulated over 1.2 s were recorded sequentially. The transition can be seen at about 50°C where there is a marked decrease in intensity of the LC diffraction pattern which is transformed into the I diffraction pattern. The mechanism of transition involves a co-existence of the initial and final phases with no evidene of intermediate structures between them. It is noteworthy that in the transition region there is a considerable amount of disorder in the phase as indicated by the intense and very broad diffraction band extending over the low-angle region.

When the isotropic phase is cooled a phase transition was observed at about 5°C to the liquid crystalline phase (LC_1). Although additional wide-angle scattering peaks are observed in the liquid crystalline phase, the mesophase structure is the same as that for the isotropic liquid phase. From this it may be concluded that the structure characterized by the single small-angle scattering peak at 7 nm is stable over relatively long periods once formed

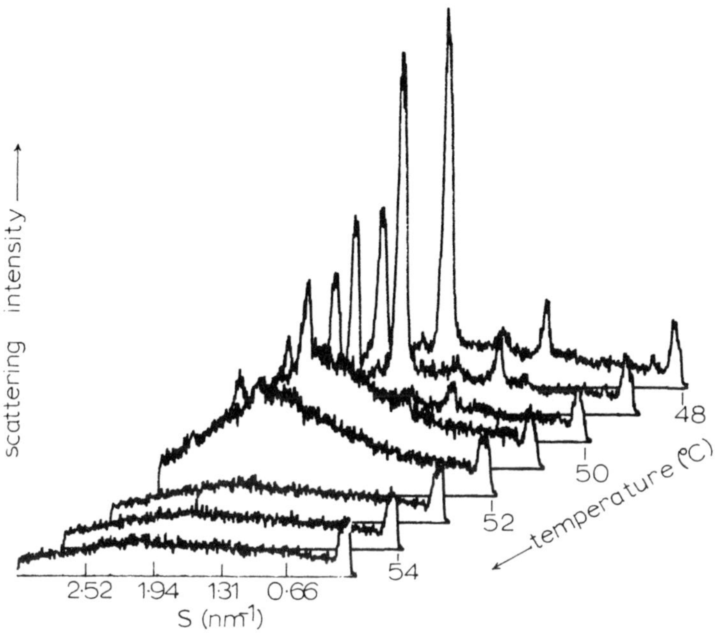

Fig. 3 - Time-resolved X-ray scattering changes during LC to I phase transition in ubiquinone-10.

from transition of the crystal to the isotropic liquid phase. Earlier studies (Quinn and Katsikas, 1985) had indicated that the phase structure represented by the wide-angle scattering peaks of the LC could be induced by prolonged storage but these experiments did not have sufficient resolution in the low-angle scattering region to establish whether the mesophase structure returns to the LC phase with unit cell spacing of approximately 6 nm.

Reheating the sample cooled from the isotropic phase from 6°C at 10°/min did not show significant changes in diffraction characteristics during heating through the complex series of thermal events centred around 20°C. The origin of these thermal events is presently unknown. There are, however, dramatic changes in the diffraction pattern at 30°C and appoximately 43°C. The first transition involves transformation from the liquid crystal phase to the LC_2 phase which involves an ordering of the molecular packing as judged by the dramatic increase in intensity of the 0.46 nm diffraction peak. The subsequent transition of the LC_2 to the I phase represents the expected order-disorder type of transition. It has been shown previously (Katsikas and Quinn, 1983) that the LC_2 phase is eventually converted to the LC phase if allowed to equilibrate isothermally at 27°C. This suggests that the reordering of the LC to LC_2 transition involves a thermally induced process during which the supercooled LC_1 requires additional energy in order to facilitate the nucleation of LC_2 phase in the isotropic mesophase (first observed when ubiquinone-10 was transformed to a liquid). It might be predicted that with prolonged equilibration, the LC_2 phase could be directly nucleated from the I phase.

The following phase transition sequence may be deduced from a combination of thermal and diffraction data:

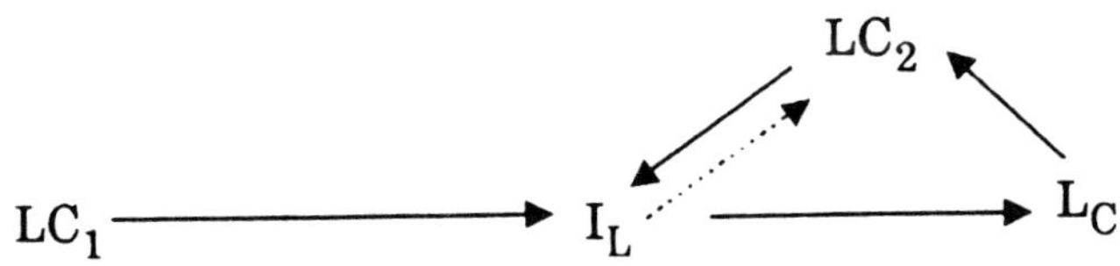

CONCLUSIONS

Ubiquinone-10 exhibits complex thermal phase behaviour with pronounced hysteresis associated with transition between phases. Time-resolved synchrotron X-ray diffraction studies have indicated that some of the phase transitions are two-state processes in which there is co-existance of the initial and final states.

REFERENCES

Katsikas, H. and Quinn, P.J., 1982, The polyisoprenoid chain length influences the interaction of ubiquinones with phospholipid bilayers. Biochimica et Biophysica Acta, 689, 363-369.

Katsikas, H. and Quinn, P.J., 1983, The thermotropic properties of coenzyme Q10 and its lower homologues. Journal of Bioenergetics and Biomembranes, 16, 67-80.

Lis, L.J. and Quinn, P.J., 1989, Real-time characterization of thermotropic liquid crystal phase transitions in coenzyme Q. Molecular Crystals and Liquid Crystals, in press.

Quinn, P.J. and Katsikas, H., 1985, Thermal characteristics of coenzyme Q and its interaction with model membrane systems. In: Coenzyme Q, edited by G. Lenaz (Chichester: John Wiley), pp 107-142.

LOCALIZATION OF UBIQUINONE IN MEMBRANES BY NMR AND NEUTRON DIFFRACTION

B.A. Cornell*, M.A. Keniry, R. Knott, A. Post, R.N. Robertson, F. Separovic, L.E. Weir, and P.W. Westerman

*Commonwealth Scientific and Industrial Research Organisation, P.O. Box 52 North Ryde, Sydney N.S.W. 2113, AUSTRALIA.

INTRODUCTION

This study addresses the location of Ubiquinone–10 (Q10) and deuterated analogues of Q10 in biological and model lipid–bilayer membranes. Two techniques have been employed; solid state, deuterium, nuclear magnetic resonance (NMR), and neutron diffraction. NMR provides information on the dynamic disorder undergone by ubiquinone in the membrane, and neutron diffraction, on the distribution and location of Q10 in relation to the surrounding lipid. The results obtained by these techniques are compared to those obtained from hydrated lipid dispersions containing alkanes, (Pope *et al.*). Alkanes are known to pool within the hydrocarbon interior of lipid dispersions, and serve as a reference when seeking the location of Q10 and its analogues. It has been shown, (Aranda & Gomez–Fernandez, 1985) that oxidation state is important in determining the effect of Q10 on the phase transition in lipid bilayers. However, the present results failed to indicate any incorporation of either the quinol or quinone form of Q10 into these structures.

Keywords; NMR, ubiquinone analogues, neutron diffraction, enzyme assays, model membranes:

METHODS

Ubiquinone analogues

The deuterated compounds shown in Figure 1 were prepared as previously described, (Cornell *et al.*, 1987).

Activity measurent

Enzyme assays were performed of the electron–transfer activities of the *E. coli* membranes with ubiquinone extracted and with membranes reconstituted with deuterated analogues of Q10. NADH oxidase activity was measured spectrophotometrically (Mackler, 1967). Cyanide inhibition experiments were also performed. The oxygen uptake with NADH as substrate was measured with an oxygen electrode (Raison & Lyons, 1970). Proton translocation of membranes in unbuffered solution was measured by the change in pH with time by a recording pH meter. The details of these assays and the membrane preparation procedures have been described previously, (Cornell *et al.*, 1987). Both Q10 and its analogues restored activity of the essentially inactive, Q10–depleted membranes to >30% of their original activity. A low–Q10 mutant experiences a 3–4 fold increase in activity by the introduction of Q10 or its deuterated analogues.

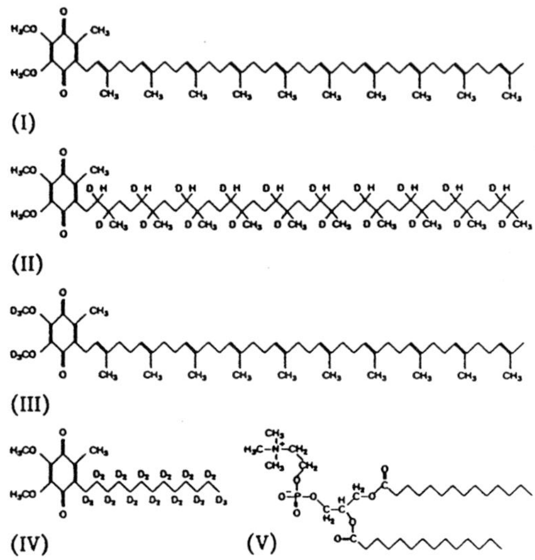

FIGURE 1: Structure of (I) ubiquinone 10, (II) the chain deuterated analogue of ubiquinone 10, (III) the benzoquinone ring deuterated analogue of ubiquinone 10, (IV) the short chain deuterated analogue of ubiquinone 10, and (V) DMPC.

NMR measurements

The samples studied by NMR were prepared as hydrated pellets of the appropriate mixtures of Q10 and membrane lipid. The model membrane systems were prepared by dissolving a known ratio of Q10 and dimyristoylphosphatidylcholine, (DMPC) in benzene–methanol or cyclohexane–methanol (95:5 v/v). These mixtures were lyophilised and hydrated with an equal weight of deuterium–depleted water. Q10 or its deuterated analogues were studied following their partitioning into *E. coli* or mitochondrial membranes from beetroot. The incorporation was performed according to the procedures of Ernster et al. (1969). The concentrations of Q10 in the reconstituted *E. coli* and mitochondrial membranes were in the range 0.05–0.4 mol % relative to the membrane lipid. Samples were washed in deuterium–depleted water.

Compound (II), in Figure 1, was also studied in the reduced state. The analogue was reduced with dithionite in ethanol and partitioned into cyclohexane (Rieske 1967). It was immediately mixed with DMPC and lyophilised giving a mixture of 2 mol %

Q10 in lipid in a sample containing 10mg of analogue. When dry,the powder was hydrated under nitrogen with an equal weight of deuterium depleted water, and then quickly flame sealed inside a glass vial. The sample appeared white, indicating complete reduction of the quinone.

Deuterium NMR was performed using a Bruker CXP–300 spectrometer operating at 46MHz and a 90 degree pulse duration of 5–6μs. A solid echo sequence with an echo delay of 40μs was used to avoid pulse breakthrough and acoustic ringing. Integrals of the spectral intensity were compared to integrals of the deuterium labelled quinones after being extracted from the sample with pentane, lyophilised, and dissolved in chloroform. This permitted a measure of the total deuterium NMR signal intensity and included any signal broadened below the level of detection in the membrane dispersion.

Neutron diffraction

In order to determine the location of the bulk of the mobile Q10 phase in relation to the lipid chains of a model membrane dispersion, a series of neutron diffraction measurements were performed on the high flux neutron source at the Brookhaven National Laboratory, New York. Samples were prepared containing 2 mol % Q10 in DMPC. This low concentration of Q10 was chosen in order to mimic the concentration found in biomembranes. Q10, deuterated at 20 sites by saturating the isoprenoid double bonds, (compound (II) in Figure 1.), was used to create a large scattering density contrast compared with that of the surrounding DMPC. Neutrons, unlike X–rays, are scattered very differently from hydrogen and deuterium, providing the basis for identifying the location of deuterium scattering centers within a diffracting medium. Samples were aligned as multilayers on glass slides and hydrated over saturated K_2SO_4 resulting in a 25% w/w of water to DMPC. Alignment of the sample on glass slides greatly enhances the intensity of the diffracted beam.

RESULTS

NMR results

To within a 5–10% precision, all of the NMR signal from the deuterated analogues (I) & (II) was visible in all systems studied, and revealed a single, featureless peak with a half–height line–width of approximately 500Hz as shown in Figure 2. This result was independent of whether the measurement was made above or below the DMPC phase transition. Reduction of the Q10 ring by dithionite yielded the same result. Measurements of the short chain analogue, revealed a resolved multiplet of quadrupolar doublets (Figure 3), anticipated for a lipid chain intercalated between the hydrocarbon chains of the DMPC bilayer. Reduction of the quinone ring caused an increase in the order of the bulk of the methylene deuterons, consistent with a change in the dynamic

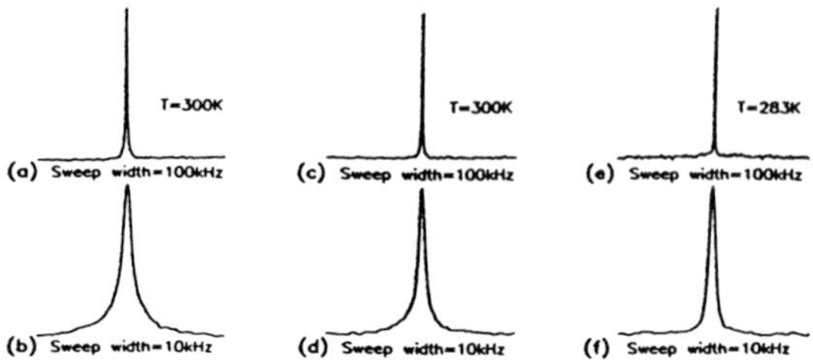

FIGURE 2: Deuterium NMR spectra of the chain–labeled Q10, (compound II), and ring labeled Q10 (compound III), dispersed in DMPC or *E. coli* membranes: (a) 0.05 mol % (II) in DMPC at 300K, sweep width 100kHz, solid echo sequence delay 40μs, repetition delay 1s, 40,000 scans; (b) same spectrum on an expanded scale of 10kHz; (c) 0.1 mol % (II) in *E. coli* at 300K and the same conditions; (d) same spectrum on an expanded scale of 10kHz; (e) 0.07 mol % (III) in *E. coli* at 283K and the same conditions; (f) same spectrum on an expanded scale of 10kHz.

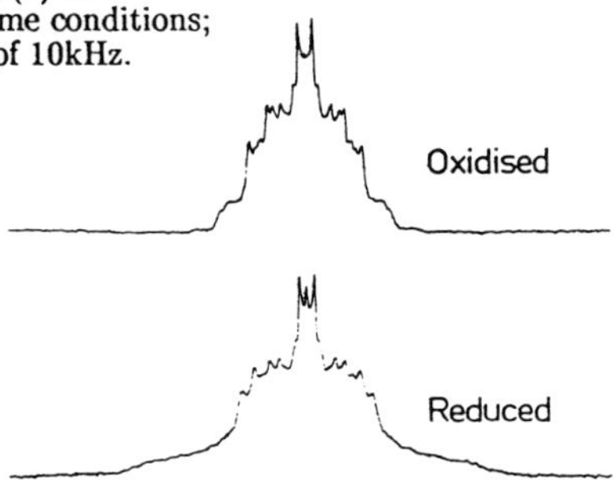

FIGURE 3: Short chain deuterated analogue of ubiquinone, (III) dispersed in DMPC in the oxidised & and reduced state. Recorded at 300K a sweep width of 100kHz, & 40,000 scans.

state of the analogue in the DMPC dispersion.

Neutron diffraction

Five diffraction orders were obtained. Phasing was determined by assuming a centrosymmetric profile and a choice of + or − assigned by the contrast variation technique. The structure factor for each order was obtained for 5 different H_2O/D_2O ratios, from each sample. An example of the dependence of the structure factor on this ratio is shown in Figure 4(a), and of the scattering amplitude profile of the membrane in Figure 4(b). Within experimental precision it was not possible to distinguish between the scattering amplitude profile of dispersions of hydrated DMPC, from those containing the native or deuterated Q10. A change in the scattering amplitude of approx. 30% would be expected if 2 mol % native Q10 was located, within a monomolecular layer at the centre of the bilayer between the terminal methyl groups. Within an experimental uncertainty of 5% in scattering amplitude, no such change was observed.

CONCLUSIONS

Both the NMR and neutron diffraction data are interpreted as showing that at the concentrations of Q10 studied here, Q10 is not oriented with the membrane lipid, but is tumbling isotropically on a timescale shorter than a μs. The lack of dependence of the motion of the Q10 analogues on the lipid phase is interpreted as showing that the Q10 is physically separate from

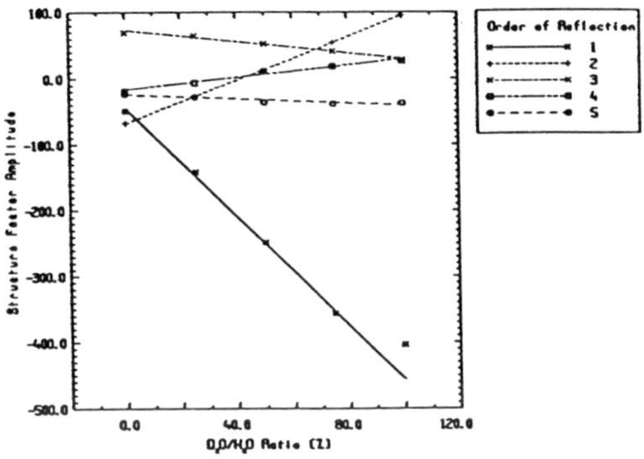

FIGURE 4(a): Dependence of neutron diffraction structure factors on the H_2O/D_2O ratio for 2 mol % ubiquinone analogue (II) in DMPC at 307K over saturated K_2SO_4.

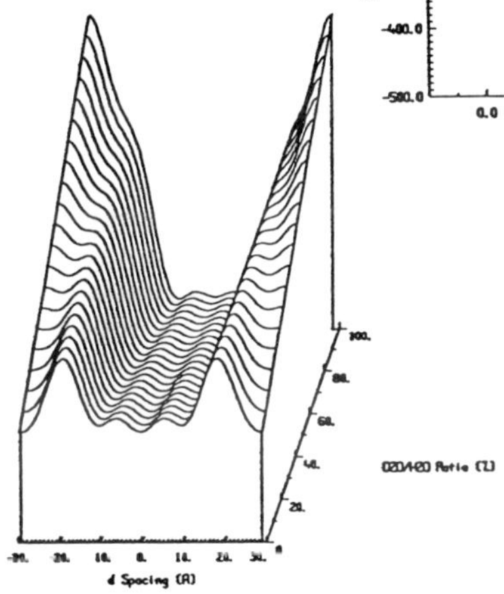

FIGURE 4(b): Scattering amplitude profile generated from this data. No significant contribution to this profile arose from ubiquinone analogue (II).

the membrane lipid. Based on the NMR measurements alone it is not possible to identify the location of Q10 with respect to the membrane. Possible sites include the centre of the membrane, between the terminal methyl groups of the lipid, Q10–rich aggregates within the aqueous environment adjacent to the membrane, or Q10–rich aggregates distributed at random throughout both of these regions. The possibility exists that a small quantity of Q10, less than 10% of the 0.05% v/v per lipid does exist within the membrane but cannot be detected by the NMR technique.

Calculations based on the present work failed to reveal a site for Q10 within fluid DMPC bilayers that correlates with either the aqueous or hydrocarbon components of the lipid. Siting Q10 within the central core of the membrane or at a defined location within the aqueous environment revealed that such positions would have easily been identified by differences in the scattering profile. Thus we conclude that the bulk of the Q10 is neither sequestered to the centre of the membrane nor to any other defined site in the dispersion. Rather, the results suggest that Q10 aggregates and is present as a separate phase that is not systematically organised relative to the lipid structure. Of course, should a small quantity (<10%) of this Q10 be organised in a regular manner relative to the lipid, this quantity may be below the level of detection.

A useful comparison for the NMR and the neutron diffraction data are the studies on the location of deuterated alkanes in phospholipid bilayer dispersions (Jacobs & White, 1984; King *et al.*, 1985; Pope *et al.*, 1989). In all studies there is clear evidence of interaction between the lipid and the alkanes leading to the conclusion that the short chain alkanes are sequestered to the centre of the bilayer between the terminal methyl groups of the hydrocarbon chains. The membrane solubility drops sharply at longer alkane chain lengths. A comparison of the concentrations and experimental conditions under which the present data were obtained indicates that Q10 neither intercalates between adjacent lipid chains nor concentrates at the centre of the hydrocarbon core consistent with the results of Stidham *et al.* (1984).

One suggestion arising from these results is that the proton and electron transporting function performed by Q10 is carried out in conjunction with another membrane molecule, and does not arise from the physical interaction of Q10 with the membrane lipid.

The authors wish to express their gratitude to Professor Benno Schoenborn for his help and advice in performing the neutron diffraction studies reported here.

REFERENCES

Aranda, F.J. & Gomez–Fernandez, J.C. (1985) *Biochim. Biophys. Acta* **820**, 19–26.

Cornell, B.A., Keniry, M.A., Post, A., Robertson, R.N., Weir, L.E. & Westerman, P.W. (1987) *Biochemistry* **26**, 7702–7.

Ernster, L., Lee, I–Y., Norling, B., & Persson, B. (1987) *Eur. J. Biochem.* **9**, 299–310.

Jacobs, R.E. & White, S.H. (1984) *J. Am. Chem. Soc.* **106**, 915–20.

King, G.I., Jacobs, R.E. & White, S.H. (1985) *Biochemistry* **24** 4637–4645.

Pope, J.M., Littlemore, L.A. & Westerman, P.W. (1989) *Biochim. Biophys. Acta* **980**, 69–76.

Mackler, B. (1967) *Methods Enzymol.* **10**, 261–3.

Raison, J.K., & Lyons, J.M. (1970) *Plant Physiol.* **45**, 382–5.

Reiske, J.S. (1967) *Methods Enzymol.* **10**, 239–45.

Stidham, M.A., McIntosh, T.J. & Siedow, J.N. (1984) *Biochim. Biophys. Acta* **767**, 423–31.

FUNDAMENTAL ASPECTS OF DIFFUSION-BASED MITOCHONDRIAL ELECTRON TRANSPORT

BRAD CHAZOTTE AND CHARLES R. HACKENBROCK

Department of Cell Biology and Anatomy, Laboratories for Cell Biology, School of Medicine, University of North Carolina, Chapel Hill, N.C. 27599-7090 USA

INTRODUCTION

This laboratory has established a fundamental role for diffusion in the process of mitochondrial electron transport (Gupte *et al.*, 1984; Chazotte and Hackenbrock, 1988, 1989; Gupte and Hackenbrock, 1988a,b) which is summarized and codified in the five postulates of the **Random Collision Model** (Hackenbrock *et al.*, 1986). Two of the postulates state primary findings that (a) all redox components are independent diffusants and (b) electron transport is a *diffusion-coupled* process, i.e. all electron transfers are preceded by one or more diffusion-based collisions. (Electron transport is defined as the process inclusive of electron transfer and the lateral diffusion of redox components.) Inherent in these findings is the fact that physical parameters, such as temperature, membrane and aqueous phase viscosities, the absence or presence of obstructions to diffusion (inclusive of obstructant size, shape, concentration, and distribution), that affect the process of diffusion in and on the inner membrane can have significant or controlling effects on the process of electron transport. This physically-related aspect of electron transport comprises the three other postulates: (c) Cytochrome *c* diffuses primarily in three dimensions; (d) Mitochondrial electron transport is a multicollisional, obstructed, long-range, diffusional process; (e) The rates of diffusion of redox components have a direct influence on the overall process of electron transport and can be rate-limiting, as in diffusion control. The analysis of such diffusion parameters must ultimately consider the limitations imposed by the actual conditions in the functional mitochondrion and the requirements of its bioenergetic function. Our studies have sought to identify and quantify the effects of such diffusion-related parameters and to understand their relationship to the bioenergetic process of electron transport in the mitochondrion.

Keywords: mitochondria, electron transport, diffusion, Random Collision Model, ubiquinone, fluorescence recovery after photobleaching

Accordingly, a primary focus of our approach has been to relate the rate of electron transport to the rate of diffusion-based collisions of appropriate redox partners over a range of physical conditions using a suitable reaction-diffusion equation for calculations (Gupte *et al*, 1984; Chazotte & Hackenbrock 1988, 1989). Various types of mitochondrial inner and model membranes were used. Electron transport rates were determined by standard kinetic methods on bulk membrane suspensions. Diffusion coefficients (Ds) of fluorescent phospholipids, immunospecifically labelled redox complexes, and a ubiquinone analogue were determined by the widely accepted technique of Fluorescence Recovery After Photobleaching (FRAP) on membranes attached to microscope slides.

UBIQUINONE DIFFUSION

Based upon our FRAP studies of a functional, fluorescent ubiquinone analogue, Q_0C_{10}NBDHA, we determined that ubiquinone diffuses at the same rate as the phospholipids in which it resides in both inner mitochondrial membranes (Wu *et al.*, 1983; Gupte *et al.*, 1984) at D ~3 x 10^{-9} cm^2/s (23 °C) and in pure dimyristoylphosphatidylcholine (DMPC) multibilayer membranes (Wu *et al.*, 1982) at D ~5 x 10^{-8} cm^2/s (30 °C) consistent with many other reports on lipoidal membrane diffusants and theoretical descriptions of membrane diffusion. The fluorescent emission maxima of our ubiquinone and the fact that its D exhibited a typical DMPC phase transition indicated that it was partitioned in the bilayer. The phospholipid-like behavior and bilayer location of the fluorescent ubiquinone was shown also by incorporating cholesterol into the DMPC membranes. Under conditions of progressively increasing cholesterol content, the ubiquinone's Ds progressively decreased above and increased below the main phase transition. This finding is in agreement with the temperature dependent diffusion study of Alecio *et al.* (1982) comparing cholesterol and phospholipid Ds. In addition, fluorescence microscopy revealed that such progressive increases in cholesterol content of DMPC membranes progressively displaced an increasing percentage of the fluorescent ubiquinone from the bilayer domain to crystal-defect regions. This latter result showed that increasing cholesterol in the inner membrane creates non-homogeneous distributions of ubiquinone in addition to affecting its D, and, as shown by Schneider *et al.* (1982), may result in the co-localization of ubiquinone with induced aggregates of integral proteins .

Our earlier finding that ubiquinone diffuses at the same rate as the phospholipids in pure lipid and mitochondrial inner membranes has been confirmed by Ferguson-Miller *et al.* (1988) who used a non-functional ubiquinone-10 with the fluorophore conjugated to the benzoquinone headgroup. We concluded that ubiquinones diffuse at the same rate as the phospholipids in which they reside, with Ds in the 3 x 10^{-9} cm^2/s range in the native inner membrane.

MULTICOLLISIONAL, OBSTRUCTED, LONG-RANGE DIFFUSION

Based upon comparisons of diffusion and electron transport rates in native and protein-diluted mitochondrial inner membranes (Chazotte and Hackenbrock, 1988) we determined that: (1) more than one collision is required to successfully carry out an electron transfer for all redox partners, i.e. electron transport is multicollisional; (2) Ds increase with decreasing inner membrane protein density, i.e. electron transport is obstructed; (3) on average the minimum straight-line distance between two redox complexes in the appropriate redox state is much greater than the short-range (<10 nm) diffusion distance measured by fluorescence quenching. The requirement for the measurement of long-range, rather than short-range, Ds is also illustrated by consideration of the 9 ms average time between ubiquinone to Complex III electron transfers in uncoupled steady-state electron transport and the Ds of ubiquinone in the $1 - 10 \times 10^{-7}$ cm^2/s range, currently estimated by fluorescence quenching (Lenaz *et al.* 1988), which we believe to be an overestimation (Chazotte and Hackenbrock, 1988, 1989). Using such high Ds, ubiquinone's calculated mean square displacement would be between 0.44 to 1.4 microns, clearly a long-range distance. Thus, the measurement of short-range diffusion is <u>irrelevant</u> to the kinetics of electron transport in the protein-containing inner membrane. We conclude that mitochondrial electron transport is fundamentally a *multicollisional, obstructed, long-range, diffusional process* and must be included in any consideration of the relationship of electron transport to diffusion.

LATERAL DIFFUSION AS A RATE-LIMITING STEP

Based upon comparisons of the temperature dependencies of diffusion and electron transport rates in accordance with Rate Process theory we determined (Chazotte and Hackenbrock, 1989) that (1) the activation energy (E_a) for the diffusion step (e.g. 12.2 kcal/mole for diffusion in the native inner memb.) alone was the most significant contribution to E_a for the overall electron transport process (e.g. 12.87 kcal/mole for succinate:cytochrome *c* oxidoreductase activity in the native inner memb.) at each protein density studied, which indicated that the diffusion and collision of ubiquinone with its redox partners is rate-limiting; (2) the E_as for electron transport activities (succinate and NADH oxidases; succinate:cytochrome *c* oxidoreductase) and for the Complex I-, Complex II-ubiquinone-Complex III diffusion steps decreased in a similar parallel fashion with decreasing protein density which indicated that as the resistance to diffusion decreased the resistance to electron transport was likewise decreased; (3) <u>no</u> significant temperature dependence was noted in Arrhenius plots for the comparison of the ratio of diffusion-based theoretical predictions to productive collisions of ubiquinone with its redox partners which indicated that the increases in the electron transport rate could be accounted for by the increases in diffusion-based collisions; (4) both multicomponent electron transport and redox component mobilities were similarly inhibited by increasing concentrations of sucrose and sucrose-like agents which indicated that electron transport rates were directly coupled to and limited by the diffusion-based collision rates. Comparison of sucrose and percoll inhibitory effects suggested that sucrose-induced dehydration of the inner membrane was largely responsible for the dramatic decrease in membrane component mobility (and consequently ubiquinone-mediated electron transport rates) beyond that theoretically expected from increased aqueous phase viscosity on the membrane surface. We concluded that the *diffusion and collision of ubiquinone with its redox partners* (Complexes I,II & III) is fundamentally *a rate-limiting step* for maximum uncoupled rates of succinate- or NADH-linked electron transport.

DIFFUSION IN THE INTACT MITOPLAST - MATRIX EFFECTS

We recently determined by FRAP that the Ds of phospholipid and ubiquinone and Complexes I, III, IV and V in matrix-containing intact condensed (300 mosM medium) or swollen (150 or 40 mosM medium) mitoplasts (Table I) were virtually the same as that determined for spherical, ultralarge, fused inner membranes where the matrix was not present (Gupte *et al.*, 1984; Chazotte and Hackenbrock, 1988). The Ds

Table I. Comparison of Ds in fused matrix-free inner membranes and matrix-containing mitoplasts[a]

DIFFUSANT	PHOSPHO-LIPID & UBIQUINONE (DiI)	COMPLEX I	COMPLEX III	COMPLEX IV	COMPLEX V
CONDITIONS					
Fused matrix-free ultralarge inner membranes					
40 mosM	39± 9	0.72±0.7	6.9±2.2	3.6±1.7	8.4±6.0
Normal matrix-containing mitoplasts					
40 mosM	59±33	0.47±0.2	3.4±2.6	2.5±1.7	7.3±3.0
150 mosM	14±12	-	3.5±2.0	2.6±5.3	6.5±3.5
300 mosM	48±43	-	4.7±3.2	2.8±1.8	2.7±1.2

a) $D \times 10^{10}$ in cm^2/s

of Complexes I, III, and IV may be at most marginally slower in mitoplasts, however, Complex V, which has the greatest penetration intothe matrix, may have a D a factor of 3-4 slower. We conclude that (1) the mitochondrial matrix does not have a significant inhibitive effect, i.e. an order of magnitude or greater, on the rate of lateral diffusion of inner membrane redox components and (2) the Ds reported by this laboratory (Gupte *et al.*, 1984; Chazotte and Hackenbrock 1988, 1989; Gupte and Hackenbrock 1988a,b) for fused, ultralarge, matrix-free inner membranes are valid for the intact functional mitoplasts containing the full complement of matrix found in the intact mitochondrion.

We conclude that any understanding of the fundamental aspects of the process of mitochondrial electron transport must consider a quintessential role for the diffusion of redox components as delineated in the **Random Collision Model of Mitochondrial Electron Transport.**

ACKNOWLEDGMENTS

We thank Dr. Tsoo King for providing isolated Complexes III and IV, Dr. C. Ian Ragan for providing isolated Complex I and Dr. Chang-An Yu for synthesizing the fluorescent ubiquinone analogue. Supported in part by NIH GM 28704 and NSF PCM 84-02569.

REFERENCES

Alecio, M.R., Golan, D.E., Veatch, W.R., and Rando, R.R., 1982, Use of a fluorescent cholesterol derivative to measure lateral mobility of cholesterol in membranes. Proceedings of the National Academy of Sciences (USA), 79, 5171-5174.

Chazotte, B. and Hackenbrock, C.R., 1988, The multicollisional, obstructed, long-range diffusional nature of mitochondrial electron transport. Journal of Biological Chemistry, 263, 14359-14367.

Chazotte, B. and Hackenbrock, C.R., 1989, Lateral diffusion as a rate-limiting step in ubiquinone-mediated mitochondrial electron transport. Journal of Biological Chemistry 264, 4978-4985.

Ferguson-Miller, S., Rajarathnam, K., Hochman, J., and Schindler, M., 1988, Is electron transfer mediated by diffusion alone? In: Integration of Mitochondrial Function, edited by J.J. Lemasters, C.R. Hackenbrock, H.V. Westerhoff and R.G. Thurmond, (New York: Plenum Press), pp. 23-32.

Gupte, S.S. and Hackenbrock, C.R., 1988a, Multidimensional diffusion modes and collision frequencies of cytochrome *c* with its redox partners. Journal of Biological Chemistry, 263, 5241-5247.

Gupte, S.S. and Hackenbrock, C.R., 1988b, The role of cytochrome *c* diffusion in mitochondrial electron transport. Journal of Biological Chemistry, 263, 5248-5253.

Gupte, S.S., Wu, E.-S., Hoechli, L., Hoechli, M., Jacobson, K., Sowers,

A.E. and Hackenbrock, C.R., 1984, Relationship between lateral diffusion, collision frequency, and electron transfer of mitochondrial inner membrane oxidation-reduction components. Proceedings of the National Academy of Sciences (USA), 81, 2606-2610.

Hackenbrock, C.R., Chazotte, B., and Gupte, S.S., 1986, The random collision model and a critical assessment of diffusion and collision in mitochondrial electron transport. Journal of Bioenergetics and Biomembranes, 18, 331-368.

Lenaz, G., Fato, R., Castelluccio, C., Degli Eposti, M., Samworth, C.M., Battino, M., and Parenti Castelli, G., 1988, Role of ubiquinone diffusion in mitochondrial electron transfer. In: Integration of Mitochondrial Function, edited by J.J. Lemasters, C.R. Hackenbrock, H.V. Westerhoff and R.G. Thurmond, (New York: Plenum Press), pp. 33-52.

Schneider, H., Höchli, M., and Hackenbrock, C.R., 1982, Relationship between the density distribution of intramembrane particles and electron transfer in the mitochondrial inner membrane as revealed by cholesterol incorporation. Journal of Cell Biology 94, 387-393.

Wu, E.-S., Yu, C.-A., and Hackenbrock, C.R., 1983, Lateral diffusion of a ubiquinone analogue in phospholipid bilayers and mitochondrial inner membrane. Biophysical Journal, 41, 374a.

PREPARATION, PROPERTIES AND RECONSTITUTION OF QP-C

By Tsoo E. King

Institute of Structural & Functional Studies, University City Science Center and Department of Biochemistry & Biophysics, University of Pennsylvania, PA 19104 USA

First I sincerely thank the organization committee for the kind invitation extended to me.

Today I have chosen this topic instead originally decided because I thought it may be more useful to extend and to discuss more about mitochondrial QP-C (ubiquinone protein-C) with the intention to clarify some of the recent confusions. The concept of Q as respiratory carrier was conceived almost at the same time when Q was claimed as an electron transfer component (*cf.* e.g.1 especially p 125). It is now practically generally agreed only Q linked with a protein is active in mitochondria or photosynthesis as an electron carrier. At present, only the nature of the protein moieties for various Q-proteins are not unanimously agreed.

We first undertook (2) the isolation of a Q-protein, QP-C, after the demonstration (3) of Q band EPR spectra of the *b*-c_1 segment at 0^0 and 77 K with prominent anisotrophy showing g = 2.0064, 20054 and 2.0021. Considerable effort was spent in developing a rather convenient assay system (2); QP-C in the starting material is depleted by controlled chymotrysin digestion (*b*-c_1-IV complex). The isolation itself with the aid of Lubrol is succeeded without much difficulty. The sample obtained can be reconstituted with the *b*-c_1-IV complex to resume about 90 to 95% of the original activity. It shows SDS-PAGE (sodium dodecyl sulfate-polyacrylamide gel electrophoresis) pure by a very rigorous means. Other results obtained demonstrate the sample is truly pure not just PAGE pure. So the amino acid sequence work is then performed(4); the sequence is show in Fig 2. It contains no cystein. Because of some reason I am not very satisfied, which I will tell you later, therefore some improvements must be found. This decision has taken us much more time than that for the original development. To make the long story short, we eventually have developed two methods. One is cholate plus antimycin and the other with cholate alone. The procedure and manipulations are of course different from the original Lubrol method. The product from original method is called QP-C(I), that from cholate plus antimycin QP-C(II), and from the cholate alone QP-C(III). The monomeric molecular weights of all these preparations in SDS-PAGE medium are found to be approximately 14 kD.

All preparations contain no detectable prosthetic groups such as heme, flavin, iron, copper, etc. Another characteristic is free of cystein or -SH group. All 3 preparations are the same as far as protein is concerned. The only peak is 278-9 nm which is decreased upon reduction. The degree of dcrease is dependnet upon which method is used for preparing the sample. The spectra of even prepared by Method II does not show any sign of antimycin at all. The amount of Q_{10} and phospholid (PL) content of the prodcuts are different varying from 0.1 to 1.0 mol Q per mol of protein and 0.4 mg to 1.0 mg PL per mg protein.

Although the SDS-PAGE gives the monomeric weight of all preparations to be about 14 kD; in the absence of a dissociating agent, such as SDS, the molecular weights of QP-C(II) and QP-C(III) are 27 kD with Stoke radius of 25.4 A in phosphate buffer, PH 7.4 containing 0.5% cholate. On the other hand, QP-C(I) in 25 mM Tris HC1 buffer with 0.5% Lubrol gives a molecular weight of 63 kD.

Now let us see some results of reconstitution activity of the QP-C preparations.

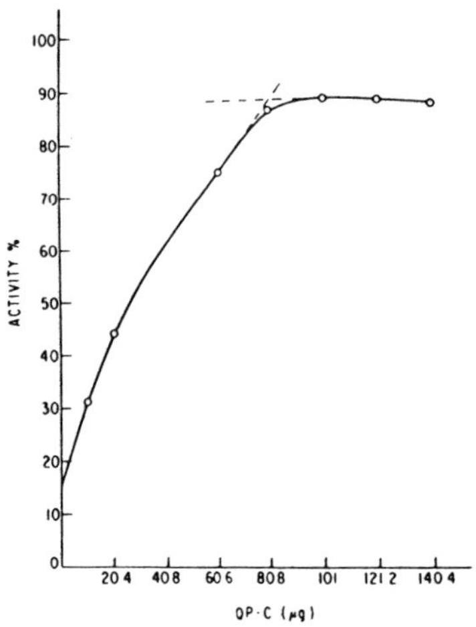

FIG. 1. RECONSTITUTION OF QP-C WITH THE CYTOCHROME bc_1-IV COMPLEX The system contained 66 *ug* of cytochrome *b*-c_1 IV complex and QP-C made by Method II as indicated in the abscissa in a total volume of 0.2 ml of 50 mM Tris-HCl buffer, pH 8.0, containing 0.67 M sucrose and 1 mM EDTA. The mixture was incubated at room temperature (22-23^0) for 10 min. After incubation, it was diluted to 1.5 ml with the same buffer. Aliquot of 10 *u*l was used for assay. The activity of the cytochrome *b*-c_1-III complex before chymotrypsin digestion was taken as 100% which was 150 *u*mol cytochrome *c* reduced per minute per mg of protein.

The cytochrome *b*-c_1-IV complex retains about 18 to 25% original quinol-*c* reductase activity. Gradual addition of QP-C increases the reductase activity, a randomly selected curve is shown in Fig. 1. Up to a certain point the activity increase is plateaued. Indeed this break point is quite sharp. Average of 20 to 30 batches of QP-C and the cytochrome *b*-c_1-IV complex have been so far tested and give a range of the recovery of 90 to 95% of the activity of the original, i.e. before the chymotrypsin digestion. The reconsitution is very sensitive to antimycin (about 1 mol antibiotic per mol of c_1) just as the activity of the reconstituted.

Another Q-protein accepts electron directly from succinate dehydrogense, namely QP-S, when prepared from "an improved" method by Drs. C.A. Yu and Linda Yu worked out in their last two years of 8 years' stay in our lab and was published (5) after they left, shows an interesting behavior in the reconstitution. At low concentration of the old QP-S, i.e. according to Yu and Yu's method, the reconstitution of QP-C with cytochrome *b*-c_1-IV complex is increased but high concentrations of old QP-S inhibited the reconstitution activity. A few years ago Yoshida *et al* (6), in our lab, isolated two fractions of cytochrome *b*, namely one is *b*-L and other *b*-s. The *b*-L possesses an apparent molecular weight of 37 kD and has no effect but only inhibition in QP-C reconstitution whereas *b*-s of molecular weight of 17 kD showed similar behavior as the old QP-S.

That is low concentration is stimulatory and high concentration inhibitory. However the high activity of the old QP-S in reconstitution has made us very suspicious. Is it just due to simply a contamination or is it due to the contamination of QP-C, which made the reconstitution more active? In other words, a synergistic effect? We did a hard job in careful analysis of the amino acid composition of a successfully separated major fraction from old Yu and Yu's QP-S preparation and eventually chemically sequenced it. All showed it is QP-C and is indeed active in reconstitution with cytochrome *b*-c_1-IV complex to form ubiqinol-*c* reductase. Triton in the preparation is inhibitory for the reconstitution of QP-C with the *b*-c_1-IV complex. Some other compounds such as the hinge protein, the Rieske protein or cytochrome c_1 does not inhibite nor activate the reconstitution. Here I must emphasize that any of QP-C preparation (I,II,III) does not stimulate nor inhibit the reconstitution of our newly developed QP-S (7) with succinate dehydrogenase to form succinate Q reductase. Likewise, the newly developed QP-S does not inhibit nor stimulate QP-C reconstitution after the Triton is removed.

We also have done some immunochemical determinations for various components, I will not belabor you the details of the methodology and only tell you that the antigen, QP-C, is in a most rigorously monodisperse form, the IgG fraction of the antibody is extensively purified, the pre-serum is always used parallel with other experimental controls. The antibody reacts with any of QP-C preparations, old QP-S, the *b*-*c*-III complex, *b*-s and weakly with the *b*-c_1-IV complex, but not *b*-L, cytochrome c_1, the Rieske protein, the hinge protein, succinate dehydrogenase, or cytochrome oxyidase.

One more point we are very much concerned with the original QP-C prepared with Lubrol method or QP-C(I). It is in reconstitution activity showing very reasonable on one hand, but on the other always exhibiting low recovery of ubisemiquinone radical in EPR spectroscopy even practically all of Lubrol in QP-C has been laboriously removed. This fact caused us great pain to develop alternative methods to prepare QP-C(II) and (III) and eventually we successfully made them. The radical recovery from QP-C (II) and QP-C (III) is in agreement of the reconstitution. EPR spectroscopy of the preparation is extremely sensitive to Lubrol and in this EPR experiments much higher concentration of the QP-C sample is required, so that QP-C(I) gives a lower EPR recovery.

But we should not forget the Q_{10} content of these 3 preparations are not the same. The protein moieties are the same, but the Q content are not. QP-C(I) is definitely in micellular form but I have no evidence to say the Q is in the original binding site. QP-C(II) has about 0.4 mol and QP-C(III) is less than 0.1 mol Q per mol of protein. On the other hand the reconstitution activity is the same. These facts prompt us of reincorborate Q to these preparations. Several methods in different media has been used. Sure enough Q is indeed incorporated to QP-C(II) and QP-C(III). Unexpectedly the QP-C(I), which contains 1.0 mol Q per mol protein also takes up Q. These additional Q cannot be separated by gel filtration chromatography but travels with the protein. Moreover, other proteins such as cytochrome *c* and bovin serum albumin, under these conditions Q is also incorporated. The amount of Q incorporated is dependent on the conditions. Naturally these artificial Q proteins do not have QP activity whatsoever. To us to base on the Q content does not seem to be very reliable.

Labeling or photolabeling experiments when done in careful and well controlled manner can throw tremendous light on the binding or other sites such as regulatory sites as well as identification of specific protein fractions or proteins. However, this may not be the case of QP-C. I do not think time permits me to discuss this matter in detail here, but I will only give you a brief summary. The labeling agent is usually an azide derivative, which is a very reactive agent as you may know. After a coupling with azido group, the product is photolyzed and covalent bond is formed. The radioactivity is usually incorporated to the side chain such as methoxy group. Yu and Yu have done most of the work, briefly they have used, for example, an azido Q derivative instead of isoprene units with ten CH_2 groups have found QP-C are 37 kD and 17 kD (8), and using azido derivative of Q_2 have claimed 17 kD to be QP-C because the decrease of the biological activity coincides with radioactive uptake (9). They commented "it greatly aides these...azido-Q derivative to unambigously identify Q binding proteins". It is formally announced the 13.4 kD of

Wang-King's QP-C definitely not QP-C but only 17 kD is QP-C. However in this year's ASBMB Meeting (San Francisco) (10) they have definitively claimed the binding of Q in QP-C is on 9.5 kD protein.

Finally from genetic view, QP-C sequences from some organisms are shown in Fig.2. You may notice that there are very high homology among them, from more than 45% in yeast (11) to 85% in human (12) tissue based on bovine heart (4). It is hardly a biological aberration.

In conclusion, to our concept a respiratory Q protein must fulfill the function as a respiratory carrier and the protein moiety is used to stabilize the Q radicals and also determines the specificity of the site or sites of QP where acts on the respiratory chain. This is in agreement of the Q cycle or even straight chain formulation. If I want to venture what is our QP-C? I would

```
                 10        20        30        40        50
1.   AGRPAVSASSRWLEGIRKWYYNAAGFNKLGLMRDD-TIHENDDVKEAIRRL
                 10        20        30        40        50
2.  MAGKQAVSASGKWLDGIRKWYYNAAGFNKLGLMRDD-TIYEDEDVKEAIRR
                           10        20        30        40        50
3.          RSKWLDGFRKWYYNAAGFNKLGLMRDD-TMHETEDVKEAIRRLPENL-YNDR
               10        20        30        40         50        60         70        80
4.MPQSFTSIARIGDYILKSPVLSKLCVPVANQFINLAGYKKLGLKEDDLIAEENPIMQTALRRLPEDESYA-RAYRIIRA
                       60        70        80         90        100        110
1.      PENL-YDDRVFRIKRALDLSMRQQILPKEQWTKYEEDKSYLEPYLKEVIRERKEREE----WAKK
                       60        70        80         90        100        110
2.     LPENL-YNDRMFRIKRALDLNLKHQILPKEQWTKYEEENFYLEPYLKEVIRERKEREE----WAKK
                       60        70         80        90        100        105
3.               MFRIKRALDLSMRHQILPKDQWTKYEEDKFYLEPYLKEVIRERKEREE----WAKK
                                        90         100       10
4.                       HOTELTHHLLPRNQWIKAQEDVPYLLPYILEAEAAAKEDELDNIEVSK
```

Fig. 2 Homology of QP-C (1. Bovin, 2. Human, 3. Rat, 4. Yeast)

speculate that our QP-C is most probably a Q_i. At any rate the report of the Q cycle greatly strengthens the Q proteins concept as well as the stabilization of Q radicals by protein and *vice versa*. Free ubisemquinone can exist only in a very small amount in mitochchdria. Without Q radical, the promotive force Q cycle can hardly exists. Many of them may remember, just a few months after announcement of Q cycle (13) practically all participants in the Bali symposium in the fall of 1975 spent virtually a whole day vehementally and sarcastically, ridiculed the existence of the Q radicals and in turn the Q cycle. Back then if one spoke of ubiquinone protein, that unfortunate chap would be laughed and looked at as a clown or science-fictionist. Of course, this chap would be kicked out from sacred lecture hall with perhaps a few laps.

There is a question, however, i.e. if this QP-C were sophistically claimed to be *not* a QP-C but a *garden variety* respiratory carrier, then the whole affair may reduce to metaphysics, or sematics or literarily a game of charade--spells c–h–a–r–a–d–e. Under such a situation, I would not have much to say.

Acknowledgement--Acknowledgement is made to Drs. N.S. Reimer, L.Q. Sun, T.Y. Wang, K.T. Yasunobu, Y.Xu in addition to those quoted in references. The work was supported by grants of NIH-HBL-12576, NIH-GM-16767 and American Cancer Society.

References

1. T.E. King, (1980) in M. Koike, T.Nagatsu, T. Okuda and T. Ozawa ed, New Horizons in Biological Chemistry, A Festschrift for Professor K. Ya) Japan Scientific Press, Tokyo, pp 121-134.
2. T.Y. Wang and T.E King (1982), Biochemical And Biophysical Research Communications 104, 591-596.
3. Y.H. Wei, C.P. Scholes, and T.E. King (1981) Biochemical And Biophysical Ressearch Communications, 99, 1411-1419.
4. S. Wakabayashi, T.Takao, Y. Shimonishi, S. Kuramitsu, H. Matsubara, T.Y. Wang, Z.P. Zhang and T.E. King (1985) Journal of Biological Chemistry 260, 337-343.
5. C.A. Yu and L. Yu (1980) Biochemistry 19, 3579-3585
6. S. Yoshida, Z.P. Zhang, and T.E. King (1982) Biochemistry International, 4,1-8.
7. Y. Xu, J.C. Salerns, Y. H. Wei, and T.E. King (1987) Biochemical And Biophysical Research Communications, 144, 315-322, and a manuscrip in preparation.
8. L. Yu and C.A. Yu (1982), Journal of Biological Chemistry 257 10215-10221.
9. L. Yu, F.D. Yang, C.A. Yu and L.Q. Gu (1985), Journal of Biological Chemistry 260, 963-973.

10. L. Yu, S. Usui, and C.A. Yu (1989) Journal of Cell Biology107 Abs. 3543.
11. M. Dehaan, A.P.G.M., Van Loon, J. Kreike, R.T.M.J Ressen, and L.A.Grivell (1984) European Journal of Biochemistry, 138, (169-177)
12. H. Suzuki, H. Yoshitoka, M. Toda, M. Nishikimi, and T. Ozawa (1988) Biochemical and Biophysical Research Communications., 156, 987-994.
13. P. Mitchell (1975), Federation of European Biological Science Letter 56, 1-12.

COMPARATIVE STUDY OF QUINONE BINDING PROTEINS IN MITOCHONDRIAL CYTOCHROME *b*-c_1 COMPLEX AND CHLOROPLAST CYTOCHROME b_6-*f* COMPLEX

Chang-An YU, Shigeyuki USUI, Liang-Bi LI, and Linda YU

Department of Biochemistry, Oklahoma State University
Stillwater, OK 74078

INTRODUCTION

While few investigators will question the essentiality of quinone in mitochondrial or photosynthetic electron transfer, opinions on how quinone mediates electron transfer and what the actual reaction mechanism of the quinone mediated electron transfer reaction is differ greatly. Basically, there are two schools of thought: one school believes that quinone serves as a mobile electron carrier, diffuses freely within the membrane, and shuttles electrons between the electron transfer complexes (Kröger and Klingenberg, 1973; Hackenbrock, *et al.*, 1986); and the other favors the existence of quinone-binding proteins (Yu and Yu, 1981) in electron transfer complexes and believes the actual redox active species of quinone or quinol is the quinone-protein complex. The free form of quinone in the membrane may or may not be in equilibrium with protein-bound quinone.

Perhaps, it should be clarified here that a quinone binding protein is considered to be a protein or a subunit of a multisubunit complex which possesses a specific quinone or quinol binding site.

The existence of QPc in ubiquinol-cytochrome *c* reductase (complex III) was established through the detection of an antimycin-sensitive ubisemiquinone radical (Yu *et al.*, 1978) and identified through photoaffinity labeling studies using arylazido-Q or azido-Q derivatives (Yu, *et al.*, 1985). Two proteins with apparent molecular weights of 37 and 17 kD were found to bind azido-Q. Both proteins have been isolated. The true molecular weights for these two proteins determined by amino acid sequences are 43 and 9.5 kD respectively. The 43 kD protein is a cytochrome *b* (subunit III) protein which has a N-terminal amino acid sequence of IVNNAFIDLP-. The first amino acid, isolecine, corresponds to the 13th amino acid residue of cytochrome *b* protein sequence deduced from gene sequencing (Anderson *et al.*, 1982). The 9.5 kD protein is subunit VII in the Swank and Munkres' gel system (1971) and is shown as subunit VI in the Weber-Osborn gel system (1969). It has an N-terminal amino acid sequence of GRQFGHLTRVRH-. C-terminal analysis by carboxylpeptidase showed an arginine as the C-terminus. This information indicates that isolated QPc is identical to the Mr=9.5 kD protein isolated and sequenced by Borchart *et al.* (1986).

In contrast to mitochondrial and bacterial cytochrome *b*-c_1 complexes, quinone binds only to subunit IV of the cytochrome b_6-*f* complex of spinach chloroplast. Cytochromes b_6 and *f* were not labeled by the azido-Q derivative after illumination. Herein we report the identification of the Q-binding site in the 9.5 kD protein of cytochrome *b*-c_1 complex and in subunit IV of the cytochrome b_6-*f* complex.

Keywords: Quinone binding protein; *b*-c_1 complex; b_6-*f* complex.

METHODS

Cytochrome b-c_1 complex (Yu, *et al.*, 1985), cytochrome b_6-f complex (Doyle, *et al.*, 1989) and azido-Q derivatives were prepared according the reported methods. Azido-Q linked cytochrome b and the 9.5 kD proteins were prepared by the method recently developed (Yu, *et al.*, 1989). SDS-PAGE was carried out under three reported conditions (Weber and Osborn, 1969; Swank and Munkres, 1971; Schägger, *et al.*,1986). Trypsin digestion of cytochrome b_6-f complex was done at 30 °C in 50 mM Tris-succ buffer, pH 7.8, containing 1% sodium cholate and 10% glycerol with a trypsin to cytochrome b_6-f complex ratio of 0.08. Peptide separation was carried out by HPLC using a C-18 column.

RESULTS

Electrophoretic Properties of Cytochrome *b* and the 9.5 kD Protein.

A procedure for the isolation of Q-binding proteins from reductase or azido-Q linked reductase has been recently reported (Yu, *et al.*, 1989). The procedure involves separation of cytochrome b and c_1 fractions by a calcium phosphate column in the presence of 2 M urea and 1.5% Triton X-100 at pH 7.8. The cytochrome b fraction is further separated into QPc-9.5 kD (QPc) and cytochrome b by a calcium phosphate column at pH 6.0. Cytochrome b, core proteins, and subunit VI are recovered in the effluent while the QPc is adsorbed to the column and eluted with 0.3 M KCl in 1% Triton X-100 containing 2 M urea. The eluate is further purified by a molecular sieving column after dissociation in 1% SDS. Cytochrome b is purified from the second calcium phosphate column effluent by Q-Sepharose fast flow column chromatography. Subunit VI is obtained as a by-product by acetone precipitation and 0.5% decanoyl-N-methylglucaminde -0.1 % sodium cholate extraction of the crude cytochrome b fraction. Since the electrophoretic mobility of these subunits depends on the SDS-PAGE condition, their identities have often been confused. To facilitate communication between investigators, we have suggested using internal protein references (Yu and Yu ,1982). Table I summarizes the relative mobilities (in reference to cytochrome c) and the band numbers of isolated cytochrome b and the 9.5 kD protein in various SDS-PAGE systems.

Table I. Relative Electrophoretic Mobility of Cyt. *b* and QPc-9.5 kD

SDS-PAGE	Cyt. *b*		QPc-9.5kD	
	Rel. Mobility	Band No	Rel. Mobility	Band No.
Weber & Osborn	0.472	III	0.841	VI
Swank & Munkres	0.320	III	1.221	VII
Schägger, *et. al.*	0.448	III	1.221	VII

The Mr=13.4 kD Protein is not a Quinone-Binding Protein.

The 13.4 kD protein, which corresponds to subunit VII in the Weber-Osborn system and subunit VI in Swank-Munkres and Schägger *et a.* system, is not labeled by azido-Q. However,this subunit has been erroneously regarded as a Q-binding protein Wang and King, 1982 Borchart *et al*, 1986) in bovine heart mitochondrial cytochrome *b-c* complex. The incorrect assignment of this protein is probably due to the abnormal electrophoretic mobilities of the QPc-9.5 kD protein and the 13.4 kD protein in the Weber-Osborn and Swank-Munkres gel systems. The amino terminal of isolated subunit VI is blocked. The molecular weight of the subunit is determined to be 13.4 kD by protein sequence analysis. Subunit VI obtained by the described method is a protein-phospholipid-Q mixture. The preparation can restore ubiquinol-cytochrome *c* reductase activity to phospholipid- and Q-depleted reductase. However, the reconstitutive activity of the subunit VI preparation is sensitive to phospholipase A_2 but not proteolytic enzyme treatment, indicating that the active component in the preparation is phospholipids and not a protein. The actual function of the 13.4 kD protein remains to be established.

Quinone Binding Site in the QPc-9.5 kD Protein (QPc)

When isolated, azido-Q linked QPc protein is treated with 2% trypsin for 3 hours and then subjected to peptide analysis by HPLC, 3 radioactive peaks are observed. If the digestion time is prolonged to 16 hours, only two radioactive peaks are obtained. Since peak II is not found in the sample digested for 16 hours with trypsin, it must be a larger fragment of another peak. The amino acid analyses of these three radioactivity-containing peaks show that they correspond to residues #48-71, 41-71, and 48-68, indicating that the azido-Q must be bound within residues #48-68. Likely binding sites are tyrosine-59 and threonine-60. The hydroxyl group of these amino acids may hydrogen bond to one of the carbonyl groups of Q molecule, and threonine-63 may provide another hydrogen bond to the other carbonyl group, since the distance between these two amino acids is about the same as that of two carbonyl groups in the benzoquinone ring. The Q-binding site would be near the surface of the matrix side if the reported topological arrangement (Borchart, *et al.*, 1986) is correct. This would fit the general idea of the Q_i site of the Q-cycle mechanism (Mitchell, 1966).

Quinone Binding Site in the Cytochrome b_6-*f* Complex

There are four subunits in the isolated, active cytochrome b_6-*f* complex. Cytochrome *f* with a molecular weight of 34 kD, cytochrome b_6 of 23 kD, iron sulfur protein of 20 kD, and subunit IV of 17 kD. Photoaffinity labeling studies show that quinone is bound to subunit IV. Cytochromes b_6 and *f*, as well as iron sulfur protein, are not involved in Q-binding.

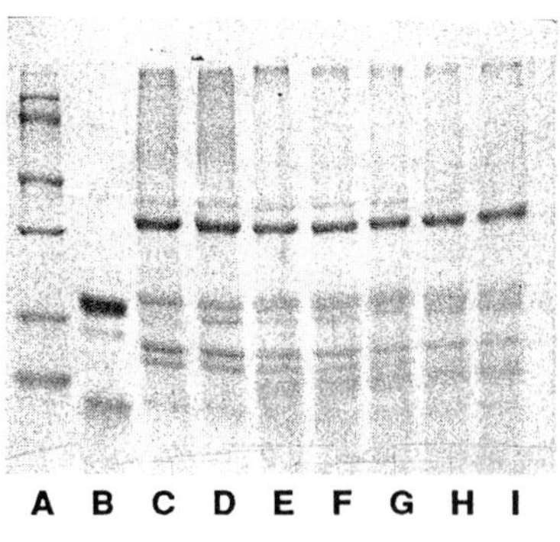

Fig.1. Effect of trypsinolysis on polypeptides of cytochrome b_6-*f* complex. Cytochrome b_6-*f* complex was digested with trypsin for 0 (D), 1 (E), 3 (F), 5 (G), 10 (H), and 15 (I) min at room temperature. The reaction was terminated by the addition of 3 molar excess of trypsin inhibitor. The samples were treated with 1 % SDS and 1% β-mercaptoethanol for 2 hrs at 37°C before being subjected to SDS-PAGE. Lanes A-C show the polypeptide patterns of the molecular weight standard proteins (phosphorylase, 97.4 kD; bovine serum albumin, 66.2 kD; ovalbumin, 42.7 kD; carbonic anhydrase, 31 kD; soybean trypsin inhibitor, 21.5 kD; and lysozyme, 14.4 kD), trypsin inhibitor, and intact cytochrome b_6-*f* complex, respectively.

The amount of azido-Q uptake by subunit IV decreases in the presence of lipids, DBMIB, UHDBT, and PQ_2H_2. This is different from the mitochondrial system. Apparently, the quinol binding sites of these two systems are not the same. Since only one protein is labeled by azido-Q, it is likely that only one Q-binding site is involved in the cytochrome b_6-*f* complex. Since the main function of the complex is to oxidize quinol, the binding site is likely to be the quinol oxidation site. The quinone reduction site predicted by the Q-cycle mechanism may either be missing or provided by another system such as ferredoxin-Q reductase as proposed by Dr. Cramer (1987).

Subunit IV, the 17 kD protein, is cleaved to a smaller fragment of about 13 to 14 kD by trypsinolysis. The trypsin inactivated sample, which still binds azido-Q upon illumination but with a slightly lower affinity, has absorption properties identical to those of the active sample, suggesting that cytochromes b_6 and *f* are not altered. The epr characteristics of the iron sulfur cluster of the trypsin-treated complex are also similar to those of the untreated sample, indicating that the iron sulfur protein is not altered. Fig. 1 shows that subunit IV is the only protein digested by trypsin. From the published amino acid sequence of the 17 kD protein (Heinemeyer *et al.*, 1984) it is clear that the only place that can be cleaved by trypsin to generate a fragment with a molecular weight of around 14 kD would be at lysine-119 or argine-125 or 126. The possible quinone-binding site has to be located before lysine-119.

To locate the trypsin cleavage site in relation to the side of the thylakoid membrane, the native and the inverted (sonicated) membranes are subjected to trypsinolysis followed by the inactivation of plastoquinol-cytochrome *c* reductase activity. Results are depicted in Fig. 2. Trypsinolysis has little inactivation effect on the native membrane but shows a significant effect on the inverted membrane, suggesting that the trypsin cleaving site is located on the lumen side of the thylakoid membrane. According to the reported amino acid sequence, the 17kD protein has three trans-membrane helices. Since the suspected trypsin cleavage site at arginine-25 or -26 and/or lysine-119 is a part of the loop which connects trans-membrane helix-II and -III, the loop must be on the lumen side. The C-terminus of the protein and the loop connecting trans-membrane helix-I and -II must be on the stroma side.

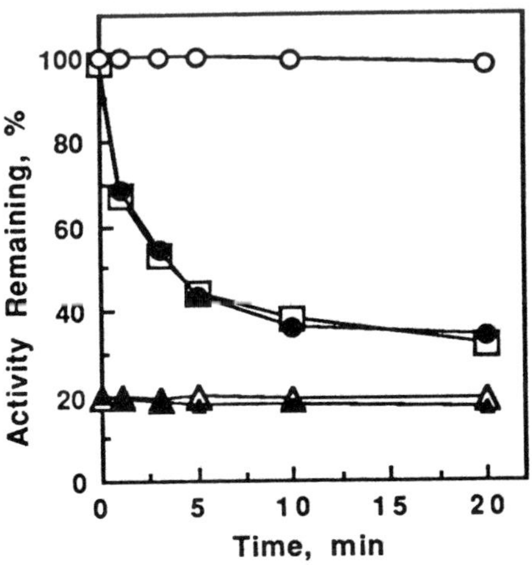

Fig. 2. Effect of trypsinolysis on the plastoquinol-cytochrome *c* reductase activity of thylakoid membranes. Thylakoid membranes as prepared (Δ, ▲), sonicated (o, •) and detergent dispersed (□) were incubated at room temperature in the presence (▲, •,□) and absence (o, Δ) of trypypsin. At the indicated times, the samples were withdrawn, mixed with trypsin inhibitor, and the activity determined after treatment with 0.5% DMG. The 100% activity represents 8.5 μmoles cytochrome *c* reduced per nmoles cytochrome *f* per hour.

If there is indeed only one Q-binding site in the cytochrome b_6-*f* complex, it would be rather difficult to explain the oxidation of plastoquinol by "Q-cycle" mechanism without some modification. Perhaps the half Q-cycle described by Cramer and his coworkers (1987) is worth exploring in the future. When cytochrome b_6-*f* complex is incorporated into phospholipid vesicles and subjected to reduction by membrane-impermeable reductants, a biphasic reduction behavior is observed; one half of cytochrome b_6 is rapidly reduced, while the other half is slowly reduced. If a lipophilic redox compound such as methyl viologen is included in the vesicles, the reduction of cytochrome b_6 becomes monophasic, suggesting that the electron transfer between the two *b* cytochromes is either very slow or does not take place. The lack of rapid electron transfer between the two cytochromes b_6 is certainly not in accord with the normal Q-cycle mechanism. Fig. 3 compares the reduction behavior of cytochrome *b* in cytochrome *b*-c_1 and b_6-*f* complexes incorporated into phospholipid vesicles using sodium dithionite as a reductant.

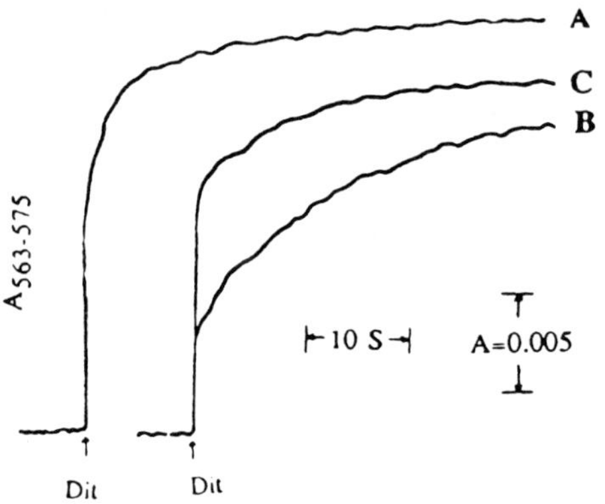

Fig. 3. Reduction of cytochrome *b* of cytochrome *b*-c_1 and b_6-*f* complexes imbedded in phospholipid vesicles. Curves A, B, and C represent cytochorme *b*-c_1 complex and b_6-*f* complex in the absence and presence of methyl viologen. Reductions are carried out by the addition of 10 mM sodium dithionite under argon atmosphere

CONCLUSION

Two proteins, cytochrome *b* and a 9.5 kD subunit, in bovine heart mitochondrial cytochrome *b*-c_1 complex are involved in quinone or quinol binding. Only one protein subunit (subunit IV) with a molecular weight of 17 kD in the cytochrome b_6-*f* complex of spinach is involved in quinone binding. Subunit IV is catalytically essential. The quinone binding site must be located before lysine-119, as removal of a part of peptide by trypsin from the C-terminal after lysine-119 causes a loss of enzyme activity, but not Q-binding. The fact that only one quinone-binding site is present in the cytochrome b_6-*f* complex, and the slow electron transfer between the two cytochrome b_6 suggests that the quinol oxidation mechanism in this complex may not be identical to that in the cytochrome *b*-c_1 complex.

ACKNOWLEDGEMENTS

This work is supported in part by grants GM30721 from the National Institutes of Health and 87-CRCR-1-2433 from the United States Department of Agriculture.

REFERENCES;

Anderson, S., de Bruijn, M. H. L., Coulson, A. R., Eperon, I. C., Sanger, F., and Young, I. G., 1982, Complete sequence of bovine mitochondrial DNA-- conserved features of the mitochondrial genome. Journal of Molecular Biology, 156, 683-717.

Borchart, U., Machleidt, W., Schägger, H., Link, T.A., and von Jagow, G., 1986, Isolation and amino acid sequence of the 9.5 kD protein of beef heart ubiquinol:cytochrome *c* reductase. FEBS Letters, 200, 81-86.

Cramer, W.A., Black, M.F., Widger, W.R., and Girvin, M.E. ,1987, Structure and function of photosynthetic cytochrome b-c_1 and b_6-f complexes. In: The Light Reactions, Amsterdam, edited by J. Barber, (Amsterdam: Elsevier Sci. Publ.), pp. 448-493.

Doyle, M.P., Li, L-B., Yu, L., and Yu, C.A., 1989, Identification of a Mr=17,000 protein as the plastoquinone-binding protein in the cytochrome b_6-f complex from spinach chloroplasts. Journal of Biological Chemistry, 264, 1387-1392.

Hackenbrock, C.R., Chazotte, B. and Gupte, S.S., 1986, The random collosion model and a critical assessment of diffusion and collision in mitochondrial electron transport. Journal of Bioenergetics and Biomembranes, 18, 331-368.

Heinemeyer, W., Alt, J. and Herrmann, R.C., 1984, Nucleotide sequence of the cluster genes for apocytochrome b_6 and subunit 4 of the cytochrome *b*/*f* complex in the spinach chromosome. Current Genetics, 8, 543-549.

Kröger, A. and Klingenberg, M., 1973, The kinetics of the redox reactions of ubiquinone related to the electron-transport activity in the respiratory chain. European Journal of Biochemistry, 34, 358-368.

Mitchell, P., 1976, Possible molecular mechanisms of the proton motive function of cytochrome systems. Journal of Theoretical Biology, 62, 327-367.

Schägger, H. , Link, T.A., Engel, W.D., and von Jagow, G., 1986, Isolation of eleven protein subunits of the b-c_1 complex from beef heart. Methods in Enzymology, 126, 224-237.

Swank, R.T., and Munkres, K.D., 1971, Molecular weight analysis of oligopeptides by electrophoresis in polyacrylamide gel with sodium sulfate. Analytical Biochemistry, 39, 462-477.

Wang, T.Y., and King, T.E., 1982, Isolation of QP-C and reconstitution of the QH_2-c reductase. Biochemical and Biophysical Research Communication, 104, 591-596.

Weber, K. and Osborn, M., 1969, The reliability of molecular weight determinations by dodecyl sulfate gel electrophoresis. Journal of Biological Chemistry, 244, 4406-4412.

Yu, C.-A., Nagaoka, S., Yu, L. and King, T.E., 1978, Evidence for the existence of a ubiquinone protein and its radical in the cytochromes b and c_1 region in the mitochondrial electron transport chain. Biochemical and Biophysical Research Communication, 82, 1070-1078.

Yu, C.-A. and Yu, L., 1981, Ubiquinone protein. Biochimica et Biophysica Acta (Review), 639, 99-128.

Yu, C.-A. and Yu, L., 1982, Ubiquinone binding protein in the cytochrome b-c_1 complex. In: Function of Quinones in Energy Conserving Systems, edited by B.L. Trumpower, New York: Academic Press, Chapter 7.

Yu, C.-A, Usui, S., and Yu, L., 1989, Ubiquinone binding protein in mitochondrial cytochrome b-c_1 complex. In: Highlights of Modern Biochemistry, Zeist, edited by A. Kotyk, J. Skoda, V. Paces and V. Kostka, (Zeist: VSP International Science Publishers), pp. 883-892.

Yu, L., Yang, F.D., and Yu, C.-A.,1985, Interaction and identification of ubiquinone-binding proteins in ubiquinol-cytochrome c reductase by azido ubiquinone derivatives. Journal of Biological Chemistry, 260, 963-973.

ROLE OF DIFFUSION IN MITOCHONDRIAL ELECTRON TRANSFER

G. Lenaz, R. Fato, M. Battino°, M. Cavazzoni, H. Rauchova, C. Castelluccio*, J. Svobodova and G. Parenti-Castelli*.

Department of Biology, University of Bologna, 40126 Bologna
*Dept. of Biochemistry, University of Bologna,40126 Bologna
°Inst. of Biochemistry, University of Ancona, 60100 Ancona
ITALY

The collisional nature of electron transfer in the mitochondrial respiratory chain is well documented (Lenaz, 1988), and both Coenzyme Q (Q) and cytochrome c (c) behave as free diffusional intermediates, although Q appears located in the hydrophobic core of the membrane and c in the intermembrane acqueous space.

The substrate-like nature of Q for its partner enzymes allows to ask whether its concentration is saturating and also whether its diffusion rate is limiting for the overall electron transfer process.

Coenzyme Q concentration is not saturating electron transfer

Under nonsaturating conditions, the rate of electron transfer across the Q-pool is a function of total Q concentration and of the maximal velocities and K_m for ubiquinone/ol of Q reductases and oxidase (Lenaz and Fato, 1986) (Table I). At physiological Q concentration, the computed NADH-c reductase activity is close to the experimental values and ca. 50% of V_{max}.

Experimentally, saturation of electron transfer is difficult to attain. We have increased the phospholipid content of rat liver mitochondria by fusing mitochondria with liposomes containing different levels of Q_{10}. The fall of NADH-c reductase observed adding only phospholipids was avoided by incorporating phospholipids plus Q; enrichment with Q levels higher than the physiological Q concentration enhanced activity above the control level (Lenaz and Fato,

Table I. Average values of the kinetic constants of Q reductases and ubiquinol oxidase in mitochondria and calculated values for Q pool function.

Activity	Vmax (nmol/s.mg)	Activity(nmol/s.mg) at normal Q	$Km(Q_{10})$ (mM in PL)
NADH - Q	45		10
succinate - Q	17		5
QH_2 - c	78		3.3
NADH - c (computed)	28.5	14.7	7.5
succinate - c (")	13.8	9.0	4.1

1986); theoretical V_{max} could not be achieved because the miscibility of long chain Qs with lipids is limited and at high Q/lipid ratios separate phases are found. Short chain Q homologs are more miscible but they inhibit NADH oxidation (Lenaz et al., 1975). Nevertheless we have been able to saturate α-glycerophospate c reductase in hamster brown adipose tissue mitochondria by adding Q_3 in vitro (Fig. 1). The strong stimulation by Q indicates that the

Fig. 1. Effect of added Q_3 on α-glycerophosphate cyt. c reductase of hamster brown adipose tissue mitochondria.

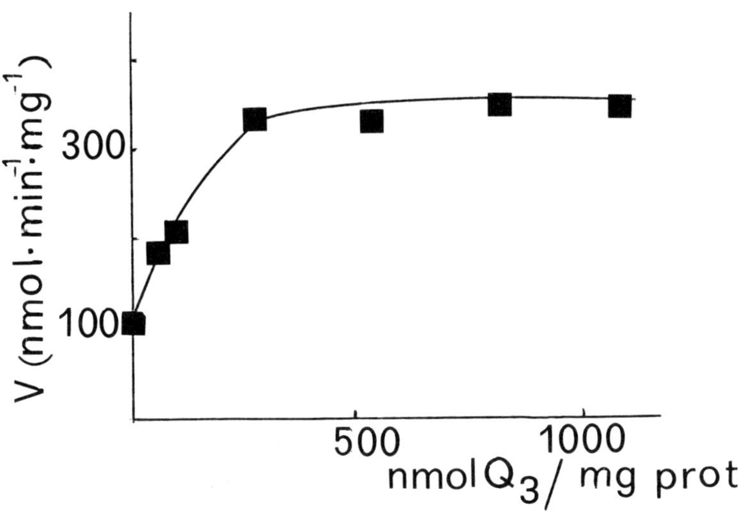

low levels of endogenous Q are far from saturating.

An important consequence emerging from these studies is that variations of Q content induce variations of electron transfer activity, at least under the conditions used experimentally. Whether these conditions reflect the in vivo situation (under coupling control in intact organelles) is unknown. Even if other control sites exist (cytochrome oxidase etc.) it may be shown from Eq. 18 of Lenaz and Fato, 1986, that at any value for ubiquinol oxidase (comprising any step below Q pool), the rate of electron transfer is a function of Q.

It is evident that this observation has deep pathological and therapeutic implications.

Coenzyme Q diffusion is not rate-limiting.

The random collision model of Hackenbrock et al., 1986 esplicitly postulates that electron transfer is a diffusion controlled process.

The lateral diffusion coefficients of Q, measured by fluorescence photobleaching are 10^{-9} cm^2/s (Hackenbrock et al., 1986) whereas those measured by fluorescence quenching are $10^{-7}cm^2/s$ (Lenaz, 1988). Calculation of the time to cover the distance between respiratory enzymes suggests that the rate of electron transfer is at the diffusion limit only at the lower coefficients. In view of this uncertainty, we have approached the problem by enzyme kinetics.

In a set of experiments we have titrated NADH-c reductase of a reconstituted fraction containing complexes I and III at different phospholipids to protein ratios corresponding to different intercomplex distances (Lenaz, 1988). The turnovers of NADH-c reductase are not affected by the distance between complexes within an experimental limit of 108 nm.

In a systematic approach to the question of the role of diffusion in mitochondrial electron transfer, we have investigated the kinetics of ubiquinol c reductase. The enzyme has a two-site ping-pong mechanism; under experimentally controlled conditions the mechanism can be simplified as in Table II; note the kmin(1) is indipendent of any rate constant at the c site, and kmin(5) is indipendent of any constant at the ubiquinol site. The kmin

Table II. Kinetic scheme of ubiquinol cyt. c reductase.

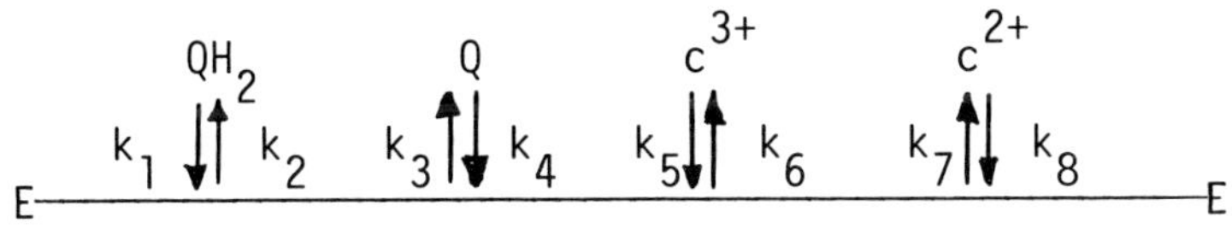

$$\frac{Kcat}{Km(QH_2)}=kmin(1)=\frac{k_1 \cdot k_3}{k_2 + k_3} \qquad \frac{Kcat}{Km(c)}=kmin(5)=\frac{k_5 \cdot k_7}{k_6 + k_7}$$

---

are of the same order of magnitude of the true second order constants k_1 and k_5, and can therefore be used for determination of a diffusion-limited step.

The rate constants for ubiquinols, as determined, are apparent rate constants, because they are expressed as concentrations in total assay medium, whereas ubiquinol concentrations in the membrane depend on their partition coefficients. True second order kmin are obtained from Km by direct kinetic determinations (Engel et al., 1980; Fato et al., 1988) and are in the range of 10^4 $M^{-1}s^{-1}$, approaching the value for ubiquinol-10 in bacterial chromatophores (Snozzi and Crofts, 1984).

It was shown by comparison of soluble and membranous bc_1 complex that the Km for ubiquinol is lower and that for c is higher in the latter enzyme (Degli Esposti and Lenaz, 1982); this is a property expected to rise from a diffusional constraint for c in the membranous enzyme.

The Arrhenius plots of soluble bc_1 complex are linear, whereas those of the membranous enzyme are biphasic with a break at 20°C. The plots are usually built by log of apparent V_{max} from ubiquinol titrations at fixed c concentrations,

---

Table III. Activation energies from Arrhenius plots of ubiquinol c reductase in SMP by ubiquinol-2 titrations at 50 μM cyt. c and cyt. c titrations at 15 μM ubiquinol-2. E_a are expressed in Kcal/mol.

	$E_a(k_{cat}/K_m)$	$E_a(V_{max})$	$E_a(K_m)$
ubiquinol-2	5.9	11.0	6.5
cyt. c	1.4	15.8	21.6

Table IV. Diffusion and reaction limited components of the association rate constants of ubiquinol-1 and ubiquinol-2 cyt. c reductase in SMP.

Parameter	Ubiquinol-1	Cyt. c	Ubiquinol-2	Cyt. c
kmin($\mu M^{-1}s^{-1}$)	6.8	32	110	33
kc($\mu M^{-1}s^{-1}$)	8	200	120	69
kdiff($\mu M^{-1}s^{-1}$)	67	38	880	64
diffusion-limited	11%	84%	12%	52%

kept relatively low because of substrate inhibition. Increasing c concentration to 50 μM, however, the plots became linear. We have then performed ubiquinol and c titrations at different temperatures and plotted the kinetic constants in form of Arrhenius and Van't Hoff plots (Table III); the temperature dependence of the Km for c was higher than that for ubiquinol. The large increase of the Km for c with temperature is therefore the reason for the biphasic Arrhenius plot when the c concentration is low.

The reason for this high Km increase may be that c diffusion is rate limiting. This suggestion is supported by the low E_a of kmin(5) compared with kmin(1). This is a strong indication that c diffusion and not ubiquinol diffusion is rate limiting.

A confirmation arises from viscosity dependence studies. We have varied the viscosity of the medium by polyethylene glycol and titrated ubiquinol c reductase with either ubiquinol-1 or -2 and c: viscosity dependence of kmin(1) was low whereas that of kmin(5) was high. By calculating the extent of diffusion control (Table IV) the reaction appears largely limited by c diffusion. We have also changed the viscosity of the membrane by cholesterol incorporation; under conditions where diffusion of Q is strongly depressed, kmin(1) is not affected.

In conclusion, electron transfer in mitochondria is a diffusion coupled process, where the concentration of Q appears to be a rate limiting step, whereas the rate of lateral diffusion appears in excess to support electron transfer maximal rates.

ACKNOWLEDGMENTS

The studies were supported by grants from CNR and Ministero della Pubblica Istruzione. Dr. Rauchova from the Czechoslovak Academy of Sciences was a recipient of a FEBS fellowship. Dr. Svobodova from Charles University, Prague, was supported by a fellowship from the Italian government. Coenzymes Q were kindly donated by Eisai Co., Tokyo, Japan.

REFERENCES

Degli Esposti M. and Lenaz G., 1982, Kinetics of ubiquinol-1 cytochrome c reductase in bovine heart mitochondria and submitochondrial particles. Biochimica Biophysica Acta, 682, 189-200.

Engel W. D., Schagger H. and Von Jagow G., 1980, Ubiquinol-cytochrome c reductase (EC1.10.22). Isolation in triton X-100 by hydroxyapatite and gel-chromatography. Structural and functional properties. Biochimica Biophysica Acta, 592, 211-222.

Fato R., Castelluccio C., Palmer G. and Lenaz G., 1988, A simple method for the determination of the kinetic constants of membrane enzymes utilizing hydrophobic substrates: ubiquinol cytochrome c reductase. Biochimica Biophysica Acta, 932, 216-222.

Hackenbrock C.R., Chazotte B. and Gupte S.S., 1986, Ubiquinone diffusion is rate limiting for electron transfer. Journal of Bioenergetics and Biomembranes, 18, 331-368.

Lenaz G., Pasquali P., Bertoli E., Parenti-Castelli G. and Folkers K., 1975, The inhibition of NADH oxidase by the lower homologs of coenzyme Q. Archives of Biochemistry and Biophysics, 169, 217-226.

Lenaz G. and Fato R., 1986, Is Ubiquinone diffusion rate-limiting for electron transfer? Journal of Bioenergetics and Biomembranes, 18, 369-401.

Lenaz G., 1988, Role of mobility of redox components in the inner mitochondrial membrane. Journal of Membrane Biology 104, 193-209.

Snozzi M. and Crofts A.R., 1984, Electron transport in chromatophores from Rhodopseudomonas Spheroides GA fused with liposomes. Biochimica Biophysica Acta, 766, 451-463.

THE MEASUREMENT OF THE LATERAL DIFFUSION PARAMETERS OF UBIQUINONES IN MODEL MEMBRANES BY TIME RESOLVED FLUORESCENCE

L. Masotti, L. Landi*, G. Sartor, A.G. Szabo**, D.T. Krajcarski**, J. Ridgeway** and M. Zuker**

Istituto di Chimica Biologica, Università degli Studi di Parma, Via Gramsci, 14, 43100 Parma, Italy, *Dipartimento di Biochimica, Università degli Studi di Bologna, Via Irnerio,48, 40100 Bologna, Italy and **Division of Biological Sciences of the National Research Council of Canada, Ottawa, K1A OR6 Ontario, Canada.

INTRODUCTION

The knolewdge of the localization and diffusion rates of lipophilyc molecules in membranes is of paramount importance for the definition of their role in complex multienzymatic processes.

Coenzyme Q is known to be a mobile electron carrier in the mitochondrial respiratory chain.

It is the aim of this work to define its localization and its diffusion rate in model systems using fluorescence quenching techniques. The fluorescence quenching constants of a variety of membrane probes that sense the membrane environment at different depths were determined and time resolved fluorescence was chosen in order to discriminate between static and dynamic (diffusion controlled) quenching.

Keywords: Coenzyme Q, fluorescence quenching, fluorescent probes, diffusion coefficients, model membranes

MATERIAL AND METHODS

The membrane fluorescent probes used were: 1,6-diphenylhexatriene (DPH) and perylene obtained from Aldrich, FRG, trimethylammonium DPH (TMA-DPH) and phosphatidylcholine DPH (DPH-PC) obtained from Molecular Probes, OR, USA. The probes were used without further purification. Phospholipids were obtained from Sigma, MO, USA. All other chemicals were analytical grade.

Model membranes were prepared by sonication of aqueous dispersions of dimiristoylphosphatidylcholine (DMPC), and egg lecithin in HEPES buffer, pH 7.2. Fluorescence quenching experiments of DPH-PC and TMA-DPH were carried out both below and above the phase transition temperature of DMPC.

Time resolved fluorescence experiments were carried out using a single photon counter fluorometer that uses as a pulsed source either an argon laser (Spectra Physics), with a half height width of 15 ps, or a nanosecond flash lamp (Edimburgh), with a half height width of 1.5 ns. The decay curves were fitted by using both the Grinvald and Steinberg (1974) and the global analysis (Knutson *et al.* 1983) approach.

RESULTS AND DISCUSSION

The probes used exhibited different efficiencies and mechanisms of quenching.

The long lifetime component of DPH was collisionally quenched (Fig. 1) with low efficiency, while the short one was quenched with a more complex mechanism, which can be explained by the change in the bilayer fluidity induced by the quencher, and by the consequent redistribution of the probe within the membrane (Landi *et al.* 1989).

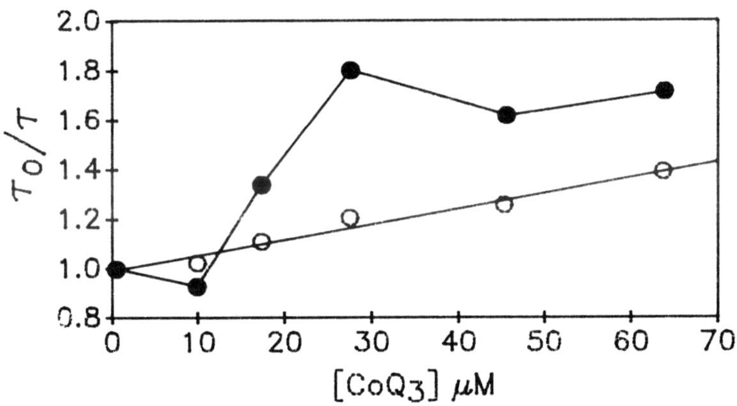

Figure 1 Stern Volmer plot of DPH fluorescence quenching by CoQ_3 in egg lecithin vesicles at 25°C. Long component (o); short component (•).

TMA-DPH showed very little quenching (data not shown).

Perylene fluorescence decay in the absence of quencher was described by a single exponential function; in the presence of CoQ_n a double exponential function was necessary to fit the decay. One component seemed to be collisionally quenched while the other one remained unquenched (6.8 ns) (Fig. 2).

DPH-PC was collisionally quenched both by CoQ_{10} and CoQ_3, above (Fig.3) and below the phase transition temperature of DMPC. The quenching efficiency was lower at 15°C than at 35°C, and in the presence of CoQ_{10} than CoQ_3.

From the results reported above CoQ homologues used seem to be localized in the inner core of the membrane with the quinone moiety, that can be assumed to be the real quencher, close to the probe.

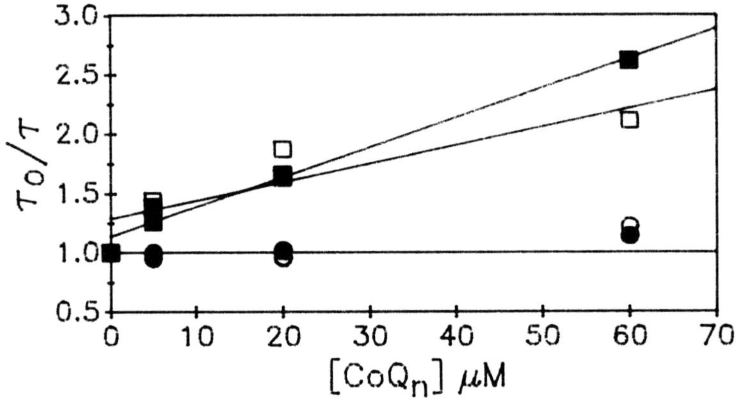

Figure 2 Stern Volmer plot of perylene fluorescence quenching by CoQ_3 (open symbols) and CoQ_{10} (filled symbols) in egg lecithin vesicles at 25°C. Long component (o), short component (□).

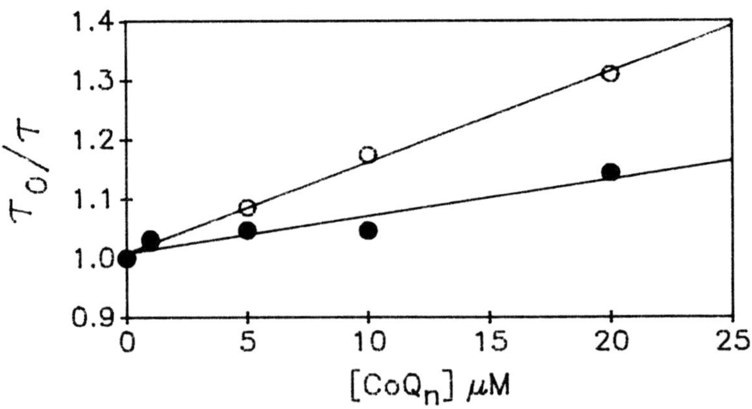

Figure 3 Stern Volmer plot of DPH-PC fluorescence quenching by CoQ_3 (o) and CoQ_{10} (•) in DMPC vesicles at 35°C.

From the slope of the Stern Volmer plots of the fluorescence quenching of DPH-PC in DMPC at 35°C the diffusion coefficient of both CoQ_3 and CoQ_{10} was determined according to Blackwell *et al.* (1987). We calculated the concentrations of quinones according to Ragan (1978) and Fato *et al.* (1986) and we found the diffusion coefficient (D) for CoQ_3 to be equal to $8.5\ 10^{-7}\ cm^2 s^{-1}$ while for CoQ_{10} D was equal to $3.3\ 10^{-7}\ cm^2 s^{-1}$. We also tried to fit our data using a non linear diffusion model based on the dependence of D on the square root of time, but the results were not satisfactory.

REFERENCES

Grinvald, A. and Steimberg, I. Z., 1974, The analysis of fluorescence decay kinetics by the method of least-squares, *Analytical Biochemistry*, 59, 583-598.

Knutson, J. R., Beechem, J. M. and Brand L., 1983, Simultaneous analysis of multiple fluorescence decay curves: a global approach, *Chemical Physics Letters*, 102, 501-507.

Landi, L., Cabrini, L., Fiorentini, D., Sechi, A.M., Sartor, G., Pasquali, P. and Masotti, L., 1989, Coenzyme Q-3 as an antioxidant: its effect on the composition and structural properties of phospholipid vesicles, *Cell Biophysics*, in press.

Ragan, C.I., 1978, The role of phospholipids in the reduction of ubiquinone analogues by the mitochondrial reduced nicotinamide-adenine dinucleotide-ubiquinone oxidoreductase complex, *Biochemical Journal*, 172, 539-547.

Fato, R., Battino, M., Degli Esposti, M., Parenti Castelli, G. and Lenaz, G., 1986, Determination of partition and lateral diffusion coefficients of ubiquinones by fluorescence quenching of n-(9-anthroyoxy) stearic acids in phospholipid vesicles and mitochondrial membranes, *Biochemistry*, 25, 3378-3390.

Blackwell, M.F., Gounaris, K., Zara, J.S.and Barber J., 1987, A method for estimating lateral diffusion coefficients in membranes from steady-state fluorescence quenching studies, *Biophysical Journal*, 51, 735-744.

AKNOWLEDGEMENTS

This work was supported by Grants from MPI Rome (LM).

CHARACTERIZATION OF UBIQUINONE AND ITS ANIONS BY ELECTROCHEMISTRY AND FTIR/UV/VIS SPECTROSCOPY: MODELS FOR QUINONES INVOLVED IN BIOLOGICAL ELECTRON TRANSFER

M. Bauscher*, E. Nabedryk+, J. Breton+ and W.G. Mäntele*

*Inst. f. Biophysik. u. Strahlenbiol., Albertstr. 23, D-7800 Freiburg FRG, +Service de Biophys., Dépt. de Biol., CEN Saclay, 91191 Gif-sur-Yvette Cedex, France

INTRODUCTION:

In the primary reactions of photosynthesis, quinones function as electron acceptors in a variety of pigment-protein complexes. In order to perform this function, a specific protein environment and binding is necessary, which allows a quinone to become an anionic semiquinone ($Q^{\cdot-}$), a dianion (Q^{2-}), or a hydroquinone with changed binding properties. In order to gain information on this interaction, Fourier transform infrared (FTIR) spectra of "model" quinones are compared with light-induced (P^+Q^-) spectra of photosynthetic reaction centres, which reflect the molecular difference between the neutral and the charge-separated state (Mäntele et al., 1985). A variety of solvents can serve to simulate a specific environment of the quinones in the protein. A comparison of bacteriochlorophyll and bacteriopheophytin spectra with P^+Q^- spectra was performed (Mäntele et al., 1988).

Keywords: Photosynthesis, Quinones, FTIR Spectroscopy

APPARATUS

For the in situ generation of quinone anions, a novel thin-layer electrochemical cell was developed.

Features of the cell are:

A high spectral transmission between 200 nm and 10000 nm, a small optical path-length of 60 µm, resulting in a fast equilibrium between the volume in the measuring beam and the electrode surface, a very small sample volume of about 10 µl, the possibility to work strictly anaerobic, a high tolerance to organic solvents, a stable and well-defined potential by the use of an external reference electrode.

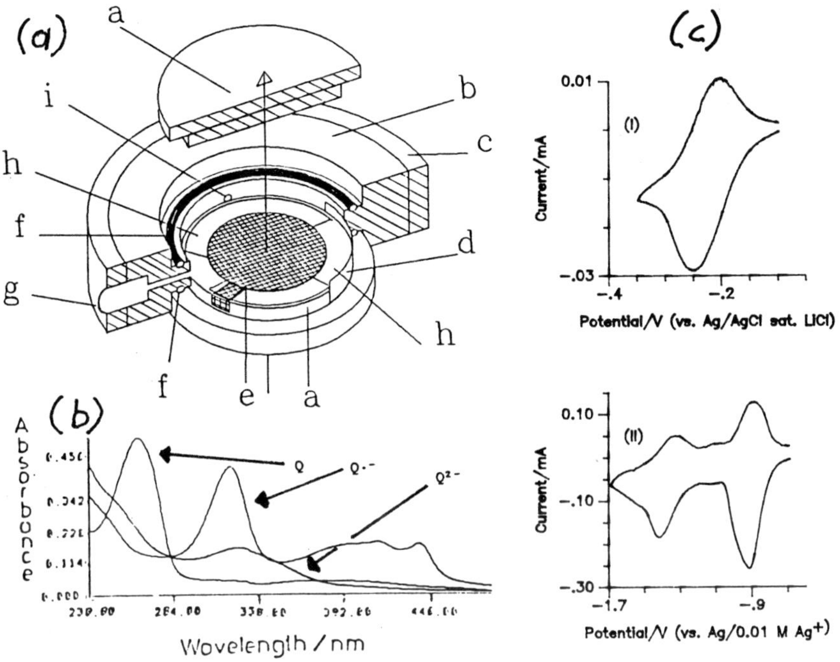

Fig. 1. (a) Spectroelectrochemical cell: a, CaF_2-windows; b, Teflon body; c, Steel ring; d, Counter electrode (Pt foil); e, Working electrode (Au grid); f, O-ring; g, Filling port; h, Teflon spacer; i, Salt bridge to the reference electrode (60 µm). (b) UV/VIS-spectra of 2,3-dimethoxy-5-methyl-1,4-benzoquinone (Q) in acetonitrile (MeCN) during a constant potential electrolysis at −1.1 V ($Q^{\cdot-}$) and −1.6 V (Q^{2-}) (vs. Ag/0.01 M Ag^+). (c) Cyclic voltammograms of Q in methanol (MeOH) (I) and MeCN (II)

RESULTS AND DISCUSSION

UV/VIS spectra and cyclic voltammograms:

The UV/VIS-spectra of Q, (fig. 1) demonstrate, by the shift of the UV-band, the formation of $Q^{\cdot-}$ (Bensasson and Land, 1973) and at a more negative potential, the occurence of Q^{2-}. Stable isosbestic points and a high reversibility show that no side reaction had occured. The cyclic voltammograms gave two waves with E_m=−0.92 V and E_m=−1.36 V (vs. Ag/0.01 M Ag^+) for $Q^{\cdot-}$ and Q^{2-} resp. in MeCN and one wave with E_m=−0.224 V (vs. Ag/AgCl sat. LiCl) in MeOH.

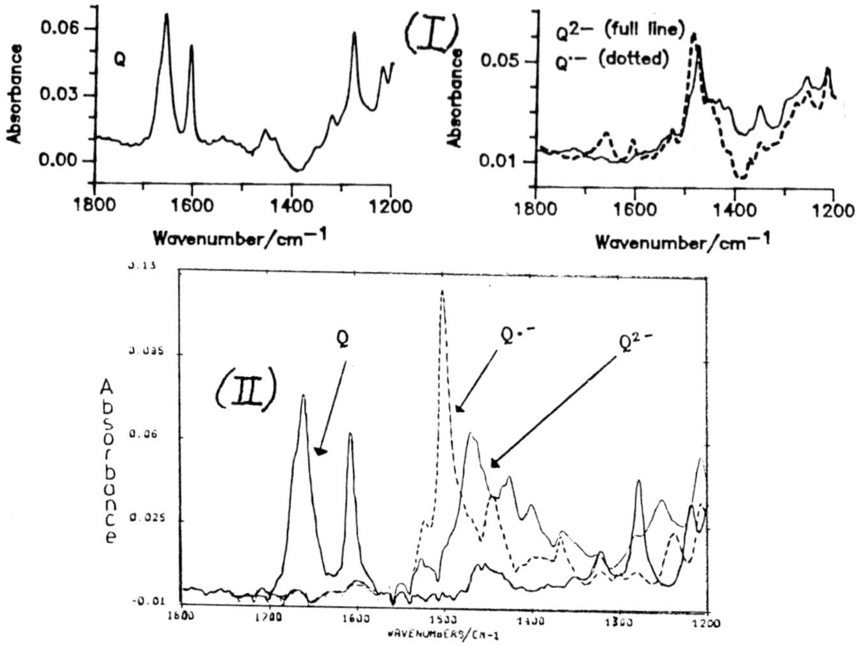

Fig. 2. (a) FTIR absorbance spectra of Q, $Q^{\cdot-}$, Q^{2-} during a constant potential electrolysis at −0.4 V (vs. Ag/AgCl sat. LiCl) in D_4−MeOH (I) and at −1.1 V ($Q^{\cdot-}$), −1.6 V (Q^{2-}) (vs. Ag/0.01 M Ag^+) in D_3−MeCN (II)

FTIR spectra of model compounds:

A pronounced difference between the FTIR spectra of Q in both solvents studied, can only be seen in case of the position of the anion bands. The spectrum (fig. 2 II), recorded at −1.1 V, in D_3−MeCN, shows a decrease of the 1660 cm^{-1} (C=O vibration) and 1608 cm^{-1} (C=C vibration) bands of the neutral quinone and an increase of a band at 1500 cm^{-1} (C⩦O vibration), which is in good agreement with the data for $Q^{\cdot-}$ obtained by Clark and Evans (1976). At −1.6 V 3 smaller bands at 1473 cm^{-1}, 1427 cm^{-1} and 1400 cm^{-1} appeared, which are typical for Q^{2-}. Clark and Evans (1976) reported no prominent bands for Q^{2-} in DMSO. The situation in D_4−MeOH is different in that $Q^{\cdot-}$ and Q^{2-} are formed subsequently at the same potential (fig. 2 I), this is consistent with CV (fig. 1)− and UV/VIS data (not shown). At first a band at 1490 cm^{-1} ($Q^{\cdot-}$, C⩦O vibration) appeared and then a new band at 1476 cm^{-1} (Q^{2-}) increased while the other one disappeared.

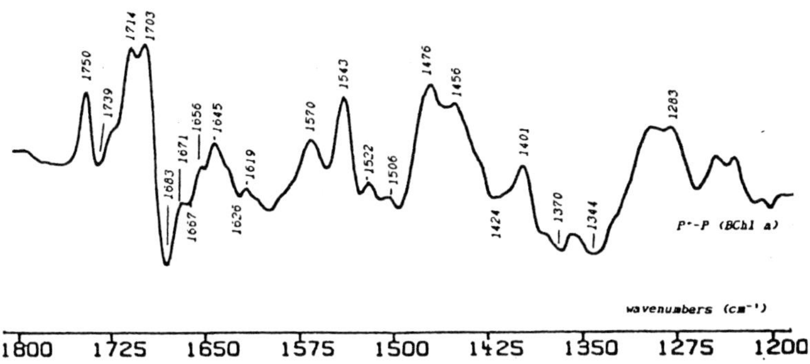

Fig. 3. Light-induced FTIR difference spectrum of Rb. sphaeroides reaction centres

FTIR spectra of reaction centres:

The bands at 1750 cm^{-1}, 1714 cm^{-1}, 1703 cm^{-1} and 1683 cm^{-1} of the $P^{+}Q^{-}$ spectrum (fig. 3) have been assigned to primary donor C=O groups (Mäntele et al., 1988). Taking the quinone "model" spectra, the C=O band at 1667 cm^{-1}, which can be assigned to quinone reduction, is remarkable small. However, the intense bands at 1476 cm^{-1} and 1456 cm^{-1}, absent in primary donor "model" spectra, are typical for a ubiquinone anion.

REFERENCES

Bensasson, R. and Land, E.J., 1973, Optical and kinetic properties of semireduced plastoquinone and ubiquinone: Electron acceptors in photosynthesis. Biochimica et Biophysica Acta, 325, 175-181.

Clark, B.R. and Evans, D.H., 1976, Infrared Studies of Quinone Radical Anions and Dianions generated by Flow-Cell Electrolysis. Journal of Electroanalytical Chemistry, 69, 181-194.

Mäntele, W., Nabedryk, E., Tavitian, B.A., Kreutz, W. and Breton, J., 1985, Light-induced fourier transform infrared (FTIR) spectroscopic investigations of the primary donor oxidation in bacterial photosynthesis. FEBS Letters, 187, 227-232.

Mäntele, W., Wollenweber, A., Nabedryk, E. and Breton, J., 1988, Infrared spectroelectrochemistry of bacteriochlorophylls and bacteriopheophytins: Implications for the binding of the pigments in the reaction center from photosynthetic bacteria. Proceedings of the Natural Acadademy of Science USA, 85, 8468-8472.

STRUCTURAL STUDIES ON THE LOCALIZATION OF UBIQUINONE IN LIPID MEMBRANES BY DIFFRACTION TECHNIQUES

Albertini G.°, Battino M.^, Bertoli E.^, Curatola G.^, Lenaz G.*, Mariani P.", Rustichelli F.", Fahmy T.§ and Zolese G.^

(°) Dip. Scienze Materiali e Terra, Università di Ancona (Italy); (^) Ist. Biochimica, Facoltà di Medicina, Università di Ancona (Italy); (*) Dip. Biologia, Università di Bologna (Italy); (") Ist. Fisica Medica, Università di Ancona (Italy; (§) Inst. Physikal. Chemie, Universität Mainz (F.R.G.).

INTRODUCTION

The Coenzyme Q_{10} (or ubiquinone-10, CoQ_{10}) is an important mobile component of mitochondria respiratory chain and is responsible for electron transport from low redox potential complexes to bc_1 complex.

The hydrophobic nature of CoQ_{10} suggests its localization inside the bilayer of cellular membranes, but till now the large number of physical studies as NMR, Calorimetry, X-ray diffraction and spectroscopic investigation did not clarify this point (see for example the book Coenzyme Q, Lenaz ed., 1985). However, in order to better understand the biological behaviour of the ubiquinone and its action in the respiratory chain, it appears to be indispensable to know the localization sites and to determine the nature of its association with the membranes. Recently, some interaction models of CoQ_{10} on liposomes systems have been proposed, including among others: (a) the localization of the coenzyme in the central part of the lipidic membrane, eventually sandwiched within the bilayer (Alonso et al., 1981; Ulrich et al., 1985; Quinn and Esfahani, 1980); (b) the arrangement in a micellar form inside the bilayer (Quinn and Katsikas, 1985); (c) the formation of a separate phase located outside the bilayer (Stidham et al., 1984). As discussed by Weir et al. (1988) other possible kinds of association of the CoQ_{10} with the membranes must be considered, as for example the adsorption on the phospholipid head-groups or the intercalation within the bilayer.

In this paper we report the preliminary results of an X-ray diffraction investigation on distearoyl phosphatidyl choline liposomes containing the CoQ_{10} or its deuterated derivative on the methoxy groups. In fact, X-ray scattering data will be correlated with neutron diffraction experiments actually in progress in our laboratory: we have deuterated in the benzoquinone cycle the Coenzyme Q in order to create a larger neutron scattering density contrast with surrounding membrane lipids.

Keywords: Ubiquinone, Diffraction techniques, Lipids, Loacalization, Model Membranes.

MATERIALS AND METHODS

The synthetic distearoyl phosphatidyl choline (DSPC) was purchased from Sigma (St. Louis, MO) and used without further purification. The deuterated CoQ_{10} was synthetized by one of us (Fahmy, unpublished): the deuteration occurred in the methoxy groups of the quinone ring with a 57% yield.

Multilamellar liposomes in excess water have been prepared containing non deuterated (H), a equimolar mixture of deuterated and non-deuterated (HD) and, finally, only deuterated (D) ubiquinone.

The DSPC and the coenzyme were mixed in $CHCl_3$/Methanol (1:3) at a CoQ_{10}/lipid molar ratio of 1:6. The mixture was dried by a nitrogen stream and the remaining solvent was removed under vacuum. The water was added in excess (the water/DSPC weight ratio was 0.75) and liposomes were obtained by leaving the samples at temperature above the melting chain transition and by vortexing several times.

The experimental X-ray diffraction techniques and procedures are described in Mariani et al. (1988).

The sample temperature was controlled within a 0.5°C using a circulation thermostat.

Neutron diffraction experiments are in progress at the Leon Brillouin Laboratory of Saclay (France) on similar samples as used for X-ray studies.

RESULTS AND DISCUSSION

The X-ray diffraction data show that the CoQ_{10} does not change the structural parameters and polymorphism of DSPC liposomes in excess water, as reported by Stidham et al. (1984) for other lipidic systems. In particular, as evicted by the presence of several small angle diffraction peaks with spacing ratio of 1:2:3:4 .... the phases are always lamellar as in DSPC pure liposomes: the patterns are not qualitatively affected by the presence of the quinone, in spite of the high concentration reached.

These results suggest unambiguously that the presence of a separate phase systematically organised relative to the lipid structure can be excluded. Moreover, the possible presence of micella phases inside or outside the lipid bilayer is not evidenced in all samples investigated. In table 1, the lamellar periodicities of the different studied samples, determined at 25°C on the gel phase, are reported: it is evident that the presence of the quinone does not change this parameter, also when the deuterated form of CoQ_{10} is considered.

Table 1. Lamellar periodicity (Å) of DSPC excess water liposomes containing CoQ_{10} (quinone/lipid 1/6 molar ratio) hydrogenated (H), deuterated (D) or an equimolar mixture of both forms (HD).

no CoQ_{10}	H	HD	D
64.9	64.8	65.2	65.0

The intensity analysis still in progress in our laboratory will give the electron density profiles of the different lamellar phases: then it will be possible to calculate other important structural parameters as, for example, the thicknesses of the lipidic and paraffinic layers and the area-per-chain at the polar-paraffinic interface.

Preliminary data obtained by small angle neutron diffraction confirm the results shown in Table 1 and above discussed, but enough analysis work is again necessary to correlate neutron experiments to X-ray data to obtain information about eventual quinone solubilization sites in the bilayer.

As a conclusion, at this point of our study, it appears to be possible to exclude the formation of a separate organized phase of ubiquinone-10 in the gel phase. Moreover our experiments have not evidenced any presence of CoQ_{10} micelles. Finally we cannot exclude a possible information on the CoQ_{10} location inside the bilayer, when the electron density estimations will be concluded.

REFERENCES

Alonso, A., Gomez-Fernandez, J.C., Aranda, F.J., Belda F.J. F. and Goni, F.M., 1981, On the Interaction of Ubiquinones with Phospholipid Bilayers. FEBS Letters, 132, 19-22.

Coenzyme Q, 1985, edited by Lenaz G. (London: J. Wiley).

Mariani, P. Luzzati, V. and Delacroix, H., 1988, Cubic Phases of Lipid-containing Systems:Structure Analysis and Biological Implications. Journal of Molecular Biology, 204, 165-189.

Quinn, P.J. and Esfahani, M.A., 1980, Ubiquinones have Surface-Active Properties Suited to Transport Electrons and Protons Across Membranes. Biochemical Journal, 185, 715-722.

Quinn, P.J. and Katsikas, H., 1985, Thermal Characteristics of Coenzyme Q and its Interaction with Model Membrane Systems. In Coenzyme Q, edited by Lenaz G. (London: John Wiley), pp. 107-130.

Stidham, M.A., McIntosh, T.J. and Siedow, J.N., 1984, On the Localization of Ubiquinonc in Phosphotidylcholine Bilayers. Biochimica et Bhiophysica Acta, 767, 423-431.

Ulrich, E.L., Girving, M.E., Cramer, W.A. and Markley, J.L. 1985, Location and Mobility of Ubiquinones of Different Chain Lengths in Artificial Membrane Vescicles. Biochemistry, 24, 2501-2508.

Weir, L.E., Cornell, B.A. and Knott, R.B., 1988, Neutron Scattering of Ubiquinone and Gramicidin A in Lipid Bilayers. Materials Science Forum, vols. 27/28, pp. 159-162.

A LINEAR DICHROISM APPROACH TO STUDIES OF UBIQUINONES IN LYOTROPIC BILAYERS

M. Battino°, I. Domini", R. Fato^, G. Lenaz^, G. Marconi* and B. Samorì"

°Istituto di Biochimica, University of Ancona - Ancona
"Dip. Chimica Organica e Dip. Chim. Ind. e Mater., University of Bologna - Bologna
^Dipartimento Biologia, University of Bologna - Bologna
*Laboratorio FRAE, CNR - Bologna
ITALY

INTRODUCTION

The localization of ubiquinones and related compounds in energy conserving membranes is still controversial (Lenaz et al., 1987), and several models have been proposed, with the extended quinone molecule either parallel or perpendicular to the membrane plane, or bent or even associated with others in a micelle-like arrangement. Much efforts has been sustained to match the appropriate physical techniques to simple model systems in order to obtain unambiguous results. Nonetheless few techniques are sensitive enough to solve the orientation and position of quinones in lipid systems at the physiological quinone to lipid ratios. Since any excess quinone might be accomodated in the lipid bilayer as a separate phase (Quinn and Katsikas, 1985), only results obtained at molar ratio of 1:50 or less should be taken with confidence. Another aspect worthy of consideration in quinone localization is its high mobility, so that, depending on the time scale of the measurement, either several orientations or a time-average of them would result.

One technique combining high sensitivity and time resolution to a wealth of extractable physical information is polarized spectrophotometry.

Keywords: Linear Dichroism, Lyotropic nematics, Liquid Crystals.

THE TECHNIQUE

Linearly anisotropic samples absorb plane-polarized light to different degrees depending on the direction of its linear polarization. The absorption intensity is in fact modulated by the dot products between the electric vector of the incident plane-polarized radiation and the electric transition moments generated within the sample oriented chromophores.

This linear anisotropy, intrinsic to the light absorption processes, is usually expressed and measured as **Linear Dichroism** (LD), i.e. as a differntial absorption of the two perpendicularly polarized components, parallel ($\parallel$) and perpendicular ($\perp$) to the sample orientation axis.

Linear Dichroism = $(I_{\parallel} - I_{\perp}) / (I_{\parallel} + I_{\perp})$

LD spectra are by us recorded by a modulated technique which increases enermously the sensitivity of the measurements up to one part in a million (Samorì, 1983; Dunlap et al., 1988). LD spectra of poorly anisotropic samples, as those provided by rod- or disc-like micelles of liquid crystalline lyotropic nematic systems, can also be recorded. A quantitative picture of the anisotropy of the orientational distribution of guest molecules within these micellar host bilayer can thus be obtained (Samorì and Mattivi,1986).

The LD signals of the transitions polarized along the i molecular axis provide the Saupe order parameter S_{ii}, which is a second rank tensor averaged over all the sample orientational distribution. These order parameters can be obtained also by NMR or ESR techniques, in general by measuring in oriented systems physical processes of rank two.

A very simple geometrical approach able to provide a stereochemical interpretation of these order parameters was recently developed. It is based upon a basic set of four limiting solubilization A-D modes (Samorì and Mattivi, 1986; Samorì, 1988). By them any orientational and localization distribution of a solute molecule in a host micellar bilayer can be described.

It must be emphasized that all the stereochemical data available from dichroic studies are always expressed in terms of relative orientations of the transition dipole moments. This information can therefore be decoded in truly stereochemical terms, i.e. in terms of relations of groups or molecules, only when the orientations and locations of the transition moments within the single chromophoric frameworks are known.

ASSIGNMENT OF THE POLARIZATIONS OF THE UV-VIS TRANSITIONS OF THE UBIQUINONE CHROMOPHORE.

The interplay of the Liquid-Crystal - Linear Dichroism (LC-LD) (Samorì et al., 1982; Samorì, 1983; Samorì, 1988) and the Magnetic Circular Dichroism (MCD) technique with M.O. Calculations can make the directions of polarization of the investigated transitions to be assigned.

The low simmetry (lower than C2v) of the ubiquinone chromophore does not allow the directions of the transition polarizations within the molecular plane to be precisely assigned by the LC-LD technique, unless independent evaluations of the solute orientational distribution are obtained (Michl and Thulstrup, 1988).

The spectrum of the ubiquinone chromophore is characterized by three bands centered at about 400, 320 and 280 nm.

We obtained experimental and theoretical evidences that the lowest energy transition is out-of-plane polarized.

On the basis of the LC-LD technique we found that the direction of polarization of the band at 320 nm has a bigger component along the long molecular axis (identified by the extended isoprenoid chain) than perpendicularly to it. The opposite holds for the transition at about 280 nm.

On the basis of these polarizations, in spite of their still very unprecise evaluation, qualitative predictions of the LD profiles expected for all possible limiting orientational distributions of ubiquinones in host bilayers, can be obtained. We thus started extracting some stereochemical information from the experimental LD spectra of ubiquinone in a bilayer.

LD SPECTRA OF UBIQUINONES IN LYOTROPIC NEMATIC SOLUTIONS

The host composition was (inpercentage by weight): Potassium laurate/decanol/water = 27.9/6.7/65.4.

We prepared solutions of a serie of ubiquinones with the isoprenoid chain of differnt length, from Q_0 to Q_{10}. The samples were oriented by a 1.5 kG magnetic field and the LD spectra recorded by the modulated technique. All the spectra were characterized by negative signals for both bands at 320 and 280 nm.

The isoprenoid chain length modifies the relative intensities of the two LD bands and the order parameters as well. This displays the important role played by the length of the chain in the ubiquinone orientational distribution settling.

The negative sign of both LD bands on the basis of the abovementioned theoretical prediction provide a picture of the ubiquinone orientational distribution which perfectly matches the localization model in lipid bilayers by Lenaz, 1988. Within this model "the ubiquinone molecule is located in the membrane midplane, with the headgroup oscillating trasversally across the membrane (see also Battino et al., 1986).

The LD data also rule out the possibility for the plane of the ubiquinone headgroup of laying preferentially parallel to the bilayers planes.

ACKNOWLEDGEMENTS

Ubiquinone homologs were a kind gift from Eisai Co., Tokyo, Japan. This work was supported in part by grants from CNR and Ministero della Pubblica Istruzione, Rome, Italy.

REFERENCES

Battino M., Fahmy T. and Lenaz G., 1986, Determination of the critical micelle concentration of short chain ubiquinones in model systems. Biochimica et Biophysica Acta, 851, 377-384.

Dunlap D., Samorì B. and Bustamante C., 1988, Why does synuscidally modulated polarization introduce a systeatic error in linear dichroism measurements? In Polarized Spectroscopy of Ordered Systems, edited by B. Samorì and E. Thulstrup (Kluwer A.P., Dordrecht) pp. 275-297.

Lenaz G., Fato R. and Mandrioli E., 1987, Localization and lateral diffusion of ubiquinone in the inner mitochondrial membrane. *Chemica Scripta*, 27, 139-144.

Lenaz G., 1988, Role of the mobility of redox components in the inner mitochondrial membrane. *Journal of Membrane Biology*, 104, 193-209.

Michl J. and Thulstrup E., 1988, Applications of IR and UV-Visible linear dichroism. In *Polarized Spectroscopy of Ordered Systems*, edited by B. Samorì and E. Tulstrup (Kluwer A.P., Dordrecht), pp. 25-57.

Quinn P.J. and Katsikas H., 1985, Thermal characteristics of Coenzyme Q and its interaction with model membrane systems. In *Coenzyme Q*, edited by G. Lenaz (New York: John Wiley & Sons), pp. 107-130.

Samorì B., Mariani P. and Spada P., 1982, The liquid crystal-linear dichroism of organic molecules by a modulation technique. *Journal of the Chemical Society, Perkin Transactions, 2*, 447-453.

Samorì B., 1983, Applications of the modulated liquid crystal-linear dichroism to spectro- and stereo-chemical problems. *Molecular Crystals and Liquid Crystals*, 98, 385-397.

Samorì B. and Mattivi F., 1986, A new LD approach for determining solute orientation in anionic micelles. *Journal of the American Chemical Society*, 108, 1679-1684.

Samorì B., 1988, Liquid crystal-Linear dichroism. In *Polarized Spectroscopy of Ordered Systems*, edited by B. Samorì and E. Thulstrup (Kluwer A.P., Dordrecht), pp. 105-133.

2. Quinones in Bioenergetics

THE CLASSICAL MOBILE CARRIER FUNCTION OF LIPOPHILIC QUINONES IN THE OSMOCHEMISTRY OF ELECTRON-DRIVEN PROTON TRANSLOCATION

Peter Mitchell

The Glynn Research Institute, Bodmin,
Cornwall, PL30 4AU, England

Key words: Osmochemistry; Mobile carrier; Mobile barrier; Protonmotive Q cycle; Solvation substitution; Nomenclature.

DEVELOPMENT OF CONCEPTS

Directness of force and energy transfer

The evolution of ideas about coupling in oxidative and photosynthetic phosphorylation provides a nice illustration of the important distinction between direct and indirect mechanisms of force and energy transfer in bioenergetics.

As shown by the flow equations below (where oxC~I means energy-rich chemical intermediate and CON~ means energy-rich conformational state of enzyme or carrier intermediate) the relatively explicit direct chemical coupling mechanism (1), was followed by the vaguely defined indirect conformational mechanism (2) described by Boyer (see Boyer, 1974).

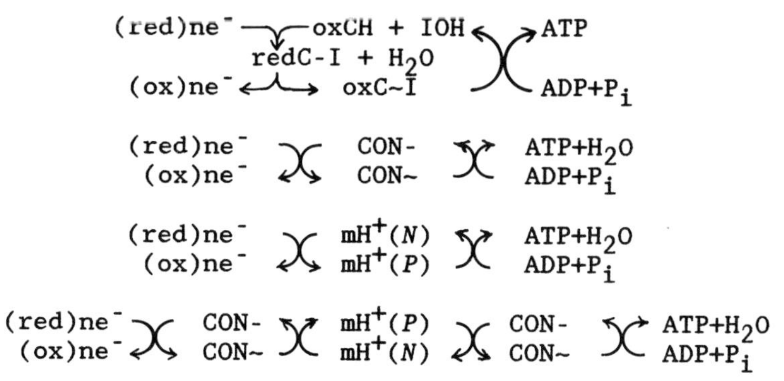

(1)

(2)

(3)

(4)

The physiological-level chemiosmotic mechanism (3) involving translocation of H^+ from negative (*N*) to positive (*P*) protonic potential was backed up by direct mechanisms of coupling between e^- and H^+ in the electron-proton symport function of hydrogen carriers, and between oxide and hydroxide groups and protons in the ADP/(ATP + P_i) antiport function of the protonmotive ATPase (Mitchell, 1981,1987).

The conformationally coupled enzyme + proton-pump type of mechanism (4) invokes two of the indirect conformational coupling interfaces introduced by Boyer for his earlier, now obsolete, conformational coupling hypothesis.

Experience has shown that the direct type of model is more strategically useful for hastening the progress of experimental research than the conformationally coupled type. To avoid misunderstanding, I must incidentally remark that conformational changes of osmoenzymes and porters that operate *in concert with* changes of ligand binding and translocation are obviously fundamentally important, especially in the mobile barrier type of mechanism (Mitchell, 1957; 1987) discussed later in this paper. But, this is quite different from Boyer's concept of energy transfer through a conformationally mobile, molecular mechanical, interface that operates *in series with* chemical or osmotic driver and driven systems, as shown in (4).

The protonmotive Q cycle

The general concept of the protonmotive Q cycle was invented at about 3 a.m. on Tuesday May 20th 1975, using a knowledge of the redox chemistry of lipophilic quinones together with a knowledge of the behaviour of the redox components of cytochrome *c* reductase, and by applying conservational topological principles to the inspection and adjustment of some protonmotive redox loop ligand-conduction diagrams (Mitchell, 1982). The fundamental Q cycle concept, as refined early-on by contributions, notably from the research groups of Garland and of King, as well as by my own further studies (see Mitchell, 1976; 1988) and by other work (see: Slater, 1983; Moss and Rich, 1987; Mitchell, 1988; Rich, 1988; De Vries *et al*., 1988; Robertson and Dutton, 1988; Drachev *et al*., 1989), has greatly assisted the progress of research on the elucidation of the protonmotive molecular mechanisms of the cyrochrome *c* and plastocyanin (PC) reductases (hereafter called *c* and PC reductase).

The interesting question arises: Why has it been possible to explain the molecular mechanism of such complicated multi-subunit quinol-oxidising osmoenzymes as the *c* and PC reductases, while the molecular mechanism of even such a simple transport protein as the lactose-proton symporter has continued to remain obscure (see Mitchell, 1987)?

The *c* and PC reductases interact with five external substrate components: reduced and oxidised cytochrome *c* or plastocyanin, hydrated protons, and the members of the QH_2/Q couple (where Q stands for ubiquinone or plastoquinone). Electrons play the part of an internal substrate between the protonic output and input centres o and i (which bind Q, QH_2

and their intermediates) and the high and low E_m haem groups, b_i and b_o, of the haem *b* doublet, respectively. Owing to this circumstance, the appropriate arrangement of the Q-binding centres i and o, in contact with the hydrocarbon domain of the membrane, and the vectorial deployment of electron-conducting and proton-conducting pathways leading to and from these centres, enables the QH_2/Q couple to act as an electron-proton symporter, coupling the flows of electrons and protons (directly coupled together as hydrogen atoms in QH_2) from centre i to centre o. It is this simple electron-proton symport function of the QH_2/Q couple, together with the return flow of electrons through the haem *b* doublet, that provides the basis of the Q cycle mechanism by which electron transfer through the *c* and PC reductases is coupled to the protonmotive action (Mitchell, 1976,1988). The coupling appears to be very tight, as would be expected, and most if not all suggestions of variable stoichiometry, have probably arisen from misinterpretations of the experimental data (Mitchell, 1988; Rich, 1988).

Mobile carrier and mobile barrier mechanisms

It is pertinent to ask: Where does the protonmotive action catalysed by the *c* and PC reductases take place? Are the components of the QH_2/Q couple conducted specifically through the osmotic barrier domain of the osmoenzyme polypeptide system, or are they conducted relatively non-specifically through the region of the hydrocarbon domain of the lipid bilayer membrane neighbouring the surface of the osmoenzyme molecules? Various lines of experimental evidence suggest that the mobile carrier function of the QH_2/Q couple is largely exercised in the hydrocarbon domain of the membrane (e.g. Chazotte and Hackenbrock, 1988; Venturoli *et al.*, 1988). It is in this respect that the mechanisms of the *c* and PC reductases, and of a few other osmoenzymes that couple the oxidation or reduction of lipophilic quinol/quinone systems to the translocation of protons or other cations, probably differ fundamentally from the great majority of osmoenzymes and solute porters, of which the substrates are strongly hydrophilic (Mitchell, 1987).

In the F_0F_1 ATPase, for example, the protonmotive mechanism may require the hydrophilic and bulky substrates ATP, ADP and inorganic phosphate (P_i) to undergo a specific ATP/(ADP + P_i) antiport process in the three catalytic domains of F_1, situated in the protonic osmotic barrier domain between the aqueous *P* domain (connected through the proton well in F_0) and the aqueous *N* domain, which contains the substrates: ATP, ADP and P_i. Such a process would require specific solvation substitution between the hydrophilic substrates and the enzyme polypeptide system. It could be more easily explained by a mobile barrier type of mechanism, in which a cleft corresponding to the catalytic domain in the osmotic barrier region of the osmoenzyme molecule opens alternately on either side, than by the classical mobile carrier type of mechanism (Mitchell, 1957,1987). The formulation of a mobile barrier mechanism would clearly require comparatively detailed information about the polypeptide structure and about its specific conformational mobility. Such detailed information is not required when translocation is effected by the relatively non-specific mobility of the hydrogen-carrying QH_2/Q couple in the fluid hydrocarbon domain of the lipid bilayer.

The question of solvation substitution

The recognition of the special mobile carrier function of lipophilic QH_2/Q systems, and the contrast with the complex solvation substitution mechanisms that may be required for the entry and exit of hydrophilic substrates in the majority of enzyme, osmoenzyme and porter proteins, leads one to ask whether the low aqueous solvation of the quinol and quinone substrates, and the hydrocarbon environment in which they reside, may give rise to some unusual characteristics of the osmoenzymes specific for lipophilic quinones.

Electronic and protonic potential profiles

As discussed in (Mitchell, 1976), the ligands in pathways conducting e^- or H^+ that traverse the osmotic barrier domain of the *c* or PC reductase molecules are subject to the electric field traversing these pathways. Conversely, an electron or proton traversing a conducting pathway generates an electric displacement and a corresponding electric field along the pathway. The membrane system acts as a capacitor. So, it is possible to estimate the proportion of unit electric charge displacement between the *N* and the *P* aqueous domains (and the equivalent electric potential change between *N* and *P*) corresponding to the movement of unit charge (i.e. e^- or H^+) between given reference points on a conducting pathway. The distance between reference points (e.g. between centre i and the iron atom of haem b_i, or between centre i and the *N* domain) may be estimated, provided we know the relative dielectric strength of the pathway; or we may estimate the 'dielectric distance' on the assumption of constant dielectric strength through the osmotic barrier domain. Distances may be estimated, either by thermodynamic poising effects of a known *P/N* electric potential difference on given redox components, or by observing the electrogenicity of transient partial actions (Mitchell, 1976, 1988; Robertson and Dutton, 1988; Drachev *et al.*, 1989). In practice, complex interpretative problems arise because electronic and protonic dissociations tend to be closely linked, not only by electrical interaction, but also by non-electrical interaction through the polypeptide system. Also, for example, the extent to which a protonic pathway known as a proton well transforms an electric potential difference to a pH difference depends on the local dielectric strength, and this may be very hard to determine independently. This symposium should provide an opportunity to discuss these difficult and important problems.

Nomenclature

The parochialisms of our nomenclature create problems of communication both inside and outside bioenergetics. As originator of the Q cycle, which has become adopted by most of my colleagues, I suggest that it would be helpful to use lower case letters o and i to identify the Q-binding centres and the neighbouring haem *b* groups. The Q-binding centres o and i can be taken to identify protonic output and input sites, respectively. In that way, they apply universally to mitochondria, chromatophores and chloroplasts. The o and i nomenclature is incidentally applicable to protonmotive osmoenzymes generally. It is fortunate that haems b_L and b_H, corresponding to haems b_o and b_i, have Lo and Hi midpoint potentials, respectively. I would also like to plead for the use of the italic capital letters *P* and *N* to identify the protonically positive and negative aqueous domains, respectively. This nomenclature is more generally applicable than the C and M nomenclature used to identify the cytochrome *c* and matrix sides of mitochondrial systems. It is especially useful, for example, in discussing the Q cycle in the closely analogous *c* and PC reductases.

REFERENCES

Boyer, P.D., 1974, Conformational coupling in biological energy transductions. B.B.A. Library, 13, 289-301.

De Vries, S., van Hoek, A.N. and Berden, J.A., 1988, The oxidation-reduction kinetics of cytochromes *b*, c_1 and *c* in initially fully reduced mitochondrial membranes are in agreement with the Q-cycle hypothesis. Biochimica et Biophysica Acta, 935, 208-216.

Chazotte, B. and Hackenbrock, R., 1988, The multi-collisional, obstructed, long-range diffusional nature of mitochondrial electron transport. The Journal of Biological Chemistry, 263, 14359-14367.

Drachev, L.A., Kaurov, B.S., Mamedov, M.D., Mulkidjanian, A.Ya., Semenov, A.Yu., Shinkarev, V.P., Skulachev, V.P., and Verkhovsky, M.I., 1989, Flash-induced electrogenic events in the photosynthetic reaction center and bc_1 complexes of *Rhodobacter sphaeroides* chromatophores. Biochimica et Biophysica Acta, 973, 189-197.

Mitchell, P., 1957, A general theory of membrane transport from studies of bacteria. Nature, 180, 134-136.

Mitchell, P., 1976, Possible molecular mechanisms of the protonmotive function of cytochrome systems. Journal of Theoretical Biology, 62, 327-367.

Mitchell, P., 1981, Biochemical mechanism of protonmotivated phosphorylation in F_0F_1 adenosine triphosphatase molecules. In: Mitochondria and Microsomes, edited by C.P. Lee *et al*. (Reading: Addison-Wesley), pp. 427-457.

Mitchell, P., 1982, Ubiquinone, bioenergetics and the still point of the turning cycle. In: Biomedical and Clinical Aspects of Coenzyme Q, Vol. 3, edited by K. Folkers and Y. Yamamura (Amsterdam: Elsevier/North-Holland),pp. 3-15.

Mitchell, P., 1987, Realistic models of transport processes. In: Integration and Control of Metabolic Processes: Pure and Applied Aspects, edited by O.L. Kon *et al*. (Cambridge University Press for ICSU Press), pp. 231-245.

Mitchell, P., 1988, Respiratory chain systems in theory and practice. In: Advances in Membrane Biochemistry and Bioenergetics, edited by C.H. Kim *et al*. (New York and London: Plenum Publishing Corporation), pp. 25-52.

Moss, D.A. and Rich, P.R., 1987, The effect of pre-reduction of cytochrome *b*-563 on the electron-transfer reactions of the cytochrome *bf* complex in higher plant chloroplasts. Biochimica et Biophysica Acta, 894, 189-197.

Rich, P.R., 1988, A critical examination of the supposed variable proton stoichiometry of the chloroplast cytochrome *bf* complex. Biochemica et Biophysica Acta, 932, 33-42.

Robertson, D.E. and Dutton, L.P., 1988, The nature and magnitude of the charge separation reactions of ubiquinol cytochrome c_2 oxidoreductase. Biochimica et Biophysica Acta, 935, 273-291.

Slater, E.C., 1983, The Q cycle, an ubiquitous mechanism of electron transfer. Trends in Biochemical Sciences, 8, 239-242.

Venturoli, G., Fernandes-Valasco, J.G., Crofts, A.R. and Melandri, B.A., 1988, The effect of the size of the quinone pool on the electrogenic reactions in the ubiquinol-cytochrome c_2 oxidoreductase of *Rhodobacter capsulatus*. Pool behaviour at the quinone reductase site. Biochimica et Biophysica Acta, 935, 258-272.

THE PHOTOSYNTHETIC REACTION CENTER AS A PARADIGM FOR ELECTRON TRANSPORT IN COMPLEX III: THE Q_A-CYCLE.

Graham Palmer

Department of Biochemistry and Cell Biology, Rice University
PO Box 1892, Houston, TX 77251-1892.

Keywords: Complex III, bc_1 complex, electron transport, quinone binding, reduction kinetics, oxidation kinetics, modified Q-cycle.

Complex III, the bc_1 complex of the mitochondrial membrane, catalyzes two fundamental reactions. The first is the transport of electrons from ubiquinol to cytochrome c and the second is the translocation of protons between two spatially distinct and physically separate domains coordinated with this electron transport. The first insights into the mechanism(s) of these processes was provided by Wikstrom and Berden (1976) who proposed that the oxidation of quinol proceeded with the concomitant reduction of two of the electron transfer prosthetic groups, currently assigned as the iron-sulfur cluster (Reiske center) and the low potential cytochrome b, b_L. This proposal was subsequently utilized by Mitchell (1976) in the first formulation of the proton-motive Q-cycle. Subsequently Trumpower (1976) revised certain details of Mitchell's scheme while more recently Crofts et al (1983), de Vries et al (1983) and Rich and Wikstrom (1986) have all volunteered extensions and refinements to the overall reaction sequence, with the scheme of Crofts having gained particular acceptance even finding its way into a recent Biochemistry textbook (Zubay, 1988).

The bc_1 complex offers a particular advantage as a system in which to probe electron and proton transfer for the structural details of the redox active components are relatively well understood[1]. Thus we (Simpkin et al 1989) have recently employed near-infrared mcd measurements to establish the ligands to the heme iron in cytochrome c_1 are histidine and methionine. Similar spectral data (Simpkin et al 1989) for cytochrome b allow the possibility of either bis-histidine or histidine-lysine coordination at the two heme centers, but as sequence comparisons make it clear that no more than 1 lysine is conserved across twenty species it follows that bis-histidine coordination is certainly present at one of the two b hemes. Furthermore recent endor data (Hoffman et al 1989) on the ^{15}N substituted Reiske center from Pseudomonas cepacia establish that the nitrogen ligation present at that cluster is provided by two histidine residues.

In probing the mechanisms of the two catalytic reactions of the bc_1 complex we have tacitly assumed that a full and complete description of the

---

[1] This situation should be contrasted with the related enzyme, cytochrome oxidase, for which there is an enormous body of kinetic data, but for which the structural properties of the redox centers is still fairly poorly defined (Palmer, 1987).

electron transfer pathway(s) will produce a realistic picture of the means whereby the transmembrane ferrying of protons is accomplished. For this approach to be reliable it is essential that experiments be conducted with well-defined enzyme, with careful control over the levels of reductant (quinol), oxidant, and other kinetic modifiers (e.g. inhibitors), and that the kinetic data be analyzed quantitatively and rigorously. Such requirements are not easily met using mitochondria or partially purified material and we have found it necessary to focus our attention to experiments using highly purified and well-characterized enzyme.

The kinetics of reduction of cytochromes b, c_1, and the endogenous Q_6 in purified yeast bc_1 by Q_1H_2 (a water-soluble substrate analog) is sufficiently rapid to support steady-state turnover at V_{max} (Kauten et al, 1987) Thus the rate of reduction of cytochrome b asymptotes to approximately 1000 s^{-1} at high concentrations of reductant while reduction rates in excess of 100 s^{-1} are observed for cytochrome c_1 and Q_6 is reduced to its steady state-level of ca 30% within the dead-time of the rapid-quenching apparatus. Pre-reduction of the high-potential components (c_1 and the Reiske Fe/S center) or addition of myxothiazol or UHDBT (center "o" or quinol oxidase inhibitors) eliminates the reduction of c_1 with little-or-no change in the rate of reduction of b or Q_6. In the presence of antimycin (a center "i" or quinone reductase inhibitor) the reduction of c_1 is unaffected while the reduction of b is decreased so that it is synchronous with c_1. The rate of reduction of Q_6 is only slightly affected. In the presence of both myxothiazol and antimycin none of the cytochromes are reduced. However the reduction of the endogenous quinone proceeded as rapidly and to a comparable extent as is observed in uninhibited enzyme. It is clear that the reduction of the endogenous Q_6 by the exogenous Q_1H_2 can proceed under conditions in which the other redox agents appear to be inactive. This fact implies that Q_6 can function as a site for the input for electrons.

The rate of oxidation of c_1 by a variety of metal-based oxidants proceeds rapidly and is kinetically simple, with electron transfer between the Fe/S cluster and c_1 proceeding at rates which appear to be faster than 2000 s^{-1} (T'sai et al, 1983; T'sai et al, 1987); however the oxidation of the cytochromes b is rate limited at ca 20-50 s^{-1}. About 50% of the endogenous Q_6H_2 is oxidized within 7 msec, the remainder being reacted within 100 msec when about half of the cytochromes b are still reduced. The oxidation of c_1 and the FeS center are insensitive to antimycin, mucidin and myxothiazol while UHDBT strongly inhibits the oxidation of c_1; apparently this latter reagent must act between the Fe/S center and c_1 and modify both of these centers. The oxidation of the b cytochromes is highly suppressed by mucidin, myxothiazol and funiculosin consistent with these inhibitors acting at the high potential components. Antimycin strongly attenuates the time course of the oxidation of the b cytochromes but the reaction is not completely inhibited. Funiculosin raises the midpoint potential of the iron-sulfur center by about 200 mvolts (T'sai et al, 1985) This reagent eliminates the reoxidation of both b centers thus localizing the FeS cluster on the direct path between the b cytochromes and c_1, rather than on a side-path in equilibrium with c_1. Funiculosin also modifies slightly the midpoint potential of the high potential b, b_H. As b_H and the iron-sulfur center appear to be on opposite sides of the membrane this is a clear example of an inhibitor having consequences which are distributed over a large part of the enzyme's structure and points to the hazards of inferring the physical location of inhibitor binding from spectral data or catalytic assays alone.

When Q_1 is used as an oxidant the reoxidation of b is multiphasic with 50% of the absorbance changes occurring extremely rapidly. This rapid phase is eliminated by antimycin but unaffected by myxothiazol.

Removal of the endogenous Q_6 has no affect on the oxidation of cytochromes b via the FeS center but yields results similar to antimycin when oxidation is initiated with Q_1. It would appear that the site for facile reaction of Q_1/Q_1H_2 is with the endogenous Q_6H_2/Q_6, most probably at center "i". Furthermore Q_6 is not required for electron transport from b_L and b_H to c_1 and the Fe/S cluster.

In attempting to find a satisfactory explanation for these data we have examined several varieties of the Q-cycle. A guiding principle in these efforts was our belief that the Q_6 which is endogenous to our preparations has a special role to play in the reaction. The bases for this belief is summarized in Table I.

Table I: Properties of the Endogenous Q_6

- Preparations of yeast bc_1 contain about 1 Q_6/c_1.
- Removal of Q_6 eliminates catalytic activity and proton-pumping capability.
- Restoration of Q_6 restores catalytic activity.
- Large Q radical observed during redox titrations.
- This radical is eliminated by antimycin.
- Q_6 undergoes very rapid reduction to Q_6H_2 in presence of reductant.
- Q_6H_2 undergoes rapid conversion to Q_6 in presence of oxidant.
- Q_6 is not displaced by a large excess of Q_1 in either fully oxidized or fully reduced bc_1.

Our first attempts to account for our kinetic results (Kauten et al, 1987) revolved around a traditional representation of the Q-cycle in which the endogenous Q_6 is able to equilibrate between two sites, one (center "i") in which it can be reduced by the external reductant (in this instance Q_1H_2) and a second site (center "o") in which it is oxidized by the combined action of the iron-sulfur center and the low potential cytochrome b. The chemical equations representing this scheme are shown in Scheme I.

Our approach has been to reproduce this system in a computer by numerical integration of the differential equations for these chemical equations. (Scheme I) The calculations yield the concentrations of the participating chemical species as a function of time and these concentrations can then be weighted by the spectral properties of the respective components so that the results of the calculations can be compared directly with experiment (full details are given in Kauten et al, 1987).

Q_1H_2	+	$Q_6(In)$	$\Leftrightarrow$	Q_1	+	$Q_6H_2(In)$
Q_1H_2	+	$Q_6(In)$	$\Leftrightarrow$	Q_1H	+	$Q_6H(In)$
$Q_6H_2(In)$	+	B_H^+	$\Leftrightarrow$	$Q_6H(In)$	+	B_H
$Q_6H(In)$	+	B_H^+	$\Leftrightarrow$	$Q_6(In)$	+	B_H
$Q_6H_2(Out)$	+	FeS^+	$\Leftrightarrow$	$Q_6H(Out)$	+	FeS
FeS	+	C_1^+	$\Leftrightarrow$	FeS^+	+	C_1
$Q_6H\ (Out)$	+	B_L^+	$\Leftrightarrow$	$Q_6(Out)$	+	B_L
B_L	+	B_H^+	$\Leftrightarrow$	B_L^+	+	B_H
		$2Q_1H$	$\Leftrightarrow$	Q_1	+	Q_1H_2
		$Q_6H_2(In)$	$\Leftrightarrow$	$Q_6H_2(Out)$		
		$Q_6(In)$	$\Leftrightarrow$	$Q_6(Out)$		

Scheme 1: The classical Q-cycle

One such comparison is given in Fig 1, top. It is clear that the version of the Q-cycle represented by Scheme I is able to provide a quantitative accounting of the temporal variation of the absorbance changes at both 553 nm (the wavelength dominated by cytochrome $\underline{c}_1$) and 562 nm (where the cytochromes $\underline{b}$ dominate the spectrum), as well as the immediate approach of of Q_6H_2 to its steady-state level.

However this representation of the Q-cycle has two obvious weaknesses. First the physical meaning of the transition of Q_6 between center "i" and center "o" is completely unclear. Wikstrom and Saraste (1984) for example, have suggested that this transition corresponds to a rotation of Q_6 about its isoprenoid backbone. Second, in scheme I, Q_6 is required to conduct electrons from the low potential centers, $\underline{b}_H$ and $\underline{b}_L$, to the high potential centers ($\underline{c}_1$ and Fe/S). However, as noted above, we find that removing the endogenous Q_6 has absolutely no effect upon the oxidation of the $\underline{b}$ cytochromes subsequent to reacting the fully reduced enzyme complex with a variety of oxidants; this result is clearly inconsistent with expectation.

The second mechanism examined is shown in Scheme II.

Q_1H_2	+	FeS^+	$\Leftrightarrow$	Q_1H	+	FeS
FeS	+	C_1^+	$\Leftrightarrow$	FeS^+	+	C_1
Q_1H	+	B_L^+	$\Leftrightarrow$	Q_1	+	B_L
B_L	+	B_H^+	$\Leftrightarrow$	B_L^+	+	B_H
B_H	+	Q_6	$\Leftrightarrow$	B_H^+	+	Q_6H
B_H	+	Q_6H	$\Leftrightarrow$	B_H^+	+	Q_6H_2
B_H	+	Q_1	$\Leftrightarrow$	B_H^+	+	Q_1H
B_H	+	Q_1H	$\Leftrightarrow$	B_H^+	+	Q_1H_2

Scheme II: A modified Crofts' cycle

It is a variation of the cycle described by Crofts et al (1983) in which the quinone reacts at both center "i" and "center "o" but does so from the medium in each case, that is, any quinol formed at center "i" must first dissociate from the enzyme and subsequently re-associate at center "o" if it is to be oxidized. Crofts' revision provides a simple and plausible scenario for the "movement" of the quinone between centers "i" and "o". The original scheme (Crofts et.al., 1983) has been modified to acknowledge the presence of the endogenous Q_6 which is placed at center "i" exclusively and must be reduced to the quinol and dissociate before Q_1 is allowed to react directly with cytochrome $\underline{b}_H$.

The best fit that I have been able to obtain is shown in Fig 1, center. It is clear that this set of equations does not provide an acceptable description of our data with all redox components becoming much more highly reduced than is found in practice.

The third and final scheme examined to date (Scheme III) is a variant of that shown in Scheme II, in which the Q_6 is once more confined to the enzyme but, by analogy to the Q_a component of the photosynthetic reaction center (PRC) of Rhodobacter sphaeroides (Allen et al, 1988), the Q_6 is firmly bound to center "i" and does not dissociate during turnover; I have dubbed this scheme the Q_a cycle. In the Q_a cycle the reaction of Q_1 with center "i" is mediated by Q_6, in the same sense that the reduction of the dissociable Q_b of the PRC is mediated by Q_a.

$$Q_1H_2 + Q_6\,(In) \leftrightarrow Q_1H + Q_6H\,(In)$$
$$Q_1H + Q_6H(In) \leftrightarrow Q_1 + Q_6H_2(In)$$
$$Q_6H_2(In) + B_H^+ \leftrightarrow Q_6H(In) + B_H$$
$$Q_6H(In) + B_H^+ \leftrightarrow Q_6(In) + B_H$$
$$Q_1H_2 + FeS^+ \leftrightarrow Q_1H(Out) + FeS$$
$$FeS + C_1^+ \leftrightarrow FeS^+ + C_1$$
$$Q_1H + B_L^+ \leftrightarrow Q_1 + B_L$$
$$B_L + B_H^+ \leftrightarrow B_L^+ + B_H$$

Scheme III: The Q_a Cycle

With this scheme a fit to the experimental data is also easily obtained (Fig. 1, bottom). Although the fits obtained with Scheme I give a slightly better fit to the initial behavior of the cytochromes b, Scheme III seems to give a somewhat better accounting of the steady-state behavior of the bound quinone. Thus, on balance, there is little to choose between the fits obtained using either Schemes I and III, with both being clearly superior to the predicted behavior of Scheme II.

The adoption of the Q_a cycle as the preferred description of the path of electron transport in yeast bc_1 is attractive for a number of reasons.

First there is the intellectual appeal of a mechanism which has features in common with a second fundamentally important Q reducing system, the photosynthetic reaction center, and which also contains a bound quinone.

Second, the presence of the bound quinone offers a clear advantage to the enzyme. One complete turnover requires that two quinols be oxidized at

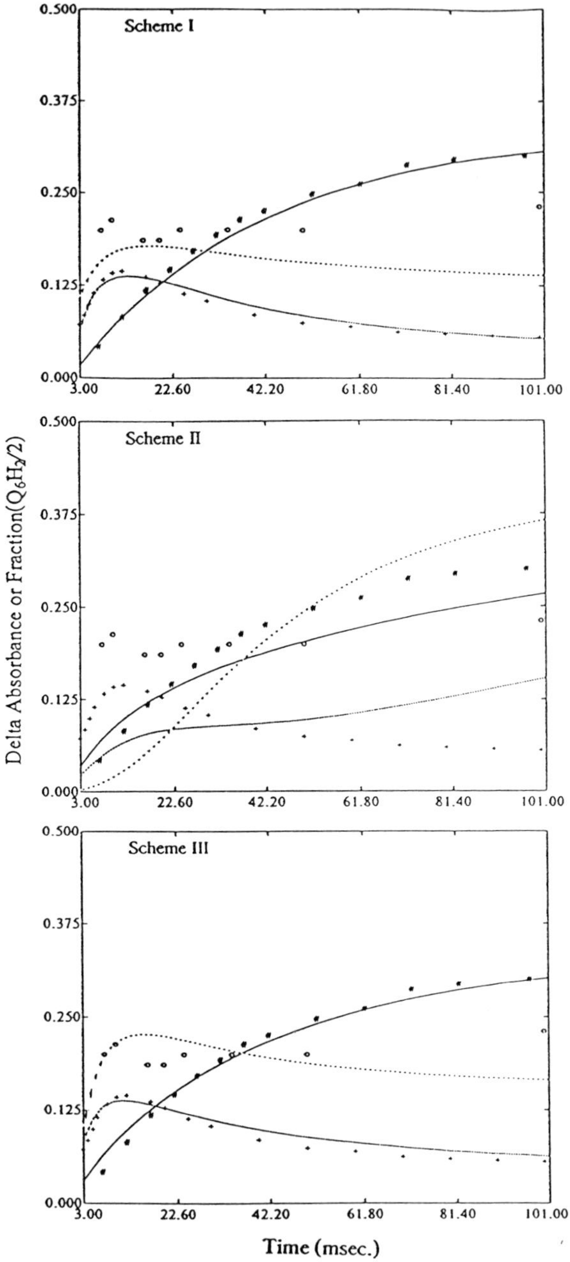

Fig. 1: Simulation of Schemes I, II and III. Calculated quantities are shown as lines and data as symbols. (_____, #), 553 nm; (.........., +), 562 nm; (. . . . . ., o), Q_6H_2.

center "i". The oxidation of the first quinol is straightforward and results in the reduction of the Fe/S center and $\underline{b}_L$ which rapidly pass the electrons onto $\underline{c}_1$ and cytochrome b_H respectively. For the second quinol to be oxidized requires the oxidation of $\underline{b}_H$, for the reduction of $\underline{b}_L$ is thermodynamically disfavored when electron transport to $\underline{b}_H$ is blocked. If were there there no endogenous quinone adjacent to $\underline{b}_H$ then reoxidation of $\underline{b}_H$ would require that center "i" be occupied by exogenous quinone. By providing a permanent quinone at center "i" the oxidation of two quinols at center "o" can proceed even though center "i" is not occupied by an exogenous quinone. Of course at this point the exogenous, pool quinone is necessary to recycle the Q_6H_2 which has been formed.

Third, the Q_a cycle provides a plausible explanation for our observations that the reaction of the reduced bc_1 complex with oxidants is unaffected by removal of the endogenous quinone. Thus it would appear that the transfer of electrons from b_L to the Fe/S center which occurs when fully reduced complex III is reacted with oxidant (e.g. cytochrome c) in the absence of exogenous quinone is a consequence of the inherent tendency for electron transfer when the free energy changes are appropriate (in this instance $\Delta G'$ is about -7 Kcal $mole^{-1}$, which is more than adequate to support electron transfer over 10Å (McLendon, 1988)) and which, in general, does not require any special geometric relationship between the redox partners provided the separation between the reactants is not excessive. The observed rates for this oxidation of $\underline{b}$, 20-50 sec^{-1}, are in fact slightly too small to support turnover at V_{max}, but as the discrepancy was only a factor of two it had previously been brushed under the rug. Because this reaction is presumably a direct transfer from $\underline{b}_L$ to the Fe/S center removal of the endogenous Q_6 need not have any effect, as we have clearly shown to be the case.

Fourth, this revision provides a simple explanation for the recent observations of Zweck et al (1989) who find that complex III from N.crassa catalyses a quinol/quinone transhydrogenase reaction which is sensitive to antimycin but not myxothiazol. This inhibitor sensitivity restricts reaction to center "i". The data exhibit a "ping-pong" kinetic patttern which most simply requires that the quinone and quinol reactants are bound to the enzyme separately with the intermediate formation of 2-electron reduced enzyme. The high-potential cytochrome b, by itself, is unable to act as a 2-electron acceptor while the potential of b_L is too low to participate; thus the proposed inclusion of Q_a at this site nicely rationalizes what is otherwise a thermodynamic dilemma.

Finally, the modification of Scheme II invoked in creating the Q_a cycle is such that it appears that the large body of data obtained with the bc_1 complex from photosynthetic bacteria (Crofts et al, 1983) is equally consistent with both formulations.

REFERENCES

Allen, J.P., Feher, G., Yeates, T.O., Komiya, H. and Rees, D.C (1988) Structure of the Reaction Center from Rhodobacter sphaeroides R-26: Protein-cofactor (quinones and Fe^{2+}) interactions. Proc. Natl. Acad. Sci. (U.S.A) **85** 8487-8491.

Crofts, A.R., Meinhardt, S.W., Jones, K.R. and Snozzi, M. (1983) The role of the quinone pool in the cyclic electron transfer chain of Rhodopseudomonas sphaeroides. A modified Q-cycle mechanism. Biochim. Biophys. Acta **723** 201-218.

DeVries, S., Albracht,S.P.J., Berden, J.A., Marres, C.A.M. and Slater, E.C. (1983) The effect of pH, Q depletion and myxothiazol on the reduction kinetics of the prosthetic groups of ubiquinol:cytochrome c oxidoreductase. Biochim. Biophys. Acta **723** 91-103.

Gurbiel, N.J., Batie, C.J., Sivaraja, N., True, A.S., Fee, J.A., Hoffman, B.M. and Ballou, D.P. (1989) Endor Spectroscopy of ^{15}N enriched phthalate dioxygenase from Pseudomonas cepacia proves that two histidines are coordinated to the 2Fe2S Reiske-type clusters. Biochemistry **28** 4861-4871.

Kauten, R., T'sai, A.L. and Palmer, G. (1987)The Kinetics of Reduction of Yeast Complex III by a Substrate Analog. J.Biol.Chem. **262** 8658-8667.

McLendon, G. (1988) Long-Distance Electron Transfer in Proteins and Model Systems. Accounts Chem. Res. **21** 160-167.

Mitchell, P (1975) The protonmotive Q-cycle: a general formulation. FEBS Lett **59** 1-3.

Palmer, G. (197) Cytochrome Oxidase: A Perspective. Pure & Appl. Chem. **59** 749-758.

Rich, P.R. and Wikstrom, M. (1986) Evidence for a mobile semiquinone in the redox cycle of the mammalian cytochrome bc_1 complex. FEBS Lett. **194** 176-182.

Trumpower, B. (1976) Evidence for a proton-motive Q-cycle mechanism of electron transfer through the cytochrome bc_1 complex. Biochem. Biophys. Research Commun. **70** 73-80.

T'sai, A.L., Olson, J.S. and Palmer, G. (1983) The oxidation of yeast Complex III. Evidence for a very rapid electron equilibration between cytochrome c_1 and the iron-sulfur center. J.Biol.Chem. **258** 2122-2125

T'sai, A.L., Kauten R. and Palmer, G. (1985) The interaction of yeast Complex III with some respiratory imhibitors. Biochim et Biophys Acta **852** 100-105.

T'sai, A.L., Olson, J.S. and Palmer, G. (1987) The Kinetics of Reoxidation of Yeast Complex III: An evaluation of the Q-Cycle. J.Biol.Chem. **262** 8677-8684.

Wikstrom, M.K.F. and Berden, J (1976) Oxidoreduction of cytochrome b in the presence of antimycin. Biochim. Biophys.Acta **283** 403-420

Wikstrom, M.K.F. and Saraste, M. (1984) The Mitochondrial Respiratory Chain. In Bioenergetics, edited by L Ernster Vol 9 of New Comprehensive Biochemistry (Amsterdam: Elsevier) pp 49-94.

Zubay, G. (1988) Biochemistry (2nd edition) Macmillan, London.

Zweck, A., Beckmann, G. and Weiss, H. (1989) The pathway of the quinol/quinone transhydrogenase reaction in ubiquinol cytochrome c reductase from Neurospora Crassa. Eur.J.Biochem. **183** 199-203.

ACKNOWLEDGEMENTS

The studies described in this manuscript was supported by grants from the NIH (GM 21337) and the Robert A Welch Foundation (C-636) and has benefitted immensely from the participation of R.Kauten, J.S. Olson and A-L. T'sai (Rice University) and M. degli Esposti and G.Lenaz (University of Bologna.). Certain of the ideas presented are the consequence of a friendly argument with Les Dutton and Tony Crofts on the Via Irnerio.

THE ROLE OF PHOSPHOLIPIDS IN THE CATALYTIC ACTIVITY OF RESPIRATORY MULTIPROTEIN COMPLEXES

G. von Jagow, Thomas A. Link and Hermann Schägger

Zentrum Biologische Chemie, Therapeutische Biochemie,
Johann Wolfgang Goethe-Universität, Frankfurt am Main
Theodor-Stern-Kai 7, Haus 25B
D - 6000 Frankfurt 70
Federal Republic of Germany

INTRODUCTION

By which particular means and to what extent did our field of research develop so rapidly during the past years?

The isolation of membrane proteins gained tremendously from a continuous examination of new **detergents** which brought us a vast number of monodisperse preparations of nearly all multiprotein complexes of the respiratory chain.

The kinetic studies which began with mitochondria or submitochondrial particles in the early 60ties and 70ties in the bioenergetic laboratories *of Philadelphia, Glynn, Wisconsin, Amsterdam, Marburg,* etc., *were later* - maybe unfortunately - overwhelmed by studies performed with isolated complexes or proteoliposomes. The phenomenological data gathered with these systems, however, did not give a conclusive picture of the bc_1 complex, which is also called ubiquinol:cytochrome reductase, or briefly the reductase.

Novel Q_o site **inhibitors**, introduced at about 1975, became a really powerful tool in the hands of red as well as green bioenergeticists and helped us to clarify many details of the electron transfer chain of the reductase.

However, our youngest and most efficient branch, was opened by colleagues from **molecular genetics**, who were originally not very interested in bioenergetics, but then joined in and enlarged our knowledge with unexpected speed.

After selection of inhibitor-resistant mutants and genetic mapping of the point mutations, they forced us to change some of our basic assumptions. Thus, for instance, we had to change the folding pattern of cytochrome b; we had to withdraw one of the presumed α-helices from the membrane. Their studies also told us the correct orientation of the two heme b centres in the membrane.

RESULTS AND DISCUSSION

Protein chemistry formed and still forms the "backbone" of bioenergetic work: a vast number of preparations has been elaborated during the past years with different detergents for different reductases of different species, - mitochondrial as well as plastidial and bacterial ones.

The authors' ambition consisted in the isolation of every single subunit of the mammalian reductase with the aim of a subsequent analysis of their primary structures, modelling of their secondary structures, and presumptive reconstitution experiments.

We were convinced that the reductase prepared by the Triton/hydroxyapatite method (Schägger *et al.*, 1986) was structurally and functionally competent. The main criterium was an unchanged ligand field of the two heme b centres, as indicated by typical absorbance maxima and half reduction potentials (Riccio *et al.*, 1977). Absorbance maxima and midpoint potentials were taken as indicators of "nativity", since regularly and instantaneously a shift of the absorbance peak connected with a redox potential shift of both heme b centres was obtained when the reductase was exposed to any kind of denaturing conditions. Thus we did not worry about a somewhat low turnover number of the catalytic activity, especially when we found out that the iron sulfur protein is in a dissociation equilibrium with the bovine complex (Engel *et al.*, 1983).

It was overlooked that the preparation method was a quite special one, because Triton X-100 in combination with chromatographic methods removed much more bound lipid as for instance cholate in combination with ammonium sulfate precipitations. Our preparation, when compared to others, has a remarkably low phospholipid content and is almost devoid of ubiquinone.

It was not before ten years later when we realised that it was not the dissociation of the iron sulfur protein that reduced catalytic activity. The reason for a low catalytic activity was neither the occurence of a conformational change of the protein oligomer, nor an incompleteness in the number of protein subunits, but a partial loss of phospholipids, as it is described in the following (Schägger *et al.*, 1989).

In order to study the action of phospholipids in detail, the reductase was delipidated by phospholipase C or A_2 in the conventional way.

Phospholipase C inactivated the electron transfer from ubihydroquinone to cytochrome c by more than 90%, which showed that diacylglycerols, which remained bound to the reductase after lipase treatment, could not substitute the natural phospholipids. Instead of turnover numbers of 900 s^{-1}, which were reached after phospholipid addition to the original preparation, turnover numbers of only 500 s^{-1} could be attained, indicating that diacylglycerols could not completely be displaced by the egg yolk phospholipids.

The response was different on phospholipase A_2 treatment. After inactivation, reactivation of only 30% could be reached, even when cardiolipin was added during replenishment.

It looks as if cardiolipin plays - like in other enzyme systems - a special "conformation preserving role". The original reductase contains 8-9 mol of cardiolipin per mol of cytochrome c_1. When those were hydrolysed by phospholipase A_2 treatment, the reductase became irreversibly damaged. Moreover, the lysophospholipids produced during phospholipase A_2 treatment may have played an additional disintegrating role.

A novel chromatographic method was developed which made sure that competing or denaturing phospholipids were not formed during inactivation of the reductase. The phospholipids were stripped off the reductase when it was bound to a hydroxyapatite column. Under these conditions, a wash with a definite volume of a buffer containing Triton did remove boundary phospholipids with the exception of cardiolipin. The method is called "Triton wash", the resulting preparation "D-1 complex".

It was shown by thin layer chromatography on silica plates that the reductase still contained the essential cardiolipin.

When the mammalian reductase was isolated by the Triton/hydroxyapatite method, its average catalytic activity varied between 100 and 600 s^{-1} depending on the amount of the retained phospholipid. An activity of 1100 s^{-1} could be reached by adding about 70 equivalents of egg yolk phospholipid.

When a gentle "Triton wash" was performed, a turnover number of 69 s^{-1} was obtained. By adding egg yolk phospholipid, again full activation could be obtained.

Phosphatidic acid and cardiolipin could not reactivate the complex. The acidic phosphatidyl inositol and phosphatidyl serine were also insufficient in their reactivating potency. The real activators were those phospholipids which form the main structural constituents of the mitochondrial membrane, i.e., phosphatidyl choline (PC) and phosphatidyl ethanolamine (PE).

On first sight it was surprising that such a high number of phospholipid molecules was required for full reactivation, that is more than 50 phospholipid molecules per monomer, corresponding to over 100 phospholipid molecules per dimer.

The question was raised wether this was due to a low affinity of the lipid molecules to the complex or due to the real existence of a high number of tight binding sites. The titration experiments speak in favor of the existence of low affinity binding sites only, but the experimental setup assures a near quantitative binding of the added phospholipids because of the phase distribution of phospholipids in the absence of Triton micelles.

Before interpretation of the experimental results, the architecture of the complex should briefly be described.

Our laboratory as well as *Capaldi's* group performed dissociation/reassociation, protease digestion and immunological studies in order to get a view of the arrangement of the 22 subunits in the dimer and obtained comparable results (Link 1988, González-Halphen *et al.*, 1988). Figure 1 depicts a cross section of the dimer divided along the twofold axis of rotational symmetry, one half of the cross section presenting the front, the other the back of the monomeric halves. The top of the section is directed to the cytosolic surface, the bottom to the matrix surface of the mitochondrial membrane.

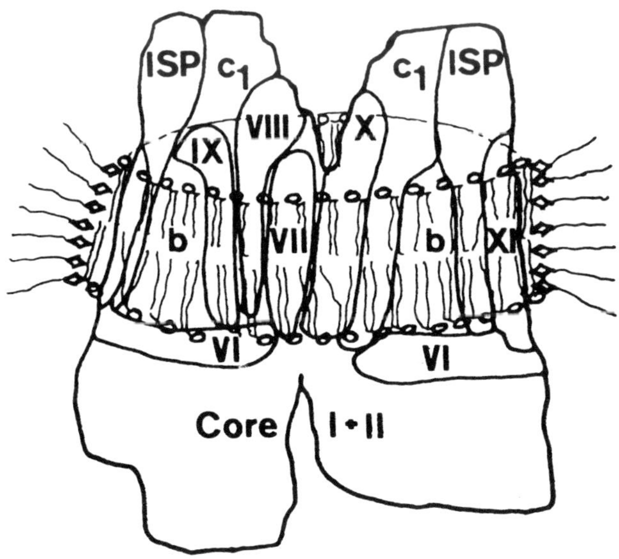

Figure 1. Topography of dimeric bc_1 complex, showing the lipid annulus.

Subunits which are intimate neighbours are marked by identical hatching. Cytochrome b forms the core of the complex, together with subunits VI and IX. The two large peptides Core I and II expose widely into the matrix space. Subunit VII is intercoiled with the two Core proteins. The primary electron acceptor at Q_o site is the iron sulfur centre, adjacent to it lies subunit XI. The electron acceptor from the iron sulfur centre is cytochrome c_1, which is intertwined with subunit VIII and X.

The question arises, why maximal catalytic activity does require such a high number of phospholipid molecules? Which part of the complex do they bind to?

One can perform an estimation of the number of phospholipid molecules which are necessary in order to surround the dimer in the form of a single bilayer annulus at that part of the protein surface, which was originally in contact with the mitochondrial membrane.

If one uses an approximate equation for a rotational ellipsoid and takes the diameters of 12 and 8 nm as given by Leonard *et al.* (1981) for the reductase dimer present in a two-dimensional crystal lattice, a circumference of 32 nm results. If we assume a chain distance of 0.4 nm for the acyl chain of a phospholipid molecule, a number of about 80 molecules of phospholipid can be estimated, which are required or can be accomodated to a full bilayer around the dimer.

We assume, that an optimal shuttle, i.e., a maximal lateral and transversal diffusion velocity of the quinone head between Q_o and Q_i centre and the various dehydrogenases, needs a complete bilayer annulus.

This annulus might not only guarantee a high hydrophobicity within the quinone reaction centres as well as an optimal diffusion of the hydrophobic quinone molecule, it may also support an optimal orientation of the headgroup of the ubiquinone molecule - necessary for a succesful collision with the respective binding site - since the headgroups of PC and PE play a crucial role for reactivation, i.e., for the interaction with quinone.

We do not interprete the linear part of our phospholipid-binding curve (not shown) as indicative for the existence of 80 specific tight binding sites. In our opinion, the observed linear phospholipid (PL) binding rather reflects a facilitated phase distribution of the added PL from the surrounding water phase to the hydrophobic phase of the protein surface, especially when one takes into account the low detergent concentration which we adjusted during lipid binding.

SUMMARY

i) The isolated, delipidated and relipidated reductase is in a monodisperse dimeric state of aggregation when handled in Triton in contrast to other preparations. Preparations of dehydrogenases performed in Triton X-100, presumably also in other detergents, should be tested for their phospholipid content. When activity is lost during preparation, relipidation may restore maximal catalytic activity.

ii) The middle part of the respiratory chain, the cytochrome c reductase, requires 20 molecules of cardiolipin and more than 100 molecules of PE and/or PC per dimer for maximal activity. This number is somewhat higher than the number of molecules required for forming a complete bilayer annulus which may cover that part of the surface of the complex which originally spanned the membrane.

REFERENCES

Engel, W.D., Michalski, C. and von Jagow, G., 1983, Reconstitution of the ubiquinol:cytochrome c reductase from a bc_1 subcomplex and the 'Rieske' iron-sulfur protein isolated by a new method. European Journal of Biochemistry, 132, 395-402.

González-Halphen, D., Lindorfer, M.A. and Capaldi, R.A., 1988, Subunit Arrangement in Beef Heart Complex III. Biochemistry, 27, 7021-7031.

Leonard, K., Wingfield, P., Arad, T. and Weiss, H., 1981, Three-dimensional structure of ubiquinol:cytochrome c reductase from *Neurospora* mitochondria determined by electron microscopy of membrane crystals. Journal of Molecular Biology, 149, 259-274.

Link, T.A., 1988, Zur Struktur des mitochondrialen bc_1-Komplexes. PhD Thesis, Ludwig-Maximilans-Universität München.

Riccio, P., Schägger, H., Engel, W.D. and von Jagow, G., 1977, bc_1-Complex from beef heart. One-step purification by hydroxyapatite chromatography in Triton X-100, polypeptide pattern and respiratory chain characteristics. Biochimica et Biophysica Acta, 459, 250-262.

Schägger, H., Link, T.A., Engel, W.D. and von Jagow, G., 1986, Isolation of the eleven protein subunits of the bc_1 complex from beef heart. Methods in Enzymology, 126, 224-237.

Schägger, H., Hagen, T., Roth, B., Brandt, U., Link, T.A. and von Jagow, G., 1989, Phospholipid Specificity of Bovine Heart bc_1 Complex. Journal of Biological Chemistry, submitted for publication.

FUNCTION, STEADY STATE TURN-OVER, AND STRUCTURE OF THE UBIQUINOL:CYTOCHROME C_2 OXIDOREDUCTASE OF *RHODOBACTER SPHAEROIDES*.

A.R. CROFTS, Z. WANG, Y. CHEN, S. MAHALINGHAM, C.-H. YUN, and R.B. GENNIS,

Biophysics Division, University of Illinois, 156 Davenport Hall, 607 S. Mathews, Urbana. IL 61801

INTRODUCTION

The modified Q-cycle (Crofts, et al. 1983) has provided a plausible explanation for the kinetics of turn-over of the photosynthetic electron transfer chain of *R. sphaeroides* following flash excitation in the uncoupled state. In this paper we report briefly on recent attempts to extend the model in three different areas. In the first of these, we have explored the temperature dependence of electron transfer to assay the activation barriers, and to provide information for a more detailed model of the mechanism. In the second, we have used the method for correction of redox changes for contamination by electrochromic absorbance changes introduces by Venturoli et al. (1986), to measure the thermodynamic profile of the electron transport chain as it approaches steady state under static head. This has provided new information on the electrogenic spans of the complex, and new insights into the control of electron transfer under coupled conditions. In the third approach, we have been using prediction algorithms, sequence comparison, site-directed mutagenesis and *pho*A fusion topological studies to probe the molecular structure of the b/c_1 complex.

Keywords: Quinol oxidation; kinetic model; photosynthesis; bc_1 complex; structural prediction

RESULTS AND DISCUSSION

Activation barriers in UQH2:cyt c2 oxidoreductase.

Values for $E_{m,7}$ of cytochromes of the UQH_2:cyt c_2 oxidoreductase of *Rb. sphaeroides* show a weak temperature dependence between 280-330 K, with coefficients of about 1 mV per degree; other components show similar dependencies, so that no significant change in E_m between components occurs over this range. Rates of electron transfer in the high potential chain (Fe.S, cyt c_1 and c_2, P870) show a weak temperature dependence, with activation energy < 10 $kJ.M^{-1}$. Oxidation of UQH_2 at the Q_Z-site showed a strong temperature dependence, with an activation energy of about 32 $kJ.M^{-1}$. The electron transfer from cyt b_L to cyt b_H was not rate determining at any temperature, and did not contribute to the energy barrier. The activation energy of 32 $kJ.M^{-1}$ for UQH_2 oxidation was the same for all states of the quinone pool. We suggest that the activation barrier is in the reaction by which UQH_2 at the catalytic site is oxidized to semiquinone. The most economical scheme would have the semiquinone intermediate at the energy level indicated by the activation barrier (Crofts and Wang, 1989). Table I (below) summarizes values from a computer model which describes the data well.

Reaction	Forward rate constant	Equilibrium constant	Notes (a,b)
c_2 oxidation	$k_0 = 3 \times 10^5\ s^{-1}$	$K_0 = 78$	
(free c_2)	$> 2.3 \times 10^7\ M^{-1}.s^{-1}$		(c)
c_1 oxidation	$k_1 = 1.2 \times 10^7\ M^{-1}.s^{-1}$	$K_1 = 23$	
FeS oxidation	$k_2 > 10^5\ s^{-1}$,	$K_2 = 0.3$	
QH_2 binding	$k_3 = 5 \times 10^6\ M^{-1}.s^{-1}$	$K_3 = 750\ M^{-1}$	(d)
$Q_Z^{\cdot}$ formation (FeS$^+$ reduction)	$k_4 > 1650\ s^{-1}$	$K_4 > \exp[-G_{act}/RT] > 1.3 \times 10^{-6}$	(d)
b_L reduction	$k_5 > k_4/K_4$	$K_5 < 3 \times 10^4$	(d)
b_H reduction	$k_6 > 100 \times k_4$	$K_6 = 255$	
b_H oxidation	$k_7 > k_4$	$K_7 = 23$	(e)
Q_Z leaving	$k_{-8} > 1000\ s^{-1}$	$K_8 = 15\ M^{-1}$	(d)

Notes. a) $E_{m,7}$ values assumed were:- $(BChl)_2$, 450 mV; cyt c_2, 340 mV; Fe.S, 290 mV; cyt c_1, 260 mV; cyt b_H, 50 mV; cyt b_L, -90 mV; Q/QH_2(pool), 90 mV. b) Stoichiometric ratios assumed were:- RC : cyt c_2 : b/c_1 complex : Q_{pool}; 1:0.7:0.4:15-30, with RC at 1 mM in membrane, cyt c_2 at 0.5 mM in aqueous phase. c) Calculated from $t_{1/2}$ = 60 us for free c_2 (P. Joliot, personal commun.). d) The equilibrium constant for the reaction at the Q_Z-site, K_Z, is given by: $K_Z = K_3 \cdot K_4 \cdot K_5 \cdot 1/K_8 = 2.2$; K_3 and K_8 are both association constants. e) Value for K_7 shown is for the two-electron reduction of Q_{pool} by 2 cyt b_H^-.

Operation of the UQH2:cyt c2 oxidoreductase in steady state.

A steady state approach has been developed by which the redox poise of the electron-transfer chain, the electron-flux and H^+-gradient could be simultaneously monitored. The poise of redox centers was measured using wavelengths previously suggested (Crofts et al. 1983, Meinhardt and Crofts, 1982, 1983) after correction for electrochromic changes (Venturoli et al. (1986). Changes were measured following flash or continuous illumination, with similar results. a) In a complete steady-state profile for the electron-transfer chain, the thermodynamically preferred pathway is consistent with the modified Q-cycle. b) The measured steady-state flux and redox poise of components of the electron transfer chain are naturally explained by kinetic parameters suggested for the modified Q-cycle when account is taken of the change of these values in response to a membrane potential. c) For the electron transfer chain at steady state, we have shown that: i) The free energy drop calculated for every electron-transfer step associated with the turnover is close to zero. ii) The UQH_2:cyt c_2 oxidoreductase complex functions as an electrogenic H^+-pump at steady state close to static head. iii) An almost symmetrical balance between E_h(Q-pool)-$E_h(b_L)$ and ($E_h(c_1)$-E_h(Q-pool)) is maintained as the chain approaches static head. iv) Near static head, the driving force for electron-transfer is close to the H^+-gradient of about 200 mV. v) Electron transfer from b_L to b_H accounts for about 0.65-0.7 of the total electrical span across the membrane. d) The degree of coupling for the chain is > 0.96 and for b/c_1 > 0.98. Results from a typical series of experiments under continuous illumination when the development of a pH-gradient is suppressed by addition of nigericin are summarized in Fig. 1. We have performed similar experiments under a wide range of conditions, and are presently exploring the role of the pH-gradient when this is allowed to develop.

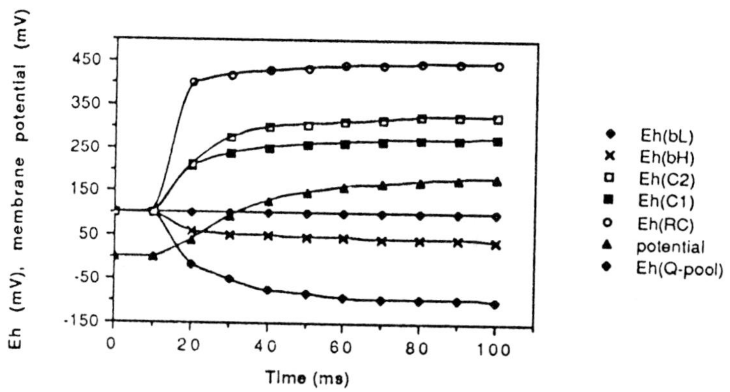

Fig. 1. Poise of reactants, and the coupled proton gradient during the approach to static head. Thermodynamic profile of the coupled electron transfer chain on continuous illumination. E_h values for each component were calculated from kinetic curves measured using appropriate wavelengths, after subtraction of contributions from other components and electrochromic changes. Development of proton gradient measured from carotenoid change, scaled to maximal change of 200 mV. Nigericin was present to collapse ΔpH. Initial E_h 100 mV.

Predicted structure for cyt b of UQH2:cyt c2 oxidoreductase.

A preliminary model for the structure of the cytochrome b subunit of the UQH_2:cyt c oxidoreductase of yeast has been predicted using a combination of methods; Chou-Fasman, Robson-Garnier and Rao-Argos prediction, hydrophobic analysis using several hydrophobic indices, and amphipathic analysis (Crofts et al., 1987). Further refinement and extension to the subunit from *R. sphaeroides* has relied on comparison of subunits from mitochondria, chloroplast and bacteria by homologous alignment, mutability analysis, site-directed mutagenesis, and comparative prediction, and by topological analysis using *pho*A fusion. A general model for bacterial and mitochondrial cyt b subunits has eight transmembrane helices, and three amphipathic helices, which are strongly conserved. The main difference from previously accepted models is in the removal of one transmembrane helix (helix IV) to an amphipathic situation. The most obvious structural consequence is that the sidedness of the sequence to the C-terminal end of helix IV has been reset, so that the liganding helices (II and V) are parallel in the new model. In line with the nomenclature adopted for the photochemical reaction center, we suggest that the putative membrane spanning helices be lettered A-H, and the amphipathic helices be

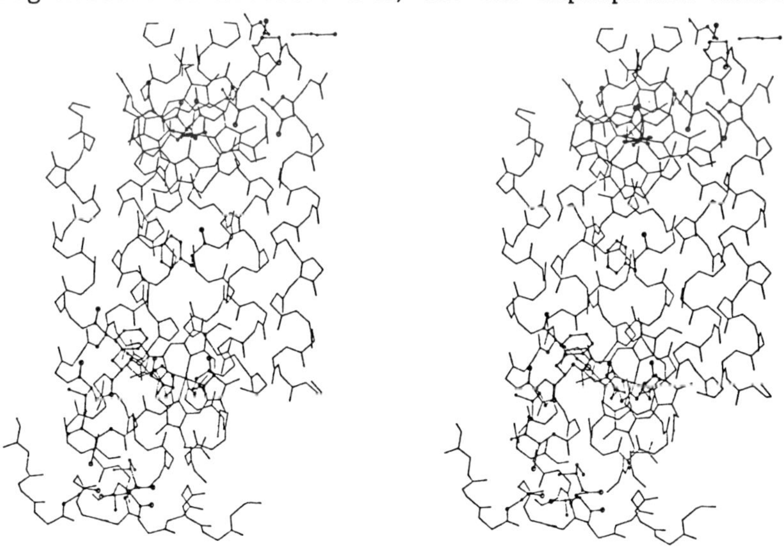

Fig. 2. Possible tertiary structure for the cyt b subunit. Stereo-pair model (can be viewed by the cross-eyed method) showing residues modified in Q_Z- (bottom left) and Q_C- (top right) site mutants, His ligands to hemes, Phe-105 as marker for helix II. Helices are (from back right, anticlockwise): A, B, C, F, D, E, with cd transverse at bottom.

lettered ab, cd, and ef to indicate their position in the sequence between the spans crossing the membrane. In the tertiary structure shown in Fig. 2, six transmembrane helices are packed about the two heme prosthetic groups to allow for the His ligands, heme-modifying mutants, inhibitor resistance lesions, data from *pho*A fusions (Yun, C.-H., Van Doren, S., Crofts, A.R. and Gennis, R., in preparation) and electrogenic spans (see data above).

Mutability moment as a method for assessing the tertiary orientation of membrane spanning helices.

Komiya et al. (1988) have noted that mutational frequency is greatest on the lipid exposed face of reaction center helices, and we have adapted this observation to help predict structure in the cyt b subunit. Using 15 aligned sequences, we have assessed the mutability, using Shannon theory to find information gain, which is proportional to binding free-energy, or "specificity" (Sarai et al., 1989). We have searched the sequence for helical repeats using the algorithm of Eisenberg (1984) to generate a value for "mutability moment" along the sequence, and an angle for the mutability vector which shows the orientation of the moment. In Fig. 3 we show profiles for hydropathy (using a modified Rao and Argos (1987) index) (dashed profile), and mutability moment.

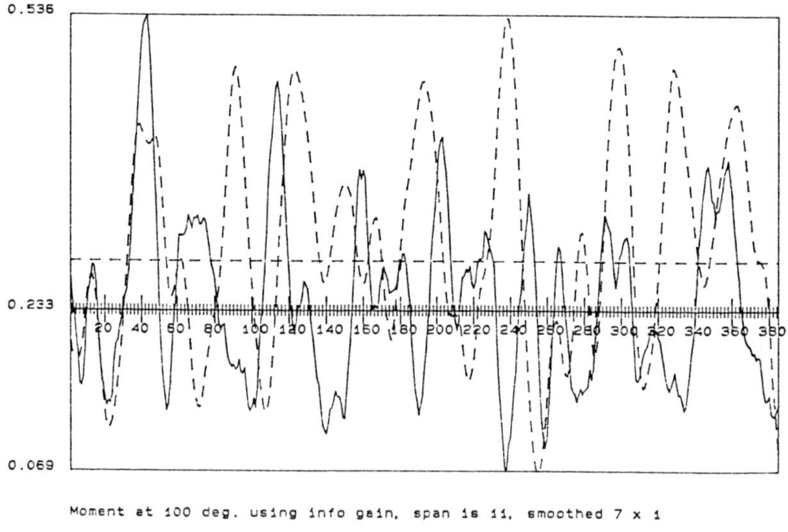

Fig. 3. Profiles for mutability moment and hydropathy. Profiles for the yeast cyt b subunit are shown; cyt b subunits from other species use the same data for information gain, and show similar profiles.

A strong mutability moment is seen for several of the helices, in particular in helices A, F, and H. Amphipathic helices ab and cd show weaker moments. Several of the helices show mutability moments at one (C) or both (D,E) of their ends, and this effect is accentuated if the span for calculation of moment is reduced, when the mutability moment for helices F and H is also seen to be greater at the ends. A similar effect is seen in profiles for amphipathy at the helical repeat, suggesting that helices may protrude beyond spans in the insulating phase indicated by hydropathy plots.

REFERENCES

Crofts, A. R., Meinhardt, S. W., Jones, K. R. and Snozzi, M. (1983). The Role of the Quinone Pool in the Cyclic Electron Transfer Chain of *Rps. sphaeroides*: A Modified Q-cycle Mechanism. Biochim. Biophys. Acta, 723, 202-218.

Crofts, A.R., Robinson, H.H., Andrews, K., Van Doren, S. and Berry, E. (1987) Catalytic sites for reduction and oxidation of quinones. In "Cytochrome Systems: Molecular Biology and Bioenergetics", Papa, S., Chance, B. and Ernster, L., eds. pp. 617-624, Plenum Publ., New York.

Crofts, A.R. and Wang, Z. (1989) How rapid are the internal electron transfer reactions of the UQH_2:cyt c_2 oxidoreductase? Photosynth. Res., in the press.

Eisenberg, D. (1984) Three-dimensional structure of membrane and surface proteins. Ann. Rev. Biochem. 53, 595-623.

Meinhardt, S.W. and Crofts, A.R. (1982). Kinetic and Thermodynamic Resolution of Cytochrome c_1 and Cytochrome c_2 from *Rps. sphaeroides*. FEBS Lett., 149, 223-227.

Meinhardt, S. W. and Crofts, A. R. (1983). The Role of Cytochrome b_{566} in the Electron Transfer Chain of *Rps. sphaeroides*. Biochim. Biophys. Acta, 723, 219-230.

Rao, J.K.M. and Argos, P. (1986) A conformational preference parameter to predict helices in integral membrane proteins. Biochim. Biophys. Acta 869, 197-214.

Sarai, A. and Takeda, Y. (1989) Information gain and specificity in the DNA sequence recognition by *Cro* and Lamda repressor. Science, submitted.

Venturoli, G., Virgili, M., Melandri, B.A. and Crofts, A.R. (1986) Kinetic measurements of electron transport in coupled chromatophores from Photosynthetic bacteria: A method of correction for electrochromic effects. FEBS Lett. 219, 477-484.

Komiya, H. Yeates, T.O., Rees, D.C, Allen, J.P. and Feher, G. (1988) Structure of the reaction center from *Rhodobacter sphaeroides* R-26 and 2.4.1: Symmetry relations and sequence comparison between different species. Proc. Natl. Acad. Sci. USA 85, 9012-9016.

THERMODYNAMIC CHARACTERIZATION OF THE QUINONE ACCEPTOR COMPLEX IN CHROMATOPHORES FROM RHODOBACTER CAPSULATUS

GIOVANNI VENTUROLI and B. ANDREA MELANDRI

Department of Biology, University of Bologna, Via Irnerio 42, 40126 Bologna, Italy

INTRODUCTION

Each reaction center (RC) of the photosynthetic bacterium *Rb. capsulatus* contains two molecules of ubiquinone-10 (UQ) bound to the apoproteins: (Q_A), permanently attached to the complex, accepts electrons from bacteriopheophytin, forming a semiquinone anion radical (Q_A^-); the second ubiquinone, (Q_B) can undergo two subsequent one-electron reductions by Q_A^-, forming the fully reduced ubiquinol. Q_B, when fully reduced or oxidized, exchanges with the large pool of ubiquinone (Q_P) dissolved in the membrane lipids, while Q_B^- remains bound to the RC (for review see Crofts and Wraight, 1983). The thermodynamics of the Q_A/Q_A^-, Q_B/Q_B^- and Q_B^-/Q_BH_2 couples has been studied by optical or ESR spectroscopy. At pH 8, in chromatophores of the related bacterium *Rb. sphaeroides*, E_m's respectively of -75, 40 and -40 mV have been reported (Rutherford and Evans, 1980). Proton binding is coupled to all three reductions, so that the E_m's are generally pH dependent. While the literature dealing with the E_m of Q_A/Q_A^- is rather abundant, the thermodynamics of Q_B is still preliminary, particularly in intact membranes. In this paper we present redox titrations which yield additional information on the thermodynamics of the quinone acceptor complex in chromatophores from *Rb. capsulatus*. By solvent extraction of Q_B the redox coupling within the two bound quinones, already observed in isolated reaction centers (Wraight, 1981), has been investigated in intact membranes.

Keywords: ubiquinone; electron transfer; bacterial photosynthesis; reaction center; herbicide

RESULTS AND DISCUSSION

Redox titration of the Q_B/Q_B^- and Q_B^-/Q_BH_2 couples

In cytochrome (cyt) c_2 rich chromatophores inhibited with myxothiazol, the RC can undergo at least two closely spaced turnovers before the electron supply for its re-reduction becomes limited. Under these conditions, the operation of the RC can be followed from the electrochromic response of carotenoids to the charge movement within the RC from cyt c_2 to the quinone acceptor complex (Wraight et al., 1978). If two single turnover flashes are fired in sequence, with a dark time between them large enough to allow a complete re-reduction of RC by cyt c_2 and electron transfer from Q_A^- to Q_B or Q_B^-, it is possible, following the amplitude of the electrochromic signal after the second flash as a function of the redox potential, to investigate the redox properties of Q_B. In Fig.1 the ratio of the absorbance changes generated by the 2nd/1st flash as a function of E_h, is presented.

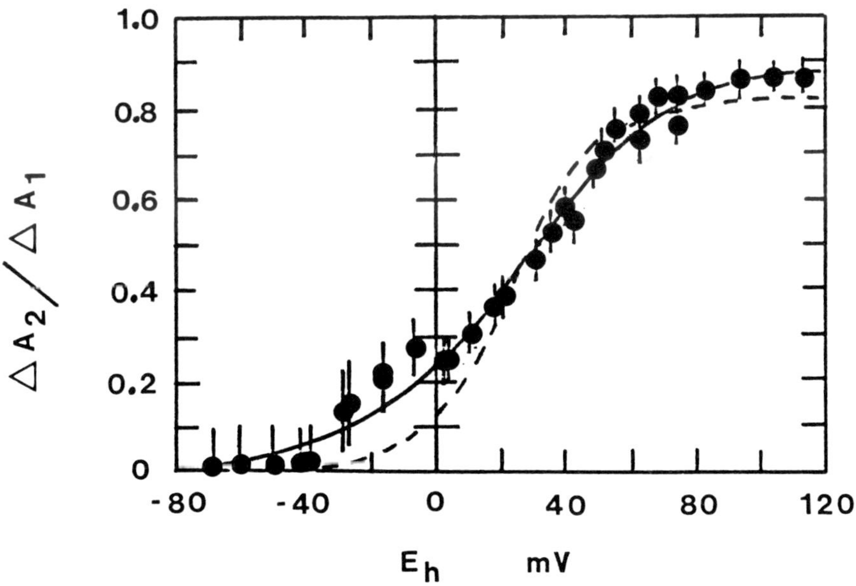

Fig.1 Redox titration of the carotenoid signal (ΔA_2) generated by the second flash in a sequence, normalized to the signal induced by the first flash (ΔA_1). Dark time between photoactivations: 5 ms. Continuous line: best fit according to eqn.1; dashed line: best fit with a single Nernstian wave (n=2).

The two states of the quinone acceptor couple compatible with two closely-spaced turnovers are Q_AQ_B and $Q_AQ_B^-$. Accordingly, the redox titration was analyzed using the equation

$$\frac{\Delta A_2}{\Delta A_1} = C \frac{1 + \exp((E_h - E_2)/26)}{1 + \exp((E_h - E_2)/26) + \exp((E_1 - E_h)/26)} \qquad (1)$$

where E_2 and E_1 are the midpoint potentials of the Q_B/Q_B^- and Q_B^-/Q_BH_2 respectively and the constant C takes into account the incomplete recovery of the system at the 2nd flash. Best fitting is obtained for E_2=77 mV and E_1=30 mV, consistent with the results of ESR equilibrium titrations. The two E_m's are compatible with a stable bound semiquinone (apparent dismutation constant of approximately 0.16).

Extraction of Q_B.
The amount of UQ-10 in chromatophores can be reduced by isoctane extraction. Following a very extensive extraction, part of the Q_B bound to the RC is lost, as can be judged from the kinetics of the back reaction when cyt c_2 is oxidized before the flash (Baccarini Melandri et al., 1982). In fact the rate of charge recombination for $Q_A^-P^+$ ($t_{1/2} \cong 50$ ms) is much faster than that for $Q_B^-P^+$ ($t_{1/2} \cong 1$s) and therefore, when the Q_B- site is partially emptied, the kinetics of the back reaction becomes biphasic. The binding constant (K_B) for the oxidized UQ to the Q_B- site should be otained by studies of the Q_B function (from the kinetics of the back reaction) at different sizes of the UQ pool. In principle, however, the simultaneous exchange of free and bound UQ can interfere kinetically with the back reaction, causing an erroneous estimation of the Q_B occupancy, particularly when the UQ pool is very reduced in size. Although, for this reason, the exact estimation of K_B is still under investigation, it is clear that the Q_B- site is progressively emptied when the number of UQ molecules present in the lipid phase is decreased below about 8 UQ/RC (approximately 16 mM in the lipid phase).

Redox coupling within the quinone acceptor system
It has been observed in chromatophores that o-phenanthroline, which displaces Q_B from the RC (Michel et al, 1986), also shifts the E_m of Q_A/Q_A^- by approximately 50 mV (Prince and Dutton,1978). A similar effect has also beeen observed in RC incorporated in phosphatidyl -choline liposomes containing UQ at a concentration non saturating the Q_B- site (Wraight, 1981). Given the coincident effets

of o-phenanthroline and of the lack of Q_B, the phenomenon suggests a redox coupling between Q_A and Q_B, rather than a distortion of the Q_B- site by the inhibitor. The possibility of extracting from the membrane the UQ pool and Q_B, allows the study of this effect in intact membranes. As shown in Fig.2, the percentage of RC in which the back reaction is slow (taken as an evaluation of the fraction of RC in which Q_B is bound) is correlated with the apparent midpoint potential of Q_A, measured from the redox titration of the amplitude of the fast electrochromic signal (phase I+II), associated with the electron transfer from cyt c_2 to the quinone acceptor complex. The shift towards more positive values of the $E_m(Q_A/Q_A^-)$ is very similar to the shift induced by sub-saturating o-phenanthroline, when the degree of inhibition is evaluated from the kinetics of the back reaction. Moreover, the reconstitution of bound Q_B with exogenous UQ-10 (Venturoli et al., 1986) reverses completely the effect of extraction.

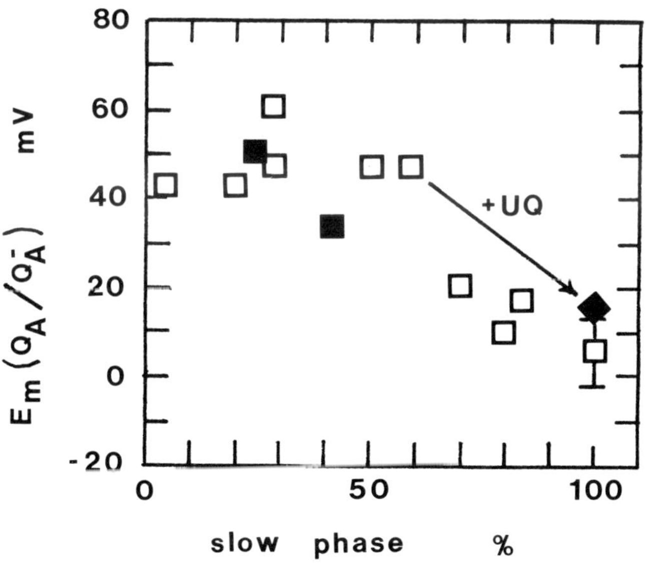

Fig.2. The dependence of $E_m(Q_A/Q_A^-)$ upon the occupancy of the Q_B-site. (□): UQ-extracted chromatophores; (◆): reconstituted with UQ-10; (■): native chromatophores in the presence of sub-saturating o-phenanthroline. The vertical bar at 100% slow phase represents the standard deviation otained for E_m measured in 5 different chromatophore preparations.

Following the analysis suggested by Wraight (1981) the data of Fig.2 can be interpreted in terms of binding constant of Q_B when Q_A is reduced (K_r) or oxidized (K_o). If the redox state of Q_A affects the binding of Q_B:

$$E_m^q(Q_A/Q_A^-) = E_m^o(Q_A/Q_A^-) + 26 \ln \frac{1 + K_r Q_P}{1 + K_o Q_P} \qquad (2)$$

where E_m^q and E_m^o are respectively the midpoint potential of Q_A in the presence or absence of bound Q_B and Q_P is the concentration of quinone in the membrane. If the amplitude of the fast-(A_f) and slow-(A_s) relaxing phases of the back reaction are taken respectively as a measure of the fraction of empty and occupied Q_B sites, substitution in eqn.2 yields

$$E_m^q(Q_A/Q_A^-) = E_m^o(Q_A/Q_A^-) + 26 \ln \frac{A_f/A_s + K_r/K_o}{A_f/A_s + 1} \qquad (3)$$

The best fitting of the data of Fig.2, replotted as $E_m^q(Q_A/Q_A^-)$ vs. A_f/A_s, yields $E_m^o(Q_A/Q_A^-)$=53 mV and K_r/K_o=0.17, i.e. the binding of oxidized UQ to the Q_B- site is 5-6 times stronger when Q_A is oxidized. This effect is consistent with the magnetic coupling of both Q_A and Q_B with the central Fe atom (Feher et al., 1978) and with the interaction of both quinones and Fe with hystidines L190 and M219 (Allen et al., 1988).

Since the E_m of Q_A/Q_A^- is equal to 6 mV, when the Q_B-site is fully occupied, the E_m of the Q_B/Q_B^- couple (77 mV) corresponds to a G of -71 meV for the electron transfer from Q_A^- to Q_B at pH 7. An independent evaluation of ΔG can be obtained from the recombination rate constants for $P^+Q_A^-$ and $P^+Q_B^-$. In fact, the back reaction for the state $P^+Q_B^-$ has been shown to proceed via the $P^+Q_A^-$ state, and the equilibrium constant for the electron transfer from Q_A^- to Q_B can be calculated from the kinetics of charge recombination in the presence and in the abbsence of o-phenantroline (Kleinfeld et al., 1984). This kinetic approach, when applied to our data, yields a ΔG value equal to -74 meV, in excellent agreement with our potentiometric measurements.

REFERENCES

Allen,J.P., Feher,G., Yeates,T.O., Komiya,H. and Rees, D.C. (1988) Structure of the reaction center from Rhodobacter sphaeroides R-26: Protein-cofactor (quinone and Fe) interactions. Proc.Natl.Acad.Sci.USA 85,8487-8491

Baccarini-Melandri,A., Gabellini,N., Melandri,B.A., Jones,K.R., Rutherford,A.W., Crofts,A.R. and Hurt,E. (1982) Differential extraction and structural specificity of specialized ubiquinone molecules in secondary electron transfer in chromatophores from Rhodopseudomonas sphaeroides, Ga. Arch. Biochem.Biophys. 216, 566-580

Crofts,A.R. and Wraight,C.A. (1983) The electrochemical domain of photosynthesis. Biochim. Biophys. Acta 726, 149-185

Kleinfeld,D., Okamura,M.Y. and Feher,G. (1984) Electron transfer in reaction centers of Rhodopseudomonas sphaeroides. I. Determination of the charge recombination pathway of $D^+Q_AQ_B^-$ and free energy and kinetic relations between $Q_A^-Q_B$ and $Q_AQ_B^-$. Biochim. Biophys. Acta 766, 126-140

Michel,H., Epp,O. and Deisenhofer,J. (1986) Pigment - protein interactions in the photosynthetic reaction center from Rhodopseudomonas viridis. The EMBO Journal 5, 2445-2451

Prince,R.C. and Dutton,P.L. (1978) Protonation and the reducing potential of the primary electron acceptor. In "The Photosynthetic Bacteria", Clayton,R.K. and Sistrom,W.R., eds. pp.439-453,Plenum Press, New York

Rutherford,A.W. and Evans,M.C.W. (1980) Direct measurement of the redox potential of the primary and secondary quinone electron acceptors in Rhodopseudomonas sphaeroides (wild type) by EPR spectrometry. FEBS Lett. 110, 257-261

Venturoli,G., Fernandez-Velasco,J.G., Crofts,A.R. and Melandri,B.A. (1986) Demonstration of a collisional interaction of ubiquinol with the ubiquinol - cytochrome c oxidoreductase complex in chromatophores from Rhodobacter sphaeroides. Biochim. Biophys. Acta 851, 340-352

Wraight,C.A. (1981) Oxidation-reduction physical chemistry of the acceptor quinone complex in bacterial photosynthetic reaction centers: evidence for a new model of herbicide activity. Israel J. Chem. 21, 348-354

Wraight,C.A., Cogdell,R.J. and Chance,B. (1978) Ion transport and electrochemical gradients in photosynthetic bacteria. In "The Photosynthetic Bacteria", Clayton,R.K. and Sistrom,W.R., eds. pp.471-511

Acknowledgements: Research partially supported by Consiglio Nazionale delle Ricerche and Ministero della Pubblica Istruzione of Italy.

Structure and Function of Fumarate Reductase, a Quinone-reactive Electron Transport Enzyme of Wolinella succinogenes

C. Körtner, F. Lauterbach, A. Kröger

Institut für Mikrobiologie der J.W.Goethe-Universität Frankfurt am Main

KEYWORDS
Fumarate reductase, Menaquinone, Cytochrome b, Electron transport, Wolinella succinogenes.

INTRODUCTION

W. succinogenes performs electron transport coupled phosphorylation with fumarate as terminal acceptor and either formate (reaction a) or H_2 (reaction b)

$$HCO_2^- + \text{fumarate} + H^+ \longrightarrow CO_2 + \text{succinate} \qquad (a)$$
$$H_2 + \text{fumarate} \longrightarrow \text{succinate} \qquad (b)$$

as donor. The ATP gain is 2 ATP/3 fumarate (Mell et al. 1986). The electron transport chain consists of fumarate reductase, MK and either formate dehydrogenase (Fig. 1) or hydrogenase. This was shown by reconstitution of the functional chains from the isolated enzymes in liposomes (Unden and Kröger 1986). In addition to the electron transport enzymes, the bacterial membrane contains an ATP synthase which looks similar to the corresponding enzyme of aerobic bacteria, mitochondria and chloroplasts (Bokranz et al. 1985). When incorporated into liposomes, the ATP synthase catalyzes the phosphorylation of ATP at the expense of an artifical $\Delta\tilde{\mu}_H$ across the liposomal membrane. The velocity of phosphorylation with the liposomes was nearly that observed with the bacterial membrane, provided that the $\Delta\tilde{\mu}_H$ was close to that generated by the bacterial electron transport (180 mV). Liposomes containing hydrogenase, MK, fumarate reductase and ATP synthase catalyzed phosphorylation

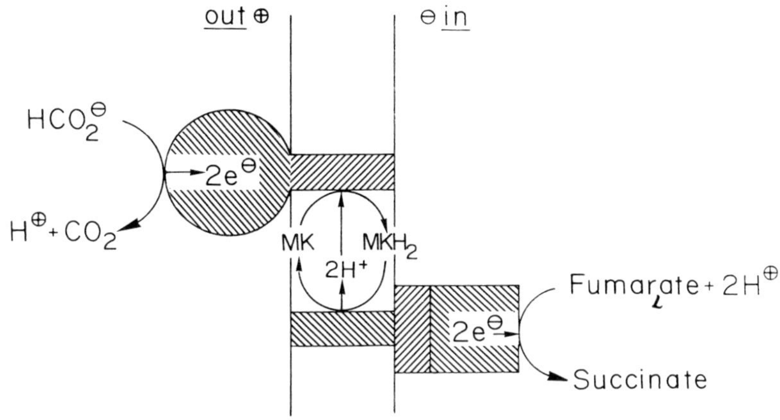

Fig. 1 Schematic drawing of the electron microscopic appearance of fumarate reductase and formate dehydrogenase which are inserted in the membrane (Unden et al. 1983). The enzymes are oriented towards opposite sides of the membrane. Fumarate reductase consists of 1 set of 3 different subunits. Their tentative location is indicated by the hatching.

driven by reaction (b). However the velocity of phosphorylation amounted to only 1% of that measured with the bacterial membrane. This discrepancy raises the question, whether during isolation one of the electron transport enzymes may have lost a constituent which is essential for energy transduction, but is not required in electron transport. To get a clue to this question, we have cloned the genes of fumarate reductase (Lauterbach et al. 1987). As the genes were found to be organized as an operon, their identity could be established by comparison with the partial sequences of the subunits of the isolated enzyme. Furthermore, information on the nature and function of the cytochrome b present in the enzyme was obtained.

THE FRD OPERON

The structural genes of the fumarate reductase of W. succinogenes are localized in an operon in the order promoter-cytb-frdA-frdB-terminator (Körtner 1989). The identity of the genes was verified by comparison of the DNA-sequences with the partial amino acid sequences of the corresponding subunits of the isolated enzyme. Cyt b consists of 269 amino acids, 60% of which are hydrophobic. The molecular weight is 31 kD. FrdA represents the bigger subunit of the enzyme which carries the fumarate site and the covalently bound FAD. The polypeptide is made up of 656 amino acids (molecular weight 74 kD) 65 % of which are hydrophilic (Lauterbach 1989). FrdB is identical with the subunit that carries the 2Fe-2S iron-sulphur cluster. The polypeptide consists of 238 amino acids (molecular weight 27 kD), 63% of which are hydrophilic. This subunit functions as a mediator in the electron transport from cytochrome b to the bigger subunit. The structural genes correspond to the constituents of the isolated enzyme, and no additional structural gene is present in the operon. This is consistent with the view that the isolated fumarate reductase is identical with the functional enzyme within the bacterial membrane.

STRUCTURE OF FUMARATE REDUCTASE

Electron microscopic inspection of lipoyomes containing the isolated fumarate reductase showed cylidrical particles (5nm in diameter and 7nm in length) protruding from the membrane surface (Unden et al. 1983) (Fig. 1). The molecular weight estimated from the dimensions (120 kD) is close to the sum of those of the two hydrophilic subunits FrdA and B (100 kD). An enzyme consisting merely of FrdA and B was earlier isolated from fumarate reductase after cleavage with guanidinium chloride (Unden and Kröger 1981). The enzyme was water-soluble in the absence of a detergent and did not contain cytochrome b. It catalyzed fumarate reduction and succinate oxidation with artificial dyes, but did not react with naphthoquinones. The molecular weight was obtained as 120 kD from its hydrodynamic properties. The molar proportion of the 2 subunits (1:1) was calculated from the distribution among the subunits of labeled amino acids, on which the bacteria were grown. The results suggest that the holoenzyme is anchored in the membrane by cytochrome b, while the hydrophilic subunits protrude into the aqueous phase. As FrdB mediates the electron transfer from cytochrome b to FrdA, it is likely that FrdB is situated on the membrane surface and that FrdA is localized on top of FrdB (Fig. 1).

The number of cytochrome b subunits per molecule of

FrdA or FrdB was measured as 1:1. The molecular weight of fumarate reductase deduced from the hydrodynamic properties suggests that the isolated active enzyme consists of 1 set of the 3 different subunits (Fig.2).

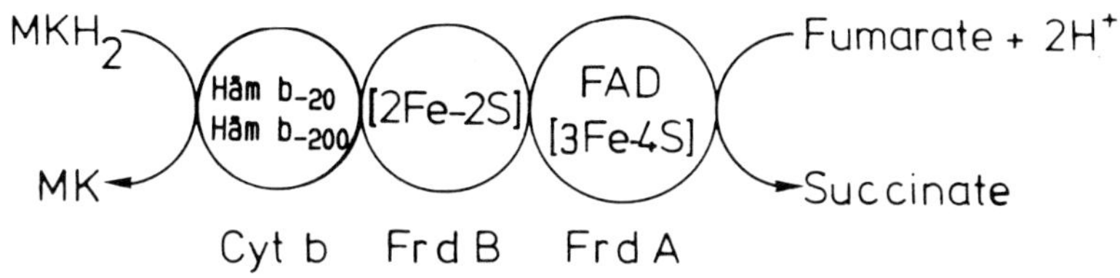

Fig. 2 Composition of fumarate reductase. The enzyme consists of 1 set of 3 different subunits, each carrying the prosthetic groups indicated (Unden and Kröger 1981).

CYTOCHROME B

The isolated fumarate reductase contains 2 heme B groups with different redox potentials (-20 mV and -200 mV). The content of each heme B is equivalent to that of FAD. To find out, whether the heme B groups belong to 1 or 2 polypeptides, the apocytochrome b was fragmented using BrCN or trypsin (Körtner 1989). Seven of these fragments were partically sequenced from the N-terminus (11-26 amino acids). The corresponding DNA-sequences were all present within the cytb-gene of the frd-operon. This strongly suggests that the enzyme contains only one type of apocytochrome b. As this polypeptide is present in only one copy per enzyme molecule, it follows that both heme B groups are bound to this subunit. Thus fumarate reductase of W. succinogenes contains a diheme-cytochrome b (Fig. 2).

A hydrophobicity plot of the amino acid sequence (Fig. 3) suggests that the cytochrome b of fumarate reductase contains 5 membrane-spanning α-helices. Each of the first 4 presumed helices contains a histidine residue in a central position. These 4 residues could possibly serve as ligands of the 2 heme B groups. A similar situation may apply to the cytochrome b present in the succinate dehydrogenase of Bacillus subtilis. This cytochrome probably carries also 2 heme B groups (Magnusson et al. 1986). Alignment of the sequences of the 2 cytochromes b shows

18% identical and 50% homologous amino acids. The homology is concentrated in the helical regions

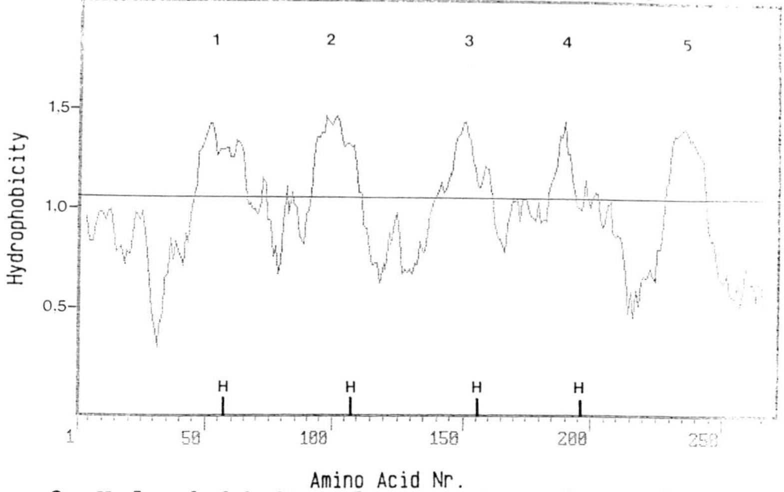

Fig. 3 Hydrophobicity plot of the amino acid sequence of the cytochrome b of fumarate reductase according to Rao and Argos. The sequence was scanned using a window of 7 amino acids. The positions of the conserved histidine residues (H) are indicated.

around the histidine residues. In contrast, no homology with the cytochrome b of the succinate dehydrogenase of E. coli is noticed.

Also the cytochrome b of the bc_1-complex does not show any homology to the Wolinella cytochrome. This argues against the view that a Q-cycle type of mechanism would operate in the electron transport from reduced MK to fumarate. Furthermore, antimycin A and HQNO do not inhibit the electron transport.

The cytochrome b of fumarate reductase functions as the direct acceptor of the electrons of reduced MK. This is concluded from the finding that the cytochrome b-deficient enzyme does not react with quinols and that quinone reactivity is recovered on reconstitution of the enzyme complex from the isolated constituents (Unden and Kroger 1981). The redox response of the high-potential cytochrome b is consistent with a function as an obligatory mediator of the electron transport chain. The redox response of the low-potential cytochrome b differs in that its reduction by formation is considerably slower, while its oxidation by fumarate is commensurate with the velocity of electron transport. This response is in line with the fact that its midpoint potential is about 100 mV more negative than of MK/MKH_2 which is clearly the donor of fumarate reductase in the electron transport from formate to fumarate. It is feasible that the function of the low-potential cytochrome b is that of a ligand to MKH_2, when it is oxidized by the high-potential cytochrome b.

ABBREVIATIONS

MK, menaquinone; MKH_2, reduced MK

HQNO, 3-n-heptyl-4-hydroxyquinoline-N-oxide

REFERENCES

Bokranz, M., Mörschel, E. and Kröger, A., 1985, Phosphorylation and phosphate-ATP exchange catalyzed by the ATP synthase isolated from Wolinella succinogenes. _Biochimica et Biophysica Acta_, 810, 322-339.

Körtner, C., _Thesis_, University of Frankfurt 1989

Lauterbach, F., Körtner, C., Tripier, D. and Unden G., 1987, Cloning and expression of the genes of two fumarate reductase subunits form Wolinella succinogenes. _European Journal of Biochemistry_, 166, 447-452.

Lauterbach, F., _Thesis_, University of Frankfurt 1989

Magnusson, K., Phillips, M.K., Guest, J.R. and Rutberg, L., 1986, Nucleotide sequence of the gene for cytochrome b_{558} of the B. subtilis succinate dehydrogenase complex. _Journal of Bacteriology_, 166, 1067-1071.

Mell, H., Wellnitz, A. and Kröger A., 1986, The electrochemical proton potential and the proton/electron ratio of the electron transport with fumarate in Wolinella succinogenes. _Biochimica et Biophysica Acta_, 852, 212-221.

Unden, G. and Kröger, A., 1986, Reconstitution of a functional electron transport chain from isolated enzymes in liposomes. _Methods in Enzymology_, 126, 387-399.

Unden, G., Mörschel, E., Bokranz, M. and Kröger, A. 1983, Structural properties of the proteoliposomes catalyzing electron transport from formate to fumarate. _Biochimica et Biophysica Acta_, 725,41-48.

Unden, G. and Kröger, A., 1981, The function of subunits of the fumarate reductase complex of Vibrio succinogenes. _European Journal of Biochemistry_, 120, 577-584.

COMPARATIVE ASPECTS OF QUINONES IN BACTERIAL ELECTRON TRANSPORT CHAINS

Giovanni Venturoli, Massimo Trotta* and Davide Zannoni

Dipartimento di Biologia, Istituto Botanico, Universita' di Bologna, Bologna (I);* Centro C.N.R. Interazione Luce Materia, Bari (I)

INTRODUCTION

Chloroflexus aurantiacus is a member of one of the earliest branching lines of the eubacteria (Gibson et al.,1985). The other genera in this grouping, the 'green non-sulphur bacteria', are *Herpetosiphon* and *Thermomicrobium*, with which *Chloroflexus* shares little apparent phenotypic resemblance. Although *Chloroflexus* contains BChls *a* and *c* in its chlorosomes, 16S rRNA sequencing studies have shown no phylogenetic relatedness to the green sulphur bacteria (Gibson et al.,1985). Indeed, in terms of reaction center photochemistry (Bruce et al.,1982), some aspects of electron transport (Zannoni and Ingledew, 1985), the light-dependent energy transducing machinary of *Chloroflexus* most resembles that of the purple non-sulphur bacteria (Venturoli and Zannoni, 1987). Unlike the Rhodospirillaceae, however, the Calvin cycle is apparently not used to fix CO_2 into cell material under autotrophic growth conditions (Holo and Siverag,1987) and a novel mechanism has been postulated. While *Chloroflexus* stains Gram negative, the composition of the cell wall is more similar to that found in Gram positive organisms (Jurgens et al.,1987). Finally, the thermophilic character of *Chloroflexus* (optimum growth at 55 °C) is seemingly rare amongst phototrophic bacteria. These characteristics indicate that *Chloroflexus* is an important organism with which to investigate the evolution of photosynthesis.

Menaquinones (MK) are widespread in photosynthetic bacteria (Zannoni and Melandri, 1985); typically, there is a distinct correlation between the presence of MK and the tolerance to oxygen in RHODOSPIRILLACEAE: menaquinones are present only in those species (*R.fulvum* , *R.molischianum* , *R.gelatinosa*) dependent for aerobic growth on microaerophilic conditions. The rest are obligate anaerobes such as *Chlorobium* , *Chromatium* and *Heliobacterium* genera. MK is normally found associated with a large amount of ubiquinone (UQ/BChl ratio of 1:2 to 1:4). *Chloroflexus* , *Chlorobium* and *Heliobacterium* are exceptions, although there are four other groups of Gram bacteria that do not contain UQ but do grow aerobically. One of these groups is the thermophilic genus *Thermus* and another is the gliding bacteria. Notably, thermophilic and gliding properties are found in *Chloroflexus* . A significant difference between *Chloroflexus* and *Chlorobium* is that the capacity of aerobic growth is present only in the former. Thus *Chloroflexus* is the only chemio-heterotrophic photosynthetic bacterium that contains MK as the sole quinone species (Hale et al.,1983). Previous work indicated that the amount of MK (as µmol/g cell dry wt.) present in light-grown *Chloroflexus* is close to that reported in similar cultures of *Chlorobium* and *Chromatium* (in:Zannoni and Melandri,1985). In these genera, a MK/BChl ratio of 1:20 is normally found. The marked difference between *Chloroflexus* and other phototrophs is that MK-10 is the predominant homologue only in the former.

Here we report on the thermodynamic properties of the MK-acceptor system of light-grown *Chloroflexus* along with a preliminary estimation of the MK/BChl and MK/RC ratios. These data are compared to those available in the literature on green and purple sulphur *plus* non-sulphur phototrophic bacteria.

Keywords : Menaquinone; acceptor quinone-complex; phototrophs; (*Chloroflexus aurantiacus*);

METHODS

Organism cultivation and membrane isolation : *C.aurantiscus* strain J-10-fl was grown in the 'Roux' medium of Pierson and Castenholz (1974) at 55°C (pH8.2) with variable incident light intensities (200-1000 W m^{-2}). Whole membrane fragments, i.e. chlorosome- containing membranes, were prepared as previously described (Zannoni and Ingledew, 1985).

Kinetic spectrophotometry and redox potentiometry : Flash excitation was provided by a xenon lamp (3.25J discharge energy), screened through two layers of Wratten 88A gelatin filter, giving a saturating (>90%) flash of 15us duration at half-maximal intensity (Venturoli and Zannoni, 1988). Redox potentiometry and data analysis were performed as previously described (Venturoli and Zannoni,1988).

High-pressure liquid chromatography (HPLC) : HPLC was used as a method for the quantitative estimation of menaquinone in the extracts. Menaquinone standards (MK-8,-9 and -10) were the kind gift of Dr.J.Bausch and Dr.H.E.Keller of F.Hoffman- LaRoche and Co., Basel, Switzerland.

Protein and pigment determination : Proteins were assayed by using the method of Lowry et al. (1951). BChl- *a* and *c* concentrations were measured in acetone methanol extracts (7:2,v/v) at 769nm and 666nm using absorption coefficients of 68.6 mM^{-1} cm^{-1} and 74 mM^{-1} cm^{-1} ,respectively.

RESULTS AND DISCUSSION

In purified plasma membranes of *Chloroflexus* the Em,8.1 of the primary quinone acceptor (Qa) has been determined to be -50mV, this value being indirectly calculated through the extent of cyt. *c554* photooxidation as a function of the ambient redox potential (Bruce et al.,1982). This latter result, however, was strongly blurred by the fact that the primary photochemistry was insensitive to O-phenantroline, an inhibitor of electron transfer between Qa and Qb. Conversely, we have recently demonstrated that whole membrane fragments from *Chloroflexus* are fully sensitive to O-orthophenantroline along with evidence that the binding of Qb is modulated by the redox and protonation state of Qa (Venturoli and Zannoni, 1988). Our measurements also indicated that Qa titrates at -210mV (pH8.2) with a pH dependence of -60mV/pH up to a pK value of 9.3 (pKa), implying that at physiological pH the equilibrium reduction of Qa is linked to a receipt of a proton. Analysis of the back reaction at high pH suggested a pKb close to 10 along with a free energy difference between Qa^{-} Qb and the $QaQb^{-}$ states of -10meV and -60meV at high and low pH, respectively. Assuming that: 1) *Chloroflexus* has an "orthodox MK-pool"; 2) Qb is in equilibrium with it, and that 3) the Qb-binding properties are similar to those of *Rhodobacter* , the predicted Em7.0 of the MK-pool would be close to -60mV, a value slightly higher than that for extracted MK (-74mV at pH7.0) (Kroger and Unden, 1985).

This is the first report on the MK/RC-BChl ratio in light-grown *Chloroflexus* . As described under MATERIALS AND METHODS, the concentrations of the RC-BChl and MK have been measured by flash spectroscopy and HPLC, respectively. Our data indicate that *Chloroflexus* contains a high amount of MK (MK/RC =100-120) supporting the concept that most of the loosely bound MK might operate as a pool. In Table 1, our results are compared to those published for some genera of phototrophs. Notably, a of $27mM^{-1}$ cm^{-1} at 542nm for the isolated *Chloroflexus* -P865 was calculated. Indeed, in the presence of ascorbate (60µM) and 1,4-naphthoquinone (20µM) a slow relaxation (>400ms) of the RC-bleaching can be observed due to electron transfer to 1,4-NQ from Qa and stabilization of P . The addition of soluble cyt. *c* (20µM) induces a relatively fast reduction (t1/2=4ms) of P^+ due to the electron donation by cyt. *c* (Zannoni et al., in preparation). Quite interestingly, the rate of charge recombination between Qa and P in the isolated RC is very similar to that observed *in vivo* in the presence of O-phenantroline. This result demonstrates that, in contrast to *Rhodobacter* species in which the RCs subunit H has a strong influence on the Qb-binding region, the lack of this subunit from the *in vitro* isolated reaction center of *Chloroflexus aurantiacus* reflects the actual nature of its functional structure *in vivo* (Shiozawa et al., 1989).

Table 1. *(Cells grown in high-light intensity)

Genera	Prenylogue	UQ/BChl	MQ/BChl	Q/RC
Chloroflexus	10	-	0.30	120
Chloroflexus *	10	-	1.20	100
Chlorobium	7	-	0.070	70
Chromatium	7	0.15	0.056	30
Rhodobacter	10	0.15	-	19
Rhodobacter*	10	1.15	-	64
Rhodospirillum	10	0.40	-	87

REFERENCES

Bruce,R.C.,Fuller,R.C. and Blankenship,R.E.,1982, Primary photochemistry in the facultatively aerobic green photosynthetic bacterium *Chloroflexus aurantiacus* Proc. Natl. Acad. Scie., 79, 6532-6536.

Gibson,J., Stackebrandt,E. and Woese,C.R., 1985, The physiology of the green photosynthetic bacteria: Lack of a close relationship between *Chlorobium* and *Chloroflexus* , System. App. Microbiol., 6,152-156.

Hale,M.B.,Blankenship,R.E. and Fuller,R.C.,1983, Menaquinone is the sole quinone in the facultatively aerobic green photosynthetic bacterium *Chloroflexus aurantiacus* , Biochim. Biophys. Acta, 723, 376-382.

Holo,H and Sirevag,R. ,1987, Autotrophic growth and CO_2 fixation of Chloroflexus aurantiacus , Arch.Microbiol., 145,173-180.

Jurgens,U.J.,Meissner,J.,Fischer,U.,Konig,W.R. and Weckesser,J. ,1987, Ornithine as a consituent of the peptidoglycan of Chloroflexus aurantiacus , diaminopimelic cid in that of Chlorobium vibrioforme f. thiosulfatophilum , Arch. Microbiol., 148,72-78.

Kroger,A. and Unden,G.,1985,The function of menaquinone in bacterial electron transport. In Coenzyme Q, edited by G.Lenaz, (Chichester: John Wiley and Sons), pp.285-300.

Lowry,O.H., Rosebrough,N.J., Farr,A.L. and Randall,R.J., 1951, Protein measurements with Folin phenol reagent, J. Biol. Chem., 193, 265-275.

Pierson,B.K. and Castenholz,R.W. A phototrophic gliding filamentous bacterium of hot springs, Chloroflexus aurantiacus . Arch.Microbiol. 143:260-265

Shiozawa,J.A.,Lottspeich,F.,Oesterheldt,D. and Feick.R., 1989, The primary structure of the Chloroflexus aurantiacus reaction-center polypeptides, Eur.J.Biochem.,180,75-84

Venturoli,G. and Zannoni,D.,1987,The mechanism of photosynthetic electron transport and energy transduction by membrane fragments from Chloroflexus .In:Green Photosynthetic Bacteria, edited by Olson,J.M. et al. ,pp.135-143.

Venturoli,G. and Zannoni,D.,1988,Oxidation-reduction thermodynamics of the acceptor quinone complex in whole- membrane fragments from Chloroflexus aurantiacus , Eur. J. Biochem., 178, 503-509.

Zannoni,D. and Ingledew,J.W.,1985,A thermodynamic analysis of the plasma membrane electron trasnsport components in photo- heterotrophycally grown cells of Chloroflexus aurantiacus . An optical and electron paramagnetic resonance study, FEBS Lett., 193,93-98.

Zannoni,D. and Melandri,B.A.,1985,Function of ubiquinone in bacteria.In Coenzyme Q, edited by G.Lenaz, (Chichester: John Wiley and Sons), pp.235-256.

PROTONMOTIVE UBIQUINOL-CYTOCHROME C OXIDOREDUCTASE OF MITOCHONDRIA. A POSSIBLE EXAMPLE OF CO-OPERATIVE ANISOTROPY OF PROTOLYTIC REDOX CATALYSIS.

S.Papa, M.Lorusso, T.Cocco, D.Boffoli and M.Lombardo

Institute of Medical Biochemistry and Chemistry
University of Bari, Bari, Italy

Everybody taking part in this Symposium will agree that the quinone cycle formality (Mitchell,1976,1988),represents a beautifully logical construction providing rational explanation of the redox and protonmotive activity of bc complexes of coupling membranes. The experimental analysis of bc complex, that the Q cycle has largely promoted, has, in fact, led to revision of its original formulation (Crofts et al. 1983; Mitchell,1988) and development of related models(double Q cycle, de Vries,1986; SQ cycle, Wikström and Krab,1986). The present difficulty in discriminating between these specific models and in explaining certain observations (Papa and Lorusso,1984; Rieske,1986) seems, however, to demand further modification of the Q cycle concept. One may wonder to what extent the original idea has to be revised or "falsified", using Karl Popper's words to make it adhere more closely to physical reality.

It is claimed (Mitchell,1976,1988) that strenght of the direct ligand conduction concept, on which the Q cycle is based, stems from redundancy of alternative models (Papa et al. 1973a and b; Papa,1976) derived from the general concept of co-operativity in macromolecules (Wyman,1968).

Our group has been developing, since the seventies (Papa et al.1973a and b; Papa, 1976,1989), proton pumping models which make use of vectorial organization of redox-linked co-operative pK shifts in cytochromes and metalloproteins (vectorial Bohr effects, Papa 1976). We would like to develop here the concept that this capacity of redox proteins, possibly in combination with direct proton transfer in primary

KEYWORDS: Cytochrome c reductase; Proton pump; bc complex.

redox catalysis, can produce mechanisms which are as equally apt as the direct ligand conduction mechanisms to rationalize the phenomenological complexity of protonmotive redox proteins of coupling membranes and develop their further experimental analysis.

Papa et al.(1973a and b) originally proposed that proton pumping in the bc_1 complex of mitochondria derived from vectorial organization of redox Bohr effects of b cytochromes (see also von Jagow and Sebald,1980; Wikström and Krab,1980) and/or the Rieske Fe-S protein (Papa et al.1975).Experimental observations indicated to us, however, that proton pumping in the bc_1 complex could not derive exclusively from redox Bohr effects (Papa et al.1981b,1982). We thus developed a proton pump model in which redox Bohr effects were conceived to be combined in series with protonmotive redox catalysis by protein bound quinones(Papa et al.1981a,1982,1989).

The model denominated Q-gated H^+ pump was derived essentially from two principles:(i) the protolytic properties of the $S\dot{Q}/QH_2$ couple which, because of the large difference in pK's of the $S\dot{Q}H$ and QH_2 species transfers $2H^+/e^-$ i.e.one H_2^+ equivalent.(ii) The general concept that proton transfer anisotropy is imposed on protolytic primary redox catalysis by co-operative pK shifts in protein environments on the opposite sides of the redox centre in the membrane and properly tuned in with the redox process (Papa,1982).

The Q-gated H^+ pump is shown in detail in Fig.1, where also the Q-cycle is presented, so as to allow for a direct comparison of the two models and appreciate the differences.

The main features of the Q-gated pump can be summarized as follows:

1. Existence in the bc complex of a ubiquinone pocket,contributed by various polypeptides, containing only a single ubiquinone molecule Q_B (Lorusso et al.1989). The Q_B pocket is delimited by the cytochrome b_{562} domain (equivalent to the <u>i</u> site of the Q cycle) in the middle (Ohnishi et al.1989) and the Fe-S protein and cytochrome b_{566} domains (<u>o</u> site in the Q cycle) at the outer side of the membrane.

2. At the steady-state Q_B is assumed to turn over between the $S\dot{Q}_B^--Q_BH_2$ state.

3. In the oxidation of one QH_2 molecule of the pool by cytochrome c, the $S\dot{Q}_B^-/Q_BH_2$ couple turns over twice. It first

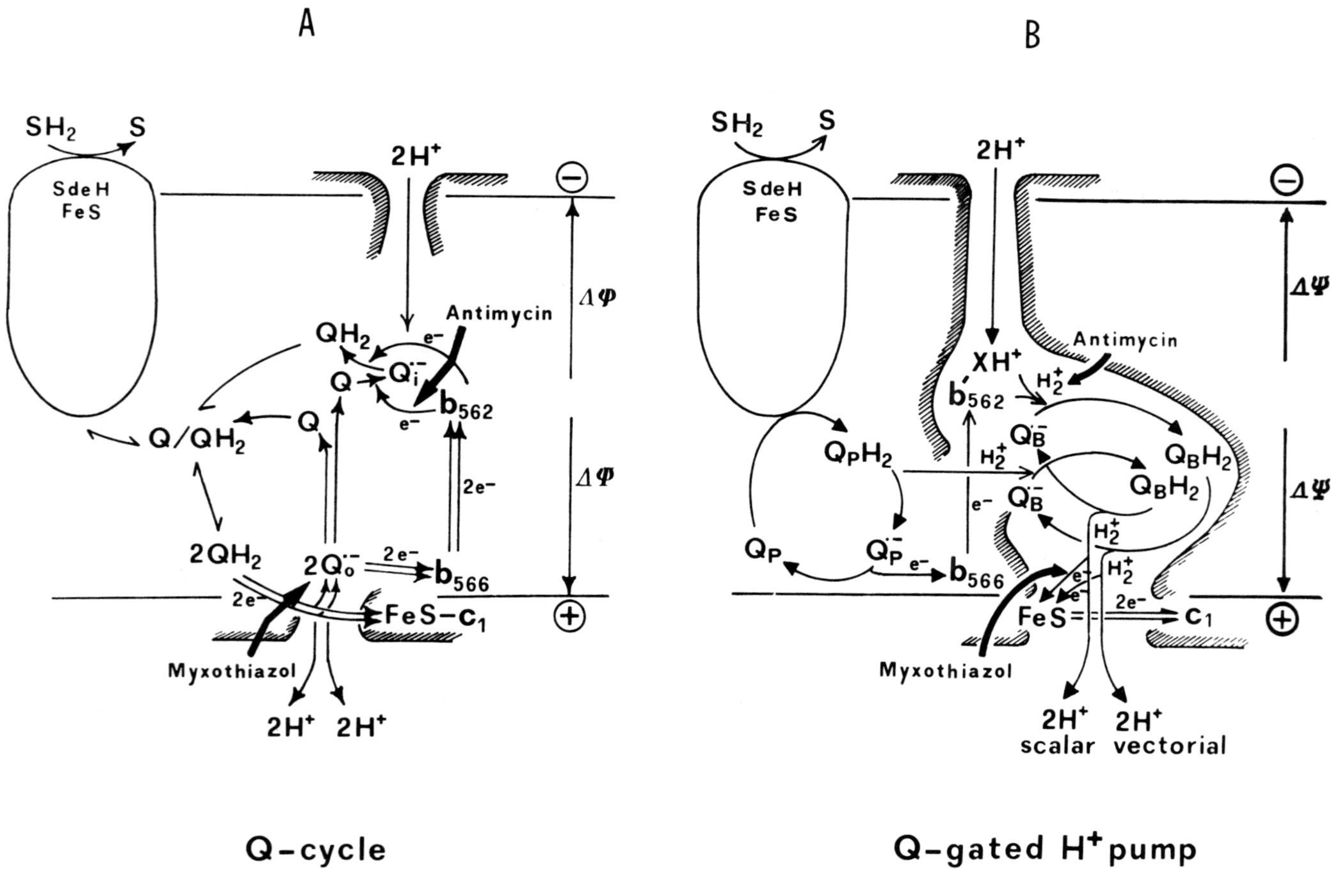

Fig.1. Models for redox and protonmotive activity of bc complex

accepts H_2^+ directly from Q_pH_2 with formation of $\dot{SQ}_p^-.SQ_p^-$, attracted at the outer positive side, reduces b_{566} which rapidly transfers the e^- to b_{562}. Q_BH_2 is oxidized by the Fe-S centre with release of 2(scalar)H^+ in the outer space. $\dot{SQ}_B^-$ stabilized by the Fe-S protein (Lorusso et al.1989), is attracted at the b_{562} domain, charged positively in the reduced state by co-operative protonation (Bohr effect, Papa et al. 1986) from the inner space. $\dot{SQ}_B^-$ is reduced by b_{562} with uptake of $2H^+$ from the inner space. Q_BH_2 is again oxidized by the Fe-S centre with release in the outer space of 2 additional (vectorial) H^+.

4. At the steady-state b_{562} exchanges e^- only with $\dot{SQ}_B$ and does not react directly with quinone of the pool. b_{566} is, instead, reduced by $\dot{SQ}_p^-$, which has a high reducing power.

5. Antimycin displaces $\dot{SQ}_B^-$ from binding sites in the Q_B pocket (Ohnishi and Trumpower,1980). Myxothiazol is assumed to displace Q_BH_2 from the Fe-S protein.

6. Simultaneous binding of antimycin and myxothiazol displaces Q_B from the pocket, thus inhibiting any reaction of Q_pH_2 with the complex. Q_B is also lost from the pocket when antimycin is added to the complex deprived of the Fe-S protein. Full reduction of Q_B to Q_BH_2 by an artificial reductant, or modification of the complex by thiol reagents, inhibit in the presence of antimycin, any reduction of its centres by Q_pH_2. These are the conditions producing the so-called "double-kill" of the complex with suppression of cytochrome b reduction by quinols (see de Vries,1986).

It should be noted that the $\dot{SQ}_B^-$ which is produced in the oxidation of Q_BH_2 by the Fe-S centre and is reduced by Q_pH_2 and by b_{562}, is assumed to be a tightly bound species which is not exchangeable with the quinone of the pool. Thus the Q-gated pump, differently from the Q cycle mechanism, and in particular its version which involve movement of SQ from centre o to centre i (diagram c of Fig.11 in Mitchell,1988; SQ cycle, Wikström and Krab,1986; see also Kröger,1976), does not involve cyclic re-reduction of quinone of the pool. It is a linear split pathway which, with regard to the original split pathway proposed by Wikström and Berden (1972), interposes the couple $\dot{SQ}_B^-/Q_BH_2$ between Q_pH_2 and the Fe-S centre.

The partial steps of electron transfer in bc complex have been the subject of intensive investigation and a large body of observations have been produced with which any mechanism proposed for the redox and protonmotive activity of the complex has to be consistent. It appears, in fact, that the Q-gated H^+ pump can provide satisfactory explanation not only for those observations which are already explained in terms of the Q cycle but also for others which seem difficult to rationalize in terms of the latter.

Studies by van Hoek et al.(1987) on the pre-steady-state kinetics of reduction of the prosthetic groups of mitochondrial bc_1 complex by duroquinol show rapid (5 ms) reduction of 1 mole of endogenous ubiquinone per c_1(8% of the total)to antimycin sensitive SQ with concomitant reduction of b cytochromes. These authors conclude "that during pre-steady-state", cytochrome b_{562} is not in equilibrium with the Q pool and that the rate of equilibration is probably determined by the rate of dissociation of the special bound quinol from centre i" This conclusion, which is considered to be consistent with the double Q cycle (de Vries,1986), seems to fit nicely with the predictions of the Q-gated pump. Consistent with this model also seems to be the finding of Linke et al.(1986) that in ubiquinone depleted bc_1 complex the rate of reduction of cytochrome c_1 by duroquinol is greatly promoted by the binding of a stoicheiometric amount of ubiquinone-10 to the complex. The enhancement of the rate of c_1 reduction,which was suppressed by antimycin A,was accompanied by a myxothiazol-sensitive depression of the pre-steady-state reduction of b cytochromes. It should, however, be noted that these observations were considered by Linke et al.(1986) to provide support for the double Q cycle.

Certain data which are difficult to be explained in terms of the Q-cycle, unless ad-hoc assumptions are made (Rich 1986; Mitchell,1988), seem, on the other hand, to be easily explained by the Q-gated pump. There is for example the observation that on initiation of electron flow by an oxidant pulse of the reduced bc_1 complex, oxidation of cytochrome c_1 is accompanied by an equally rapid oxidation of both cytochromes b_{562} and b_{566} (see Fig.2 of Papa et al.1981a),also when the b cytochromes are fully reduced by dithionite(Rich and Wikström, 1986). This is unexpected in the Q-cycle where the oxidant of b_{562} is the quinone produced by SQ-reduction of b_{566} at center o. In the Q-gated pump $S\dot{Q}_B^-$ produced in the oxidation of Q_BH_2 by the Fe-S center, is, instead,the direct oxidant of b_{562}. Also in the SQ cycle (Wikström and Krab, 1986), the SQ produced by oxidation of QH_2 by the Fe-S center is the oxidant of b_{562}, but in this mechanism,at difference with the Q-gated pump, QH_2 and SQ exchange with the Q-pool.

The transient oxidation of b cytochromes observed in their pre-steady-state reduction by quinols (Jin et al.1981) and the inhibition by antimycin of reduction of cytochrome c_1 (Degli Esposti et al.1982;Degli Esposti and Lenaz,1982; Jin et al.1981) require additional assumptions to be explained in terms of cyclic mechanisms (Rich,1983,1986).The same observations can,on the other hand,be directly explained by the Q-gated pump where the $S\dot{Q}_B^-/Q_BH_2$ couple mediates antimycin-sensitive electron flow from b_{562} to the Fe-S center and c_1.

A central feature of the Q-gated pump is that electron flow from b_{562} to FeS-center is coupled to electrogenic vectorial translocation of $2H^+/e^-$ from the inner to the outer space.This was supported by our finding (Papa et al.1983) that $\Delta\psi$ collapse by valinomycin promoted e^- flow not only from b_{566} to b_{562} but also from the latter to c_1. This observation confirmed by others (Glaser et al.1984), is however considered to be equally consistent with the SQ cycle, and, on the basis of further extensions, also with the Q cycle (Mitchell,1988).

Evidence has been presented indicating that cytochrome b (Yu and Yu,1982),the 14 kDa polypeptide (T.E.King,this volume) and the 11 kDa polypeptide (Yu et al.this volume) contribute to the binding of ubiquinone in the bc complex of mitochondria. Our group has shown that proteolytic cleavage of the Fe-S protein of isolated bc_1 complex to a fragment which was released from the complex, resulted in destabilization of the antimycin sensitive semiquinone radical and inhibition of redox activity (Lorusso et al.1989).Direct aminoacid sequencing of the proteolytic fragment has revealed that it resulted from the digestion of an N-terminal segment of 63 residues (Fig.2). It should be recalled that the Fe-S center is associated to the carboxyl-region of the protein (Gatti et al.1989). The cleaved N-terminal segment has an abundance of polar residues, 40 out of 63, and in mitoplasts is susceptible to proteolytic cleavage at various points(Gonzales-Halphen et al.1988).Thus it represents an amphiphatic anchor

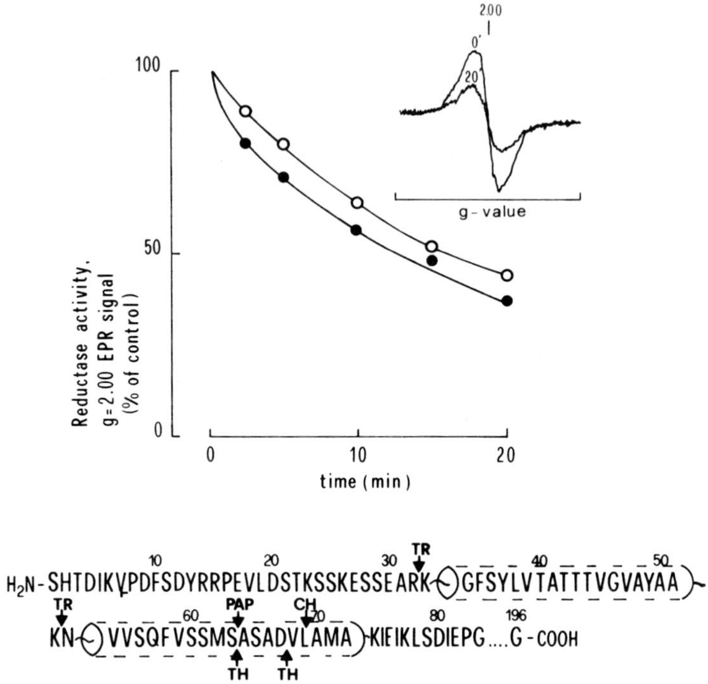

Fig.2. Papain digestion of the N-terminal region ot the Rieske Fe-S protein of bovine-heart bc_1 complex,destabilization of endogenous semiquinone and inhibition of redox activity.
The isolated bc_1 complex was digested with papain. The proteolytic fragment of the Fe-S protein, apparent Mr 19000 from SDS PAGE, was electrotransferred onto immobilon sheets and sequenced. This showed that the N-terminal portion was cleaved off up to residue 63 (T.Cocco, M.Lombardo,G.Tedeschi,S.Ronchi,M.Lorusso and S.Papa, in preparation). The NH_2 terminal sequence (Schägger et al. 1987) and the sites of proteolytic cleavage by papain (PAP), trypsin(TR),chymotrypsin(CH) and thermolysin(TH) (see Gonzales-Halphen et al.1988), are shown at the bottom. For details see Lorusso et al.1989.

which attaches the Fe-S protein to a region of the complex situated rather superficially at the C side of the membrane. The observation that the Fe-S protein contributes to stabilize the same semiquinone species destabilized by antimycin supports the idea that there is in the complex only one compact ubiquinone pocket with a single quinone molecule/c_1 whose benzenoid head oscillates in the "turning over" enzyme between the reduction site(s) and oxidation site in the Fe-S protein. The N-terminal 63 residues anchor of the Fe-S protein, with 21 hydroxyl-residues, 9 positively charged residues, 9 acidic residues (Fig.2) can provide binding sites for the quinone and constitute a H^+ conducting pathway from the quinone pocket to the outer space.

There is evidence that the conformation of the bc_1 complex is affected by redox state of the prosthetic groups. Reduction of the high redox potential centers (Fe-S center and c_1), which was reported to stabilize the complex against dissociation by guanidinium salts into b and c_1 moieties (Baum et al.1967) resulted in a "squeezing out" of hydrophylic N-terminal segments of Core protein II and the 14 kDa protein out of the M side of the membrane (Gonzales-Halphen et al.1988;Lorusso et al.1989) and of the Rieske Fe-S protein out of the C side (Fig.3). These observations indicate that the conformation of the enzyme in the "turning over" state may differ substantially from those of the oxidized and the reduced enzyme (a situation then analogous to that described for cytochrome c oxidase, Brunori et al.1979). If this were the case, observations obtained in the pre-steady-state reduction and oxidation of the complex, respectively in the oxidized and in the reduced state may not be directly applicable to the "turning-over" enzyme.

The possibility envisaged in the Q-gated pump, that segments of constituent proteins may provide asymmetric proton conduction from the M aqueous phase to the quinone pocket for protonation of Q_B^{2-} to Q_BH_2 is supported by the observation that proteolytic digestion of the isolated bc_1 complex under conditions leading to cleavage of the N-terminal segment of the 14 kDa protein results in decoupling of the proton pumping activity (Fig.3 and 4).

CONCLUSIONS

Two essential features of the Q-gated pump are represented by the proposed existence in the bc complex of a single tightly bound ubiquinone species not exchangeable with the Q pool and by <u>direct</u> transfer of reducing equivalents between this spe-

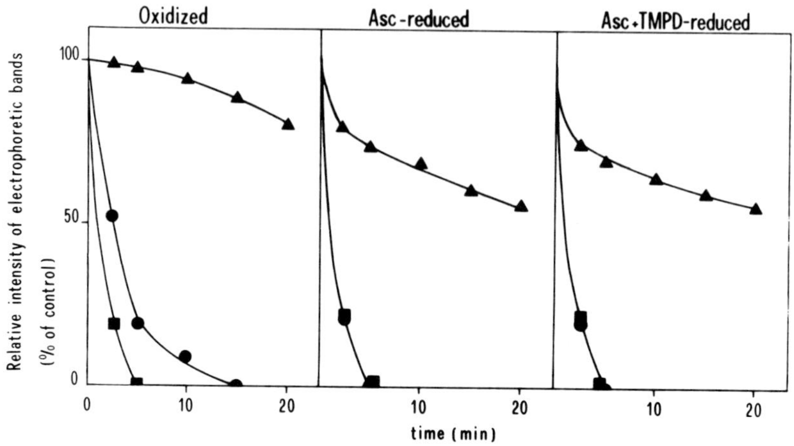

FeS-protein: H_2N-S....SMSASADVLAMAKIEIK....G-COOH (residues 1, 61, 71, 196; arrow after SMS)

14kDa : H_2N-AGRPAVSASSRWLEGIRKWYYNAAGF....K-COOH (residues 1, 11, 21, 110; arrow after AGRPAVS)

Core II : H_2N-SLKVAPKVKATEAPAGVPPHPQDLEFRR....-COOH (residues 1, 11, 21; arrow after SLKVAPKVK)

Fig.3. Densitometric analysis and aminoacid sequencing of polypeptides of papain-digested soluble bc_1 complex. (Lorusso et al.1989; T.Cocco,M.Lombardo,G.Tedeschi,S.Ronchi, M.Lorusso and S.Papa, in preparation); ▲—▲ Fe-S protein; ●—● 14 kDa protein; ■—■ Core protein II. The N-terminal sequences reported are from Schägger et al. 1987; Wakabayashi et al.1985 and Gonzales-Halphen et al. 1988, respectively.The arrows indicate the cleavage sites by papain.

cies and quinone of the pool. These predictions find precedents in the photoreaction center of bacteria (Breton and Vermiglio,1988). Both the reaction center and bc complex will contain a single tightly bound quinone species never exchanged with the Q pool. This is Q_A in the reaction center and what we call Q_B (from bound) in the bc complex. However,whilst Q_A "turns over" in the $Q_A/S\dot{Q}_A$ state, Q_B of bc complex "turns over" in the $S\dot{Q}_B^-/Q_BH_2$ state. The different behaviour will be determined by the protein environment and the bonds that the quinone species exchange with specific residues.

In both systems there will occur a direct transfer of redu-

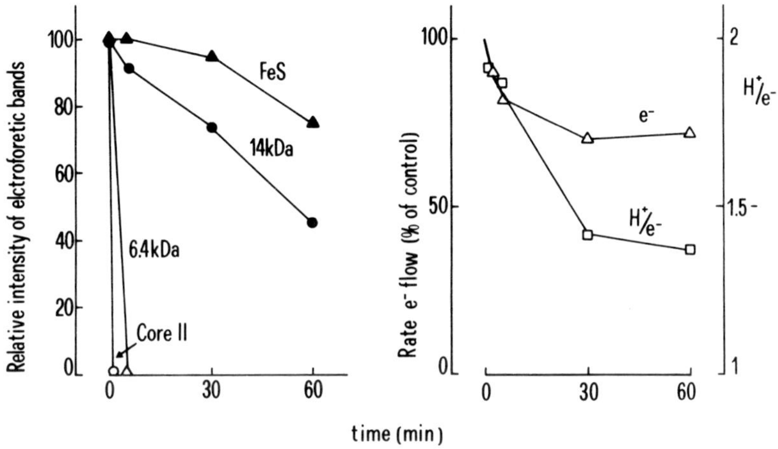

Fig.4. Trypsin digestion of polypeptides of bc_1 complex, inhibition of cytochrome c reductase activity and decoupling of proton pumping in the reconstituted complex. Oxidized bc_1 complex (60 μM c_1) was incubated with trypsin at a concentration of 1 mg trypsin/200 nmol complex for the intervals indicated, digestion being blocked by addition of the trypsin inhibitor.

cing equivalents between the tightly bound quinone species and quinone of the pool. In the reaction center the exchangeable quinone is made to react by the specific domain in the protein as a two e^- gate reduction system according to the sequence:

$$1 \text{ flash} \quad \underset{*}{\dot{Q}}_A \underset{*}{Q}_B \longrightarrow \underset{*}{Q}_A \underset{*}{\dot{Q}}_B$$

$$2 \text{ flash} \quad \underset{*}{\dot{Q}}_A \underset{*}{\dot{Q}}_B \xrightarrow[\uparrow \, 2H^+]{} \underset{*}{Q}_A \underset{*}{Q}_B H_2$$

$$\underset{*}{Q}_A \underset{*}{Q}_B H_2 \longrightarrow \underset{*}{Q}_A + Q_B H_2$$

In the bc complex quinol of the pool will be oxidized to Q in two one e^- transfer steps according to the sequence:

$$Q_p H_2 + \underset{*}{\dot{Q}}_B^- \longrightarrow \underset{*}{Q}_B H_2 \underset{*}{\dot{Q}}_p^-$$

(the asterisk denotes protein bound quinone species).

Differences in the binding parameters of the various quinone species make the reaction center (or the NADH and succinate dehydrogenase of mitochondria) a Q-reductase and the bc complex a Q-oxidase. In the reaction center $KdQ < KdQH_2$, in the bc complex $Kd\ QH_2 < KdQ$ and $Kd\ \dot{Q}_B$ in the reaction center lower than $Kd\dot{Q}_p$ in the bc complex.

REFERENCES

Baum,H., Rieske,J.S., Silman,H.I. and Lipton,S.H.,1967, On the mechanism of electron transfer in complex III of the electron transfer chain.*Proc.Natl.Acad.Sci.USA.USA* 57,798-805

Breton,J. and Vermiglio,A.,1988,*The photosynthetic Bacterial Reaction Center,Structure and Dynamics*, Plenum, N.Y.

Brunori,M., Colosimo,A.,Rainoni,G.,Wilson,M.T. and Antonini,E. 1979, Functional intermediates of cytochrome oxidase. *J. Biol.Chem.*,254, 10769-10775.

Crofts, A.L.,Meinhardt,S.W., Jones, K.R. and Snozzi,M.1983, The role of the quinone pool in the cyclic electron transfer chain of Rhodopseudomonas Sphaeroides. A modified Q-cycle mechanism. *Biochim.Biophys.Acta*, 723, 202-218.

Degli Esposti, M. and Lenaz,G.,1982,Effect of antimycin on the rapid reduction of cytochrome c, in the bc_1 region of the mitochondrial respiratory chain, *FEBS Lett.*,142, 49-53.

Degli Esposti,M., Lenaz,G., Izzo,G. and Papa,S.,1982,Kinetics and sidedness of ubiquinol-cytochrome c reductase in beef-heart mitochondria,*FEBS Lett.*,146, 101-105.

de Vries,S.,1986, The pathway of electron transfer in the dimeric QH_2-cytochrome c oxidoreductase. *J.Bioenergetics Biomembranes*, 18, 195-224.

Gatti,D.L., Meinhardt,S.W., Ohnishi,T. and Tzagoloff,A.,1989, Structure and function of the mitochondrial bc_1 complex. A mutational analysis of the yeast Rieske iron-sulfur protein. *J.Mol.Biol*., 205, 421-435.

Glaser,A.G.,.and Crofts,A.R.,1984, A new electrogenic step in the ubiquinol-cytochrome c_2 oxidoreductase complex of Rhodopseudomonas Sphaeroides.*Biochim.Biophys.Acta*, 766,322-330.

Gonzales-Halphen,D.,Lindorfer,M.A. and Capaldi,R.A.,1988,Subunit arrangement in beef-heart complex III.*Biochemistry*,27 7021-7031.

Jin,Y.Z., Tang,H.L.,Li,S.L. and Tsou,C.L.,1981, The triphasic reduction of cytochrome b in succinate-cytochrome c reductase. Biochim.Biophys.Acta, 637, 551-554.

King, T.E., This volume.

Kröger,A.,1976, The interaction of radicals of ubiquinone in mitochondrial electron transport. FEBS Lett.,65,278-280.

Linke,P.,Bechmann, G.,Gothe,A. and Weiss,H.,1986, Dimeric ubiquinol cytochrome c reductase of Neurospora mitochondria contains one co-operative ubiquinone-reduction centre. Eur.J.Biochem, 158, 615-621.

Lorusso,M.,Cocco,T.,Boffoli,D.,Gatti,D.,Meinhardt,S.,Ohnishi,T. and Papa,S.,1989, Effect of papain digestion on polypeptide subunits and electron transfer pathways in mitochondrial bc_1 complex. Eur.J.Biochem.,179, 535-540.

Mitchell,P.,1976, Possible molecular mechanism of the protonmotive function of cytochrome system. J.Theor.Biol.,62,327-367.

Mitchell,P.,1988,Respiratory chain systems in theory and practice. In:Biochemistry and Bioenergegics, Kim.C.H.,Tedeschi,H.,Diwan, J.J. and Salerno,J.C.(Eds.),Plenum Publ.N.Y.,pp.25-52.

Ohnishi,T., Schägger,H.,Meinhardt,S.W.,LoBrutto,R.,Link,T.A.and von Jagow,G.,1989, Spatial organization of the redox active centers in bovine heart ubiquinol-cytochrome c oxidoreductase. J.Biol.Chem.,264, 735-744.

Ohnishi, T. and Trumpower,B.L.,1980,Differential effects of antimycin on ubisemiquinone bound in different environments in isolated succinate cytochrome c reductase complex. J. Biol.Chem., 255, 3278-3284.

Papa,S., 1976, Proton translocation reactions in the respiratory chains.Biochim.Biophys.Acta,456, 39-84.

Papa,S.,1982,Molecular mechanism of proton translocation by the cytochrome system and the ATPase of mitochondria.Role of proteins. J.Bioenergetics Biomembranes,14, 69-86.

Papa,S.,1989, Redox-linked proton transfer translocation in coupling membranes. In:Highlights of Modern Biochemistry,Kotik, A.,Skoda,J.,Paces,V. and Kostra,V.(Eds.),VSP International Science Publishers, Zeist, pp.781-796

Papa,S.,Guerrieri,F.,Lorusso,M.,Izzo,G.,Boffoli,D. and Maida,I. 1981b, Redox Bohr effects in the cytochrome system of mitochondria and their role in oxido-reductions and proton translocation. In:Vectorial Reactions in Electron and Ion Transport in Mitochondria and Bacteria, Palmieri,F.,Quagliariello E.,Siliprandi,N.,and Slater,E.C. (Eds.), Elsevier,Amsterdam, pp.57-69.

Papa,S., Guerrieri,F.,Lorusso,M.,Izzo,G. and Capuano,F.,1982, The proton translocation function of the ubiquinone-cytochrome-c-oxidoreductase of mitochondria. In:Function of Quinone in Energy Conserving Systems, Trumpower,BL.(Ed.), Academic Press N.Y.,pp.527-539.

Papa,S., Guerrieri,F.,Simone,S.,and Lorusso,M.,1973a,Mechanism of the respiration-driven proton translocation in mitochondria.In:Mechanism in Bioenergetics. Azzone,G.F.,Ernster, L.,Papa,S.,Quagliariello,E.,and Siliprandi,N.(Eds.), Academic Press;New York,pp.451-472.

Papa,S.,Guerrieri,F.,Lorusso,M. and Simone,S.,1973b,Proton translocation and energy transduction in mitochondria,Biochemie, 55, 703-716.

Papa,S.and Lorusso,M.,1984, The cytochrome chain of mitochondria:electron transfer reactions and transmembrane proton translocation. In:Biomembranes, Burton,R.M. and Carvalho-Guerra,F. (Eds.),Plenum Publ.N.Y.,pp.257-290.

Papa,S.,Lorusso,M.,Boffoli,D. and Bellomo,E.,1983,Redox-linked proton translocation in the bc_1 complex from beef-heart reconstituted in phospholipid vesicles. General characteristics and control of electron flow by $\Delta\mu H^+$. Eur.J.Biochem.,137, 405-412.

Papa,S.,Lorusso,M.,and Guerrieri,F.,1975,Mechanism of respiration driven proton translocation in the inner mitochondrial membrane.Analysis of proton translocation associated with oxidation of endogenous ubiquinol. Biochim.Biophis.Acta, 387, 425-435.

Papa,S.,Lorusso,M.,Izzo,G.,and Capuano,F.,1981a,Control of electron transfer in the cytochrome system of mitochondria by pH, transmembrane pH gradient and electrical potential. The cytochrome bc segment. Biochem.J.,194,395-406.

Rieske,J.S.,1986,Experimental observations on the structure and function of mitochondrial complex III that are unresolved by the protonmotive ubiquinone-cycle hypothesis. J.Bioenergetics Biomembranes,18, 235-257.

Rich,P.,1983, Electron Transfer through the isolated mitochondrial cytochrome bc_1 complex.Biochim.Biophys.Acta,722,271-2

Rich,P.,1986,A perspective on Q-cycles,J.Bioenergetics Biomembranes, 18, 145-156.

Rich,P. and Wikström,M.K.F.,1986,Evidence for a mobile semiquinone in the redox cycle of the mammalian cytochrome c_1 complex, FEBS Lett.,194, 176-182.

Schägger,H., Borchart,U.,Machleidt,W.,Link,T.A. and von Jagow, G.,1987, Isolation and aminoacid sequence of the 'Rieske' iron sulfur protein of beef-heart ubiquinol:cytochrome c reductase. FEBS Lett.,219, 161-169.

Van Hoek,A.N., Van Gaalen,M.C.M., de Vries,S. and Berden,J.A., 1987, Pre-steady-state reduction kinetics of QH_2 cytochrome c oxidoreductase and the Q pool, evidence for a special quinone not in rapid equilibrium with the Q pool.Biochim.Biophys Acta, 892, 152-161.

Von Jagow, G.,and Sebald,W.,1980, b-Type cytochromes,Ann.Rev. Biochem.,49, 281-314.

Wakabayashi,S., Takeo,T.,Shimonishi,Y.,Kuramitsu,S.,Matsubara, H.,Wang,T., Zhang,Z.,King,T.E.,1985,Complete aminoacid sequence of the ubiquinone binding protein (Q_p-C), a protein similar to the 14,000- Dalton subunit of the yeast ubiquinol-cytochrome c reductase complex. J.Biol.Chem., 260,337-343.

Wikström, M.K.F. and Berden,J.,1972,Oxidoreductase of cytochrome c in the presence of antimycin.Biochim.Biophys.Acta,283,403-4

Wikström,M.K.F. and Krab,K.,1980, Respiration-linked H^+ transloc tion in mitochondria:stoicheiometry and mechanisms. Cur.Top. Bioenerg.,10, 51-101.

Wikström,M.K.F. and Krab,K.,1986, The semiquinone cycle.A hypothesis of electron transfer and proton translocation in cytochrome bc-type complexes.J.Bioenergetics Biomembranes,18,181-

Wyman,J.,1968,Regulation in macromolecules as illustrated by hae globin. Quant.Rev.Biophys.1, 35-80.

Yu,C.A., Usui,S., Bili,L.,and Yu,L., This Volume.

Yu,L., and Yu.C.A.,1982, The interaction of Arylazide ubiquinone derivative with mitochondrial ubiquinol-cytochrome c reductas J.Biol.Chem.,257,10215-10221.

PARTITION COEFFICIENTS OF QUINONES AND QUINOLS AND THEIR RELATION TO BIOCHEMICAL REACTIVITY

Peter R. Rich

Glynn Research Institute
Bodmin
Cornwall PL30 4AU
U.K.

INTRODUCTION

Ubiquinone reacts physiologically with a diverse range of redox enzymes. In many cases, analogues can be made to substitute for the natural compound. It is of interest to probe the binding sites on the protein for their quinone specificity with a range of differently substituted quinone rings. A difficulty of interpretation in this approach is that a change of the quinone ring substituents often changes two factors in addition to the steric one, viz. the thermodynamics of the redox couples and the partition coefficients of the chemical species. The long polyisoprenyl side chain of the naturally occurring quinones gives the overall molecules partition coefficients which favour the apolar phase by many orders of magnitude, and so makes them reside essentially exclusively in the lipid bilayer (from data in Hansch and Leo (1979), each isoprenyl unit may be calculated to increase the partition coefficient in octanol\water by almost 2.5 orders of magnitude). However, it is the microscopic physical properties of the head group itself which determine the biological activity and specificity. In this report, some major effects of ring substituents on quinone headgroup partition coefficients are described and attention is drawn to the differences in partition coefficients between the two major freely diffusing redox forms, the quinone, Q, and the quinol, QH_2.

Keywords: quinones; quinols; hydroquinones; partition coefficients; bc complex; quinone specificity.

METHODS

Beef heart mitochondrial succinate-cytochrome *c* reductase was prepared by the method of Yu *et al.* (1974) and was stored at 77°K until required.

Partition coefficients were measured at 23°C by shaking the compound in the appropriate two-phase system until no change in distribution occurred. Concentrations were then measured spectrophotometrically after separation of the two phases, and assuming an equal absorption coefficient for the compound in both phases.

Rate constants for quinol reaction with succinate-cytochrome *c* reductase were determined from the rate of cytochrome *c* reduction at pH 7 and 23°C in 50 mM potassium phosphate, 2 mM EDTA and 1 mM KCN, under conditions of high [cytochrome *c*] and low [quinol] so that rates became pseudo-first order in quinol. Second order rate constants were calculated from: k = Rate/([quinol]•[bc complex]).

RESULTS

Figure 1 summarises the effects of methyl group substitution of the quinone ring on partition coefficient between water and either the nonpolar, hydrogen bonding solvent octanol or the nonpolar, non-hydrogen-bonding solvent cyclohexane.

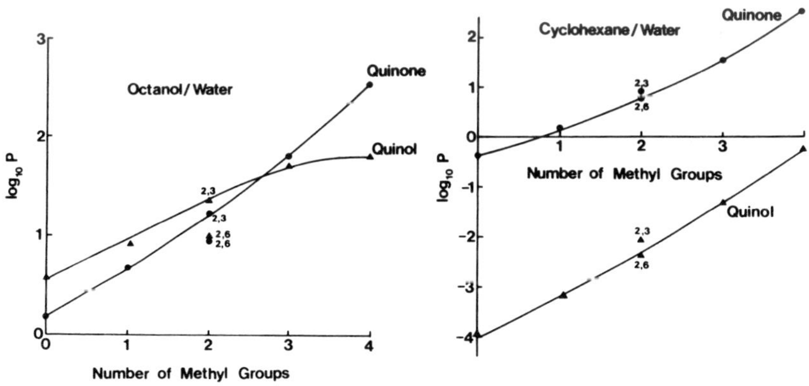

<u>Figure 1. Partition coefficients of methyl-substituted *para*-benzoquinones/ols.</u> The points for the 2,3-dimethyl- or the trimethyl- derivative may be considered reasonable models of the headgroup of the naturally-occurring plastoquinone.

The results are much as predicted by calculation by the FRAGMENT method (Hansch and Leo, 1979). The quinone and quinol partition rather similarly in octanol/water, but the quinol partitions much more strongly into the water phase in cyclohexane/water. The behaviour can be rationalised in terms of two major factors, the polarity of the molecules and their propensity to form hydrogen bonds. The quinol, compared to the quinone, is of roughly similar polarity but can form stronger hydrogen bonds with either water or with octanol (but not with cyclohexane). Hence quinol, in comparison to quinone, partitions more into water in the cyclohexane/water system, but in octanol/water is rather similar in behaviour to the quinone.

Addition of each methyl group causes further partitioning into the hydrophobic phase in all cases as expected. Addition of methoxy, rather than methyl, groups in general is predicted to cause little change in partition coefficient compared to the parent compound. This is roughly borne out experimentally (figure 2) for partitioning of both quinone and quinol in octanol/water and of quinone in cyclohexane/water. However, the partition coefficients of the methoxy-quinols in cyclohexane/water change by roughly an order of magnitude

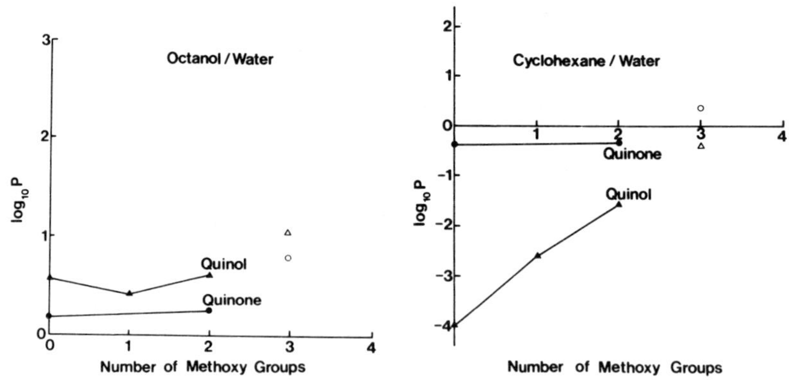

Figure 2. Partition coefficients of methoxy- and 2,3-dimethoxy- *para*-benzoquinones/ols. Also included are points for the quinone (○) and quinol (△) forms of ubiquinone-0 (2,3-dimethoxy-5-methyl-*p*-benzoquinone).

for each group added for the mono- and 2,3-disubstituted compounds. It is suggested that this is caused by intramolecular hydrogen bonding between the methoxy- and hydroxy- substituents on the quinol ring which can occur in a non-hydrogen-bonding solvent.

RELEVANCE TO BIOLOGICAL SYSTEMS

Also included in figure 2 are the measured partition coefficients for the headgroup of the naturally occurring

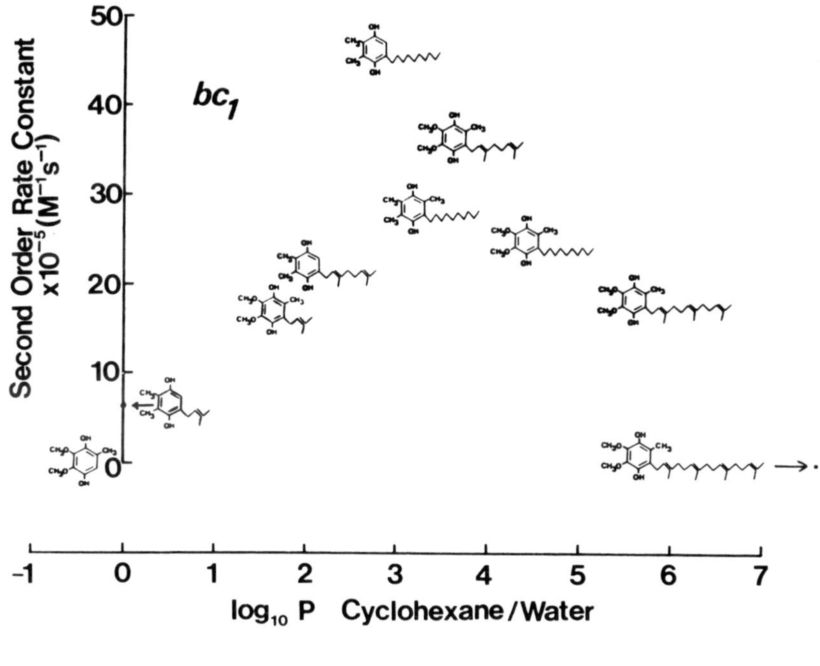

Figure 3. Second order rate constants for oxidation of quinols by beef heart mitochondrial *bc*$_1$ complex. Rate constants for quinol oxidation were determined with beef heart mitochondrial succinate-cytochrome *c* oxidoreductase. Unmeasurable (>10^3) partition coefficients of the quinols in cyclohexane\water were estimated by adding an offset for the hydrophobic side chain to the measured values of the parent compounds (isoprenyl = 10^2; *n*-decyl = $10^{4.8}$). These offsets were obtained by difference from measured values and are slightly lower than calculated values.

ubiquinone. The values are very much as expected from the above discussion. An important point to note is that the methoxy groups will have the effect of bringing closer together the partition coefficients of the quinone and quinol forms in a non-hydrogen bonding solvent, as would be the case in the interior of the mitochondrial lipid bilayer. Hence the distribution of headgroups within the membrane can be similar for both redox forms.

A major factor for binding of the headgroups into a proteinaceous quinone binding site will presumably be the formation of hydrogen bonds, as is the case with the known structures of the Q_A and Q_B sites in bacterial reaction centres (Allen *et al*., 1987). A function of the very long hydrophobic tail of the natural quinones will be to counteract these binding forces with the entropic forces associated with re-entry into the fluid membrane interior. In this way rapid exchange with the binding site can occur, as has been shown for the Q_B site. In the case of the quinol form, intramolecular hydrogen bonding should also aid re-entry into the quinone pool.

When attempting to determine steric specificity of a quinone binding site, it is important to consider the influence of partition coefficient. For example, a plot of rate constant for reaction of quinol with the mitochondrial bc_1 quinol oxidation site *versus* $\log_{10}P$ (figure 3) suggests that partition coefficient is a major factor in reactivity determination, with steric specificity for sidechain or for methoxy- or methyl- substituents playing little role. A peak in the profile at around $\log_{10}P$ of 3 is seen. The fall in rate constant at higher values probably represents an artefact caused by the very limited aqueous solubility of the quinols involved.

The low steric specificity of the site for the substituents on the quinone ring is unusual in biology. A similar low steric specificity of the Q_A site of bacterial reaction centres has been found by Gunner *et al*. (1986). Presumably, the results indicate some flexibility of the binding sites and the dominance in the binding forces of hydrogen bonding to the quinone moieties. Such low specificity is not necessarily expected to be the case for all quinone binding sites. Indeed, in analogous experiments with the chloroplast cytochrome *bf* complex a limited specificity for plastoquinone-like quinones was found (unpublished data). Succinate dehydrogenase has also been reported to exhibit quinone specificity (Chen *et al*., 1986). However, in both of these cases a true steric specificity, rather than a thermodynamic or partitioning mismatch, remains to be established.

ACKNOWLEDGEMENTS

This work is supported by the Glynn Research Foundation Ltd and was carried out with the able technical assistance R. Harper and S.A. Madgwick.

Allen, J.P., Feher, G., Yeates, T.O., Komiya, H. and Rees, D.C., 1987, Structure of the reaction center from Rhodobacter sphaeroides R-26: The cofactors. Proceedings of the National Academy of Sciences USA, 84, 5730-5734.

Chen, M., Liu, B.-L., Gu, L.-Q. and Zhu, Q.-S., 1986, The effect of ring substituents on the mechanism of interaction of exogenous quinones with the mitochondrial respiratory chain. Biochimica et Biophysica Acta, 851, 469-474.

Gunner, M.R., Robertson, D.E. and Dutton, P.L., 1986, Kinetic studies on the reaction centre protein from Rhodopseudomonas sphaeroides: the temperature dependence and free energy dependence of electron transfer between various quinones in the Q_A site and the oxidized bacteriochlorophyll dimer. Journal of Physical Chemistry, 90, 3783-3795.

Hansch, C. and Leo, A., 1979, Substituent Constants for Correlation Analysis in Chemistry and Biology, John Wiley, New York.

Yu, C.A., Yu, L. and King, T.E., 1974, Soluble cytochrome b-c_1 complex and the reconstitution of succinate-cytochrome *c* reductase. Journal of Biological Chemistry, 249, 4905-4910.

THE MYXOTHIAZOL-SENSITIVE, ANTIMYCIN A-INSENSITIVE RESPIRATORY PATHWAY OF *Candida parapsilosis*

Martine GUERIN, Souad ZNIBER and Nadine CAMOUGRAND

Institut de Biochimie Cellulaire et Neurochimie du CNRS, 1 rue Camille Saint-Saëns, 33077 Bordeaux cedex, France

INTRODUCTION

C.parapsilosis CBS 7154 is a strictly aerobe yeast, lacking fermentative pathway (Caubet *et al.*, 1988) but able to grow on high glucose concentrations as well as on a non-fermentescible carbon source in the presence of antimycin A or of drugs acting on mitoribosomal protein synthesis (Camougrand *et al.*, 1986), due to the peculiar organisation of its respiratory chain (Guérin *et al.*, 1989). Indeed, besides the classical respiratory chain, *C.parapsilosis* develops a second electron transfer route, different from the known alternative pathways. Entry of electrons occurs through dehydrogenases specific for NADH and NADPH respectively, which are amytal- and mersalyl-sensitive, and located on the external side of the inner membrane (Camougrand *et al.*, 1988). Electrons are transferred to a specific pool of cytochrome c and then, either to an alternative oxidase (cytochrome 590) or to cytochrome c oxidase (Guérin and Camougrand, 1986).

What is the intermediate step between NAD(P)H dehydrogenases and cytochrome c ?

Key words : alternative pathway - Rieske's protein - myxothiazol -

RESULTS

First, mitochondria were progressively depleted from endogenous quinones and, at each step, antimycin A-insensitivity of residual respiration with NADH was measured. Then, organic extract was reincorporated in mitochondria and same measurements were taken. From results reported in Table 1, it can be seen that : (i) Lyophilisation of mitochondria decreased the oxidation rate but did not affect functioning of antimycin-insensitive pathway. (ii) After one pentane extraction, 60% of respiratory capacity was lost but the remaining respiration was fully antimycin A-sensitive. Reincorporation of pentane extract led to 63% recovery of initial respiration which was partly antimycin A-insensitive again. (iii) After two pentane extractions, 80% of respiration was lost, but reincorporation of organic extract led to a respiration recovery partly antimycin A-insensitive.

Table 1. Mitochondrial respiration and antimycin A-sensitivity after quinone extraction.

Experimental steps	Respiration rate nat O $min^{-1}.mg^{-1}$	%	Antimycin sensitivity
Control	273		65 %
After lyophilisation	219	100	53 %
First extraction	92	42	100 %
Extract I + supernatant I	140	63	64 %
Second extraction	41	19	100 %
Extract II + supernatant II	115	52	50 %

In a second set of experiments, effect of myxothiazol, another inhibitor of bc1 complex was assayed (Figure 1). This compound acts at another site than antimycin A, and inhibits cytochrome b reduction by blocking oxidation of ubiquinol at the level of the iron-sulfur protein (Von Jagow et al., 1984). As a control experiment, inhibition of oxoglutarate oxidation as a function of myxothiazol concentration was measured : only one inhibitory site was evidenced with Ki similar to that measured with rat liver mitochondria (not shown). On the opposite, when NADH was the substrate, two inhibitory sites were found with half-inhibition constants equal to 0.025 µmole and 0.7 µmole myxothiazol/ mg protein respectively. In the presence of antimycin A, only the low affinity site was effective whereas pentane- treated mitochondria were sensitive to low myxothiazol concentrations.

Figure 1. Effect of myxothiazol on respiration rate of C.parapsilosis mitochondria. Mitochondria were suspended in a 0.65 M mannitol, 10 mM Tris-maleate pH 6.7, 0.36 mM EGTA. Respiration was started by addition of substrate and then, increments of myxothiazol were added. Average of 2 experiments. □—□ NADH 2 mM; ■—■ mitochondria were preincubated with 0.2 μg antimycin A before NADH addition; ◐—◐ mitochondria were submitted to one extraction with pentane; ▲—▲ substrate: 2-oxoglutarate.

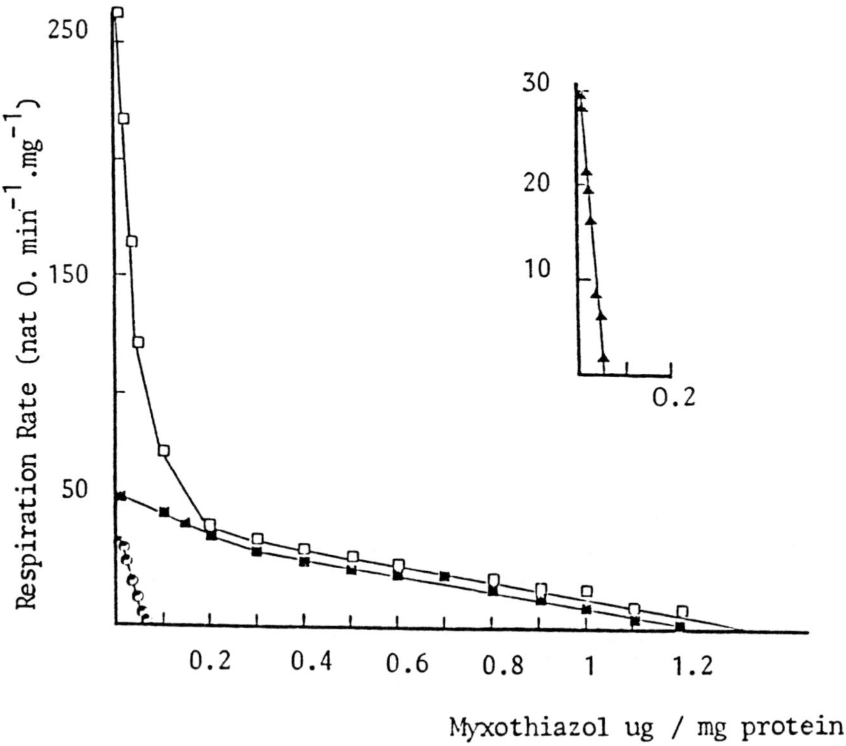

Preliminary analysis of CoQ content of pentane extract was performed by chromatography on silicagel plate : only one quinone species (Q_9) was evidenced.

CONCLUSION

From these results, it was hypothesized that the second respiratory chain of *C.parapsilosis* branched the main chain at the level of Rieske's protein. Two quinone pools should be present in *C.parapsilosis*, one related to Coenzyme Q-cytochrome b cycle and the other to the second respiratory pathway. The quinone pool related to this second chain could be able to bind Rieske's protein either to a site requiring more myxothiazol to be displaced than the site implicated in the main chain, or to an unknown Q-binding protein.

REFERENCES

Camougrand, N., Cheyrou, A., Henry, M.F. and Guérin, M., 1988, The alternative respiratory pathway of the yeast *C.parapsilosis*. Oxidation of exogenous NAD(P)H. *Journal of General microbiology*, 134, 3195-3204.

Camougrand, N., Velours, G. and Guérin, M., 1986, Resistance of *C.parapsilosis* to drugs. *Biology of the Cell*, 58, 71-78.

Caubet, R., Guérin, B. and Guérin, M., 1988, Comparative studies on the glycolytic and hexose monophosphate pathways in *C.parapsilosis* and *S.cerevisiae*. *Archives of Microbiology*, 149, 324-329.

Guérin, M. and Camougrand, N., 1986, The alternative oxidase of *C.parapsilosis*. *European Journal of Biochemistry*, 159, 519-524.

Guérin, M., Camougrand, N., Caubet, R., Zniber, S., Velours, G., Manon, S., Guélin, E. and Cheyrou, A., 1989, The second respiratory chain of *C.parapsilosis* : a comprehensive study. *Biochimie*, in press.

Von Jagow, G., Ljungdahl, P.O., Graf, P., Ohnishi, T. and Trumpower, B.L., 1984, An inhibitor of mitochondrial respiration which binds to cytochrome b and displaces quinone from the iron-sulfur protein of the cytochrome bc1 complex. *The Journal of Biological Chemistry*, 259, 6318-6326.

SOLUBLE NAD(P)H-QUINONE REDUCTASE IN PLANT CELLS

Franca Guerrini, Paolo Pupillo and **Vincenzo Valenti**

Department of Biology, Laboratory of Plant Physiology
via Irnerio n. 42, I-40126 Bologna

INTRODUCTION

Plant cells contain different NADH- and NADPH-dependent quinone reductases and diaphorases. Most of the activity of etiolated tissues is found in high speed supernatants but some is associated to mitochondria and microsomes, essentially plasma membrane (Pupillo et al., 1986; Luster and Buckhout, 1988; Valenti et al., 1989). Microsomal NAD(P)H-quinone reductase of etiolated zucchini hypocotyls has been solubilized and purified (Guerrini et al., 1987).

The major supernatant NAD(P)H-quinone reductase (NAD(P)H-diaphorase) from etiolated plant tissue catalyzes the oxidation of NADPH or NADH using various redox dyes and quinones as acceptors. We are now describing some properties of the partially purified enzyme from maize (*Zea mays* L.), to be compared to those of DT-diaphorases from animal sources.

METHODS

NAD(P)H-quinone reductase was obtained from the 100K g supernatant of etiolated maize coleoptiles (usually 100 g). The usual purification procedure, outlined in Table 1, consists of an ammonium sulfate step (40-55% saturation) followed by acetone fractionation (50% v/v, -20 °C). The resuspended precipitate is applied to a Sephadex G-200 column (2.6 x 60 cm) equilibrated with 50 mM Tris-HCl containing 50 uM $NADP^+$, 10 mM 2-mercaptoethanol, 100 mM NaCl (pH 7.5). The main peak is dialyzed, concentrated and applied to a Blue Ultrogel column. Enzyme elution is developed by a linear NaCl gradient as shown before for the microsomal enzyme (Guerrini et al., 1987).

RESULTS

In preliminary experiments, ammonium sulfate preparations from maize supernatants were investigated by gel filtration. The main NADPH- and NADH-dependent quinone reductase had a mol. wt. of 140 kDa but minor peaks were detectable. The mol. wt. of the main peak did not change with inclusion of substrates in buffer or exclusion of reductant (this being essential for full activity). The enzyme from etiolated zucchini hypocotyls also showed a major 140 kDa band. In contrast, the dominant species in extracts from green leaves is a protein with about 40 kDa. At variance from the microsomal enzyme, soluble NAD(P)H-quinone reductase from dark-grown maize is fairly stable to cold acetone, and some activity is further recovered during the subsequent gel filtration.

Therefore, a suitable combination of purification procedures was chosen (Table 1).

Table 1. Summary of the purification procedure from 100 g etiolated maize coleoptiles.

Step	NADPH-DQR	NADH-DQR	Protein	Purification
	(umol min^{-1})		(mg)	
S100K	91.3	61.4	55	1
$(NH_4)_2SO_4$	72.4	38.0	35.6	1.22
Acetone	31.3	14.2	17.1	1.10
Sephadex G-200	43.0	24.3	4.90	8.78
Blue-U I	12.1	6.7	0.11	66.5
Blue-U II	5.9	3.3	0.08	44.6

Only the 140 kDa peak from Sephadex G-200 was used for further purification. Blue Ultrogel peak I (the principal form, eluting at 280 mM NaCl) and peak II (eluting at 350 mM NaCl) were collected. Specific activities were 50-150 umol min^{-1} mg^{-1} protein using duroquinone as acceptor (see Table 1), representing a purification factor from supernatant 100K g of 50 or more. However, both peaks still contained several bands in SDS-PAGE. Two major polypeptides with mol. wt. 35 and 51 kDa respectively exhibited some correlation with activity profiles of the Blue Ultrogel peaks.

Peak I was used for kinetic studies. Initial velocity responses with NADPH or NADH as donor and ubiquinone-1 (or other quinones) as acceptors were nonlinear in double reciprocal plots, suggesting negative cooperativity. Biphasicity was especially sharp with reduced pyridine nucleotide, with Hill coefficients down to 0.68 (NADPH varied) and 0.50 (NADH varied). Velocity was higher with NADPH than NADH as donor. Extrapolation of the plots yields apparent Km values of approximately 1 uM for both NADPH and NADH in the low-velocity, high-affinity portion, and 25 uM for NADPH (40 uM for NADH) in the high-velocity, low-affinity range. Limiting Km for ubiquinone-1 was 120 uM with NADPH and 40 uM with NADH as donor (Fig. 1).

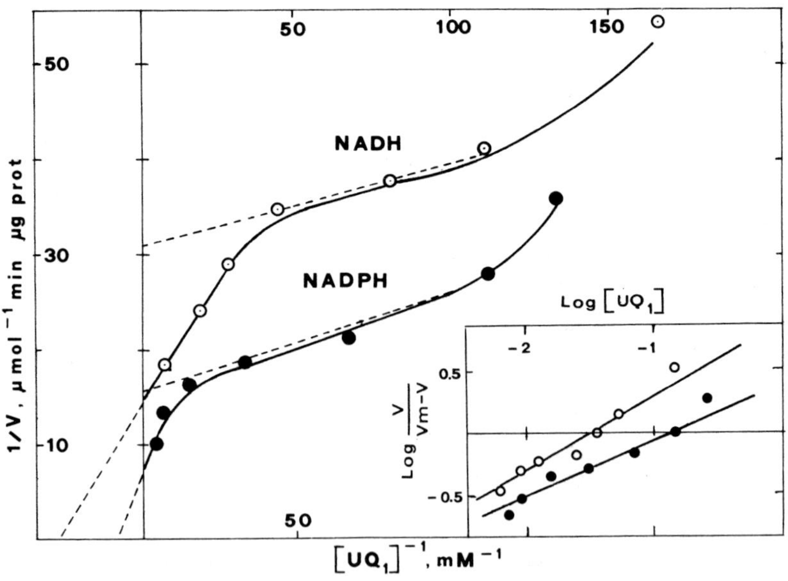

Fig. 1. Dependence of enzyme activity on ubiquinone-1 concentrations. Inset: Determination of Hill coefficients.

Hill coefficients were 0.44 (NADPH donor) and 0.58 (NADH donor). Activity was inhibited by dicoumarol with a Ki of 50 uM (NADH as donor).

CONCLUSIONS

Soluble NAD(P)H-quinone reductase of etiolated plant tissues clearly resembles the microsomal NAD(P)H-diaphorase (Guerrini et al., 1987) in some important properties including molecular weight (140 kDa), dependence on reductants, and kinetics. These features do not match those of DT-diaphorase preparations from animal tissues of Lind et al. (1982) and Prochaska and Talalay (1988) with respect either to molecular size or dicoumarol sensitivity (two order of magnitude higher in the latter case). However, it may be expected that further research may reveal the existence of animal DT-diaphorase types more closely resembling this major plant diaphorase.

REFERENCES

Guerrini, F., Valenti, V. and Pupillo P., 1987, Solubilization and purification of NAD(P)H dehydrogenase of *Cucurbita* microsomes. *Plant Physiology*, 85, 828-834.

Lind, C., Hochstein, P. and Ernster, L., 1982, DT-diaphorase as a quinone reductase: a cellular control device against semiquinone and superoxide radical formation. *Archives of Biochemistry and Biophysics*, 216, 178-185.

Luster, D.G. and Buckhout, T.J., 1988, Characterization and partial purification of multiple electron transport activities in plasma membranes from maize (*Zea mays*) roots. *Physiologia Plantarum*, 73, 339-347.

Prochaska, H.J. and Talalay, P., 1987, Crystallization and characterization of NAD(P)H : quinone reductase from murine liver. *Chemica Scripta*, 27A, 43-47.

Pupillo, P., Valenti, V., De Luca, L. and Hertel, 1986, Kinetic characterization of reduced pyridine nucleotide dehydrogenase (duroquinone-dependent) in *Cucurbita* microsomes, *Plant Physiology*, 80, 384-389.

Valenti, V., Guerrini, F. and Pupillo, P., 1989, NAD(P)H-duroquinone reductase in the plant plasma membrane, *Journal of Experimental Botany*, in press.

MUTANTS OF ESCHERICHIA COLI FUMARATE REDUCTASE ANCHOR PEPTIDES WHICH EFFECT ELECTRON TRANSFER WITH QUINONES

[1]Gary Cecchini, [2]David J. Westenberg, [1]Brian A. C. Ackrell, and [2]Robert P. Gunsalus

[1]Molecular Biology (151S), VA Medical Center, and Department of Biochemistry & Biophysics, University of California at San Francisco, 94121, USA, and [2]Department of Microbiology, University of California at Los Angeles, 90024, USA

INTRODUCTION

The fumarate reductase (FRD) of *Escherichia coli* is a four subunit (FrdA, FrdB, FrdC and FrdD) integral membrane protein complex of the anaerobic respiratory chain. Electrons are transferred from any of several primary dehydrogenases to FRD via menaquinol so that the cell is able to generate energy using fumarate as the terminal electron acceptor (Kröger, 1978). The subunits of FRD are arranged into two domains: the catalytic domain composed of the flavin and iron-sulfur-containing subunits FrdA and FrdB (Cole *et al*, 1985, and Johnson *et al*, 1988) respectively, and the hydrophobic membrane anchor domain composed of FrdC and FrdD (Lemire *et al*, 1982). The FrdC and FrdD subunits have extensive hydrophobic regions and each is predicted to contain three membrane-spanning helices (Cole *et al*, 1985). The subunits are also required for interaction with menaquinol and ubiquinone (Cecchini *et al*, 1986a). Specific mutations in FrdC (Weiner *et al*, 1986) and FrdD (Cecchini *et al*, 1986b) have been shown to affect activity with quinones. These subunits thus have analogous functions to the small subunits of mammalian succinate-ubiquinone oxidoreductase (Complex II), which stabilize the radical intermediate(s) generated in the two-step reduction of ubiquinone (Ruzicka *et al*,1975). In this report we describe the properties of a second *frdC* mutant defective in electron transfer either to or from quinones. The roles of FrdC and FrdD in electron transfer with quinones are discussed.

Keywords: fumarate reductase, menaquinone, Complex II, quinone mutant, anaerobic respiration.

METHODS

Mutagenesis for FrdCD mutations

Mutations in *frdCD* independent of the *frdAB* genes were obtained by transforming a plasmid containing *frdC*$^{+}$*D*$^{+}$ into the *E. coli mutD* mutator strain KD1087 (Degnen and Cox, 1974), followed by isolation of independent pools of transformed cells. The mutagenized plasmid pool was screened for mutations by transforming the *frd* deletion strain DW12, containing *frdA*$^{+}$*B*$^{+}$, and lack of anaerobic growth on glycerol-fumarate medium. Strains with a Frd^{-} phenotype were isolated and DNA sequence analyses of the resulting mutations was determined by standard sequencing methods.

Enzyme assays

The oxidation of reduced benzyl viologen (BV) and menaquinol-6 (MQ-6), and the reduction of phenazine methosulfate (PMS) and the ubiquinone analogue 2,3-dimethoxy-5-methyl-6-pentyl-1,4 benzoquinone (DPB) were determined as previously described (Cecchini *et al*,1986a).

RESULTS

Previous approaches for the isolation of *frdCD* mutants have been relatively inefficient due to the small size of these genes compared to *frdAB*. The *mutD* mutagenesis protocol described above resulted in the isolation of a number of *frdCD* mutants. One of these was found to be a single base change that resulted in the substitution of a lysine residue for the glutamate residue at position-30 of the FrdC peptide. Hydropathy plots suggest that this residue is located at the juxtaposition of the membrane and cytoplasmic surface in the first transmembrane domain of the FrdC peptide. The effect of the substitution on enzymatic properties is shown in Table 1. The mutation prevented *E. coli* from growing on anaerobic glycerol-fumarate medium (data not shown) consistent with an inability of the enzyme to react with either menaquinol or ubiquinone. However, BV oxidase and PMS reductase activities were normal. The enzyme complex assembled normally in the membrane, as evidenced by similar levels of covalently bound FAD in the membranes of normal and mutant cells (data not shown). The properties of this mutant are similar to those described by Weiner *et al* (1986) for a His-82 to Arg-82 substitution, a residue located at the cytoplasmic surface of the second membrane spanning domain of FrdC.

The effects that the various FrdC and FrdD mutations thus far described have on catalytic activity are summarized in Table 2. Whereas, mutations in FrdC

Table 1. Catalytic activites for fumarate reduction and succinate oxidation for wild-type and mutant membranes.

Strain and Genotype	Enzyme activity: μmol/min/mg protein BV	MQ-6	PMS	DPB
DW12 $frdA^{+}B^{+}C^{+}D^{+}$	11.3	10.5	2.7	1.8
DW12 $frdA^{+}B^{+}C^{-}D^{+}$	10.3	0.0	2.6	0.3

interfere with both menaquinol oxidation and ubiquinone reduction that in FrdD affects the oxidation of menaquinol and benzyl viologen, but not the ability of the FRD complex to reduce the ubiquinone analogue DPB or PMS.

Table 2. Summary of properties of FrdC and FrdD mutants.

	Succinate Acceptor Oxidoreductase		Fumarate Reductase	
	PMS	DPB	BV	MQ
FrdC Glu_{30}→Lys_{30}	+	-	+	-
[a]FrdC His_{82}→Arg_{82}	ND	ND	+	-
[b]FrdD point mutation	+	+	-	-

[a]Weiner *et al*, 1986. [b]Cecchini *et al*, 1986b.

DISCUSSION

Identification of mutations which affect MQ-oxidase, reduced BV oxidase, and ubiquinone reductase activities differently is taken as evidence for two separate binding sites for quinones. This would be consistent with Rich's (1981) suggestion that reduction of the oxidized and radical forms of quinones occurs at distinct sites. As an alternative to a scheme (Cammack *et al*, 1986) in which these sites are serviced by a branched electron transport chain constituted by the low and high redox potential components of the enzyme, respectively, it is proposed that the mechansim of Q reduction is similar to that in reaction centers of photosynthetic bacteria. Thus, Fe-S Center 3 of FRD would donate both electrons to a quinone pair (Q_AQ_B) assumed to be present by analogy to that identified in Complex II. In the model, the primary acceptor Q_A transfers successive electrons to Q_B to form fully reduced quinone, which then exchanges with Q of the membrane pool. The effects of the FrdC mutations might thus be to destabilize $Q_A^{\bullet}$. This would preclude electron transfer from both Center 3 and a MQH_2 at the Q_B site, and thus both DPB reductase and MQ oxidase activities, yet permit reduction of Q_A by low potential donors such as reduced BV (-359 mV). A possible explanation for the FrdD mutation is that ability to stabilize protonated forms of quinone is lost (effectively raising pK_a and potential of the $Q/Q^{\bullet}$ couple) such that electron transfer from Center 3 but not the reverse reaction is permitted. This would allow DPB reduction, but not oxidation of MQH_2 and reduced BV.

This work was supported by the VA, NSF, and NIH.

REFERENCES

Cammack, R., Crowe, B.A., and Cook, N.D., 1986. Biochemical Society Transactions, 14, 1207-1208.

Cecchini, G., Ackrell, B.A.C., and Gunsalus, R.P., 1986, Journal of Biological Chemistry, 261, 1808-1814.

Cecchini, G., Thompson, C.R., Ackrell, B.A.C., Westenberg, D.J., Dean, N., and Gunsalus, R.P., 1986, Proceedings National Academy of Sciences USA, 83, 8898-8902.

Cole, S.T., Condon, C., Lemire, B.D., and Weiner, J.H., 1985, Biochimica et Biophysica Acta, 811, 381-403.

Degnen, C.E., and Cox, E.C., 1974, Journal of Bacteriology, 117, 477-487.

Johnson, M.K., Kowal, A.T., Morningstar, J.E., Oliver, M.E., Whittaker, K., Gunsalus, R.P., Ackrell, B.A.C., and Cecchini, G., 1988, Journal of Biological Chemistry, 263, 14732-14738.

Kröger, A., 1978, Biochimica et Biophysica Acta, 505, 129-145.

Lemire, B.D., Robinson, J.J., and Weiner, J.H., 1982, Journal of Bacteriology, 152, 1126-1131.

Rich, P.R., 1981, FEBS Letters, 130, 173-178.

Ruzicka, F.J., Beinert, H., Schepler, K.L., Dunham, W.R., and Sands, R.H., 1975, Proceedings National Academy of Sciences USA, 72, 2886-2890.

Weiner, J.H., Cammack, R., Cole, S.T., Condon, C., Honore, H., Lemire, B.D., and Shaw, G., 1986, Proceedings National Academy of Sciences USA, 83, 2056-2060.

THE REACTION BETWEEN CYTOCHROME *B* AND UBIQUINONE-UBIQUINOL AT CENTRE I OF UBIQUINOL:CYTOCHROME *C* OXIDOREDUCTASE

J.A. Berden and P. Nieboer

E.C. Slater Institute for Biochemical Research, University of Amsterdam, Plantage Muidergracht 12, 1018 TV Amsterdam, The Netherlands

INTRODUCTION

It is a long standing knowledge that ubiquinol or other quinols can reduce cytochrome *b* of the ubiquinol:cytochrome *c* oxidoreductase via two different pathways. Since specific inhibitors of each pathway are available, it is possible to study just one pathway under conditions that the other is not active. The pathway *via* centre o, an integral part of the proposed Q-cycle mechanism (Mitchell, 1976) for electron transfer from ubiquinol to cytochrome *c* *via* the oxidoreductase, can be inhibited with compounds like myxothiazol, while the other pathway, *via* centre i, can be inhibited specifically with antimycin. This second pathway is involved in the Q-cycle mechanism of electron transfer as well, but only in the opposite direction, i.e. the oxidation of cytochrome *b*.

Since antimycin at stoicheiometric concentrations inhibits both the reduction of cytochrome *b* in the presence of myxothiazol *and* the steady-state electron transfer, concomitant with the induction of an increased reduction level of cytochrome *b*, it has always been assumed that oxidation of cytochrome *b* as part of the steady-state electron transfer is the same reaction, but in opposite direction, as the reduction of cytochrome *b* by quinol (in the presence of myxothiazol).

In the present study we will show that the system is more complex and that steady-state electron transfer can still be fully operative under conditions in which the reduction of cytochrome *b* by quinol is inhibited. Some possible consequences of this finding for the mechanism of electron transfer through the bc_1 complex will be indicated.

Keywords: dimeric ubiquinol:cytochrome *c* oxidoreductase, centre i, cytochrome *b*, antimycin-binding sites, DCCD, fluorescamine.

MATERIALS AND METHODS

Submitochondrial particles from bovine heart and from yeast (*S. cerevisiae*) were prepared according to the usual procedures. Purified ubiquinol:cytochrome *c* oxidoreductase was isolated from bovine heart according to Rieske (1976). Ubiquinol-2 was prepared in our laboratory by

A.F. Hartog. 2,5-dihydroxy-3-undecyl-1,4-benzoquinone (embelline) was purchased from ICN, dicyclo-hexylcarbodiimide (DCCD) and fluorescamine were obtained from Sigma and antimycin A and myxothiazol from Boehringer. The redox states of cytochrome b and cytochrome c were measured with an Aminco DW2 spectrophotometer at the appropriate wavelength pairs (562 minus 577 and 550 minus 540 nm, respectively).

RESULTS AND DISCUSSION

Studies with the Q-analogue embelline, using both submitochondrial particles and isolated ubiquinol:cytochrome c oxidoreductase, showed a peculiar phenomenon: about 4-5 μM embelline inhibited, in the presence of myxothiazol, the reduction of cytochrome b by quinol, but steady-state electron transfer to cytochrome c (in the absence of myxothiazol) was only slightly affected. Also no increased reduction of cytochrome b in the presence of both substrate and oxidant could be detected. Only at about ten-fold higher concentrations embelline inhibited the steady-state reduction of cytochrome c. At those concentrations, however, the pre-steady-state reduction of cytochrome c_1 was inhibited as well, due to inhibition of centre o. Also the pre-steady-state reduction kinetics of cytochrome b in the absence of myxothiazol indicated that at 5 μM embelline cytochrome b cannot be reduced any more via centre i, but it can still be oxidized via this pathway.

A second finding, only obtained with yeast particles and not with bovine heart particles, was that the inhibition of the reduction of cytochrome b by ubiquinol-2 or duroquinol titrates non-linearly with the concentration of antimycin (Fig.1). The titration curve suggests that reduction of cytochrome b via centre i is inhibited with only one antimycin/two cytochrome c_1. Since the enzyme is a dimer (Wingfield et al., 1979 and De Vries et al., 1982), the data suggest that in yeast particles antimycin binds with a higher affinity to the first protomer of the dimer than to the second and that binding to one protomer is enough to inhibit the reduction of cytochrome b. Binding to both protomers, however, remains necessary for inhibition of the oxidation of cytochrome b, induction of oxidant-induced reduction of cytochrome b and inhibition of steady-state electron transfer. A similar result as described here for yeast particles is obtained by Bechmann et al. (1988) with particles from Neurospora crassa.

Our conclusions about two different antimycin-binding sites in the dimeric bc_1 complex, belonging to two different electron transfer pathways, are confirmed by experiments with DCCD and fluorescamine. Incubation of yeast particles for one hour with 50 μM DCCD resulted in complete inhibition of cytochrome b reduction by quinol (in the presence of myxothiazol) without any effect on the steady state reduction of cytochrome c. Exactly the same result was obtained with 100 μM fluorescamine. If after incubation with DCCD or fluorescamine the steady-state reduction of cytochrome c was titated with antimycin, full inhibition was obtained with about half the usual concentration of antimycin (Fig. 2). This result is in agreement with earlier data of Clejan and Beattie (1983) and ourselves. Van Hoek (1989) in our laboratory has shown with direct-binding experiments that incubation of particles with DCCD results in a lowering of the

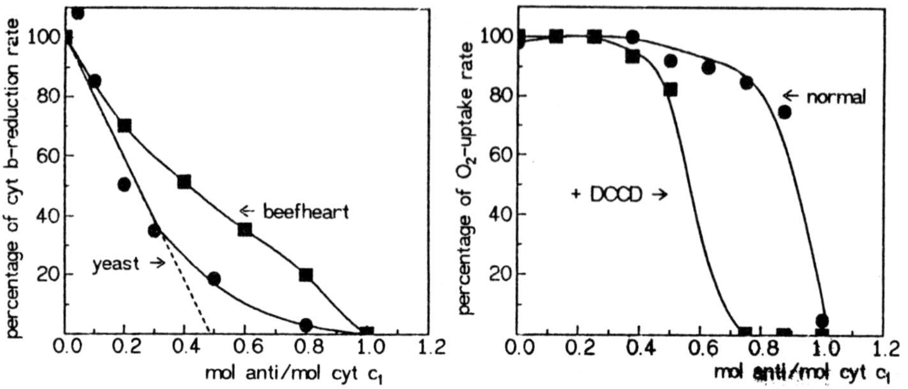

Fig. 1. Titration with antimycin of the cytochrome b reduction in submitochondrial particles (5 mg protein/ml) by duroquinol in the presence of 5 μM myxothiazol.

Fig. 2. Titration with antimycin of the oxygen uptake by submitochondrial particles in the presence of 10 mM succinate. The medium is a sorbitol/phosphate buffer at pH 7.4.

concentration of antimycin-binding sites with maximally 50%. Both DCCD and fluorescamine, apparently, occupy one of the two antimycin-binding sites of the bc_1 dimer, thereby inhibiting the reduction of cytochrome b *via* centre i, but not the oxidation of cytochrome b *via* this site or the steady-state electron transfer. Since it is known that both DCCD and fluorescamine lower the experimental proton/electron stoicheiometry of the bc_1 complex (Degli Esposti and Lenaz, 1985), one may speculate that the antimycin-binding site occupied by these inhibitors, is involved in proton translocation. Bechmann et al. (1988) have suggested that proton translocation is a property of the bc_1 dimer and that DCCD induces the monomeric form which is active in electron transfer, but without coupled proton translocation and energy conservation. This proposal dismisses the data of Brand et al. (1985) who showed that in the presence of DCCD energy conservation is still intact and also the fact that state 4 respiration is not stimulated by DCCD.

We like to propose as a working hypothesis that one of the ubiquinone- and antimycin-binding sites is involved in the interaction with exogenous quinol and the other only in the interaction with the bound endogenous quinol/quinone. For the reduction of cytochrome b by an exogenous quinol, therefore, are both sites required, while during steady-state electron transfer the endogenous quinone, after its reduction by cytochrome b, can reduce centre o again without involvement of an exogenous quinone and therefore only one of the two antimycin-binding sites is required for steady-state electron transfer. This proposal fits with the finding (Linke et al., 1986) that for reconstitution of electron transfer only one ubiquinone per bc_1 dimer is required. An explanation for the finding that in the presence of DCCD or fluorescamine (when interaction with exogenous quinol/quinone at centre i is inhibited) no proton extrusion towards the cytoplasmic side of the membrane can be detected, although energy is still conserved, could be that in proton uptake at centre i also the interaction site with exogenous quinol/quinone is involved. The rate-limiting step in state 4 respiration, then, is not the backflow of protons from the C-side to the M-side of the membrane, but the uptake of protons by ubiquinone, concomitant with its reduction, from the C-side of the membrane. ATP synthesis in state 3 respiration occurs as described in the collision hypothesis (Slater et al., 1985), and the only additional assumption has to be made that, upon an energy-transducing collision between the bc_1 complex and the ATPsynthase, protons, produced at centre o, can move from the C-side to centre i. In this way a Q-cycle mechanism of electron transfer, substantiated by many data, can be combined with the reported effects of DCCD.

REFERENCES

Bechmann, G., Zweck, A. and Weiss, H., 1988, The regulation of dimeric mitochondrial ubiquinol:cytochrome c reductase by the electrical membrane potential. EBEC Reports, 5, 82.

Brand, M.D., Al-Shawi, M.K., Brown, G.G. and Price, B.D., 1985, Thermodynamic and steady-state kinetic investigation of the effect of dicyclohexylcarbodiimide on H^+ translocation by the mitochondrial cytochrome bc_1 complex. Biochemical Journal, 225, 407 - 411.

Clejan, L. and Beattie, D.S., 1983, Dicyclohexylcarbodiimide blocks proton ejection and affects antimycin binding but not electron transport in complex III from yeast mitochondria. Journal of Biological Chemistry, 258, 14271 - 14275.

Degli Esposti, M. and Lenaz, G., 1985, A clarification of the effects of DCCD on the electron transfer and antimycin binding of the mitochondrial bc_1 complex. Journal of Bioenergetics and Biomembranes, 17, 109 - 121.

De Vries, S., Albracht, S.P.J., Berden, J.A. and Slater, E.C., 1982, The pathway of electrons through QH_2:cytochrome c oxidoreductase studied by pre-steady-state kinetics. Biochimica et Biophysica Acta, 681, 41 - 53.

Linke, P., Bechmann, G., Gothe, A. and Weiss, H., 1986, Dimeric ubiquinol: cytochrome c reductase of Neuraspora crassa mitochondria contains one cooperative ubiquinone-reduction site. European Journal of Biochemistry, 158, 615 - 621.

Mitchell, P., 1976, Possible molecular mechanisms of the protonmotive function of cytochrome systems. Jornal of Theoretical Biology, 62, 327 - 367.

Rieske, J.S., 1976, Preparation and properties of reduced coenzyme Q-cytochrome c reductase. Methods in Enzymology, 10, 239 - 245.

Slater, E.C., Berden, J.A. and Herweijer, M.A., 1985, A hypothesis for the mechanism of respiratory-chain phosphorylation not involving the electrochemical gradient of protons as obligatory intermediate. Biochimica et Biophysica Acta, 811, 217 - 231.

Van Hoek, A., 1989, On the role and function of quinone reaction domains of the quinol:cytochrome c oxidoreductase. Ph. D. Thesis, University of Amsterdam.

Wingfield, P., Arad, T., Leonard, K. and Weiss, H., 1979, Membrane crystals of ubiquinone:cytochrome c reductase from Neurospora crassa. Nature, 280, 696 - 697.

The Interaction Between Bovine Heart Mitochondrial Complex III and the Non-Heme Iron Protein.

Diego González-Halphen

Departamento de Bioenergética, Instituto de Fisiología Celular, Universidad Nacional Autónoma de México, Apdo. Postal 70-243; 04510 México D.F.

INTRODUCTION

The ubiquinol-cytochrome c oxidoreductase (Complex III) is formed by eleven different polypeptides (Schägger et al.,1986), three of which are associated with prosthetic redox groups. In a recent model (González-Halphen et al., 1988), a topological arrangement for the eleven subunits has been proposed. It has been shown that the non-heme iron protein (subunit V) may be dissociated from the rest of the complex by relatively mild procedures in a reconstitutively active form (Trumpower and Edwards, 1979). As established by chemical labeling, protease sensitivity, and functional studies, the bulk of the non-heme iron protein faces the intermembrane space. In addition, the knowledge of the primary amino acid sequence of the mitochondrial non-heme iron protein (Schägger et al., 1987) has led to the formulation of structural models, which are based on hydropathy plots and predict two hydrophobic regions, i.e. one between residues 30 and 70, and another near the putative site of binding of the iron-sulfur moiety. In these models the non-heme iron protein is shown either as a monotopic or a polytopic membrane protein with one or two membrane spanning regions respectively. Here, we briefly discuss some of the experimental data that attempt to identify the possible site of interaction for the non-heme iron protein in complex III.

Keywords: Complex III, non-heme iron protein, cytochrome b.

METHODS

Complex III was purified as described by Rieske (1967). Limited proteolysis of the complex with trypsin and sucrose gradient centrifugation were carried out as described by

González-Halphen et al.,(1988), except that detergents were omitted during the centrifugation step.

RESULTS AND CONCLUSIONS

Limited proteolysis of complex III with trypsin inhibits its activity by cleavage of the non-heme iron protein (Ball et al., 1977). The trypsin cleavage sites were identified in the primary amino acid sequence of the protein (González-Halphen et al., 1988), i.e., one after arginine 32 which generates a fragment named V' and a second one after lysine 52, which yields fragment V". Sucrose gradient centrifugation of trypsinized complex III renders a pellet which contains both the non-trypsinized subunit V and the V' fragment, still associated to the rest of the complex; fragment V" is recovered in the supernatant as a soluble polypeptide. The same result is obtained when the sucrose gradients are made in the absence of detergents, suggesting that fragment V" behaves as a soluble protein (Figure 1). Therefore, the non-heme iron protein binds to complex III by a single hydrophobic stretch of 18 residues flanked by lysine 33 and lysine 52 (Figure 2, bottom part).

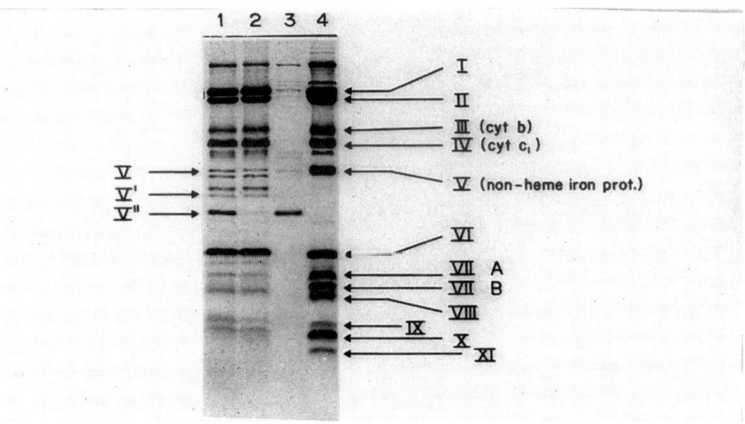

Fig. 1. Lane 1: Trypsinized complex III. Lane 2: Pellet obtained after sucrose gradient centrifugation of trypsinized complex III. Lane 3: Supernatant. Lane 4: Control complex III. (Modified from Biochemistry 27:7021).

Gutweniger et al., (1981) carried out labeling studies of complex III with azidophosphatidylcholine derivatives bearing the cross-linking photoreactive group at two different levels of the fatty acid chain. Although large labeling of cytochrome b was found in the protein patterns, no significant radioactivity was found associated with the non-heme iron protein. The same results were obtained with other membrane intercalated probes such as [^{125}I TID]; no labeling of the non-heme iron protein was found (Gonzãlez-Halphen et al., 1988). In addition, amphipathic arylazido derivatives of phosphatidylethanolamine which label the protein at the lipid head group-water interface of the membrane (Aggeler et al., 1987) do not react at any significant extent with subunit V. Thus, a relatively large variety of hydrophobic probes, whose photoreactive groups are localized at different depths of the bilayer (Fig. 2 shows the chemical structures) fail to react covalently with the non-heme iron protein.

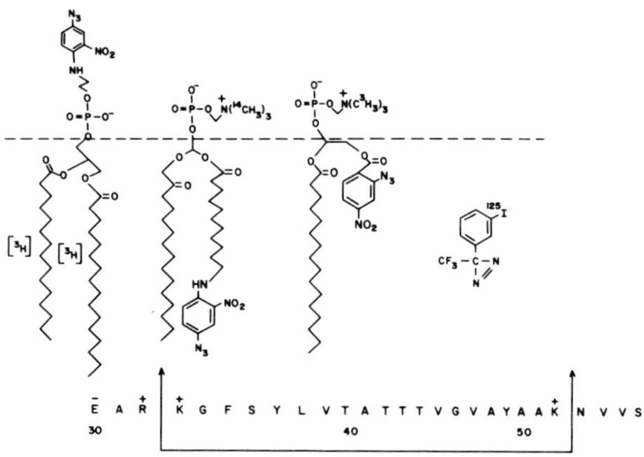

Fig. 2. Upper part: hydrophobic probes used in the labeling of complex III, from left to right: arylazidophosphatidylethanolamine, arylazidophosphatidylcholine derivatives, and [^{125}I]TID. Bottom part: the putative hydrophobic stretch of the non-heme iron protein and the cleavage sites of trypsin.

These results suggest either a shielding of the putative transmembrane stretch by some other membrane-imbedded subunit of complex III or a peripheral location of the non-heme iron protein with no direct contact to the membrane. The former possibility seems improbable, since studies on the biogenesis of complex III suggest that the non-heme iron protein is the last subunit to assemble with the nascent complex (Crivellone et al., 1988). The suggestion that the non-heme iron protein may be peripherally located is more plausible; in fact, this would explain why it is easily released from the rest of the complex as well as the observation that both the NH_2- and COOH-terminal regions of this protein are located towards the intermembrane space in intact mitochondria (Hartl et al., 1986). We would like to suggest that the non-heme iron protein binds to cytochrome b through hydrophobic protein-protein interactions. Indeed, the recent EPR studies by Ohnishi et al., (1989) in which the iron-sulfur cluster and the cytochrome b566 heme are found to be in close proximity, support this idea. In this respect, it is noteworthy that the topological model for cytochrome b proposed by Saraste, (1984) was modified by Crofts et al.(1989); one transmembrane helix (helix IV) was relocalized to an amphipathic location, based on the experimental results obtained by di Rago and Colson (1988) with inhibitor-resistant mutants of cytochrome b. In a more speculative line of thought, it is conceivable that helix IV of cytochrome b (which would be exposed towards the mitochondrial intermembrane space) interacts with the 18 amino acid hydrophobic stretch of the non-heme iron protein. (Supported by a grant from Fundación Mexicana para la Salud and by a travel grant A128CCOE890235 from CONACyT, México).

REFERENCES

Aggeler, R., Zhang, Y.Z., and Capaldi, R.A. (1987) Biochemistry, 26, 7107-7113.

Ball, M. B., Bell, R.L. and Capaldi, R.A. (1977) FEBS Letters 83, 99-102.

Crivellone, M.D., Wu, M., and Tzagoloff, A. (1988) Journal of Biological Chemistry, 263, 14323-14333.

Crofts, A.R., Wang, Z., Chen, C.H., Yun, S., Mahalingham, S., and Gennis, R.B. (1989) Abstracts. Biochemistry, Bioenergetics and Clinical Applications of Ubiquinone, Bologna, 18-22 May.

di Rago J.P., and Colson A.M. (1988) Journal of Biological Chemistry, 263, 12564-12570.

González-Halphen, D., Lindorfer, M.A., and Capaldi, R.A. (1988) Biochemistry, 27, 7021-7031.

Gutweniger, H., Bisson, R. and Montecucco, C. (1981) Journal of Biological Chemistry, 256, 11132-11136.

Hartl, F., Schmidt, B., Wachter, E., Weiss, H., and Neupert, W. (1986) Cell, 47, 939-951.

Ohnishi, T., Schägger, H., Meinhardt, S.W., LoBrutto, R., Link, T.A. and von Jagow, G. (1989) Journal of Biological Chemistry, 264, 735-744.

Rieske, J.S. (1967) Methods Enzymology, 10, 239-245.

Saraste, M. (1984) FEBS Letters, 166, 367-372.

Schägger, H., Link, T.A., Engel, W.D. and von Jagow, G. (1986) Methods Enzymology, 126, 224-237.

Schägger, H., Borchart, U., Machleidt, W., Link, T.A. and von Jagow, G. (1987) FEBS Letters, 219, 161-168.

Trumpower, B.L., and Edwards, C.A., (1979) Journal of Biological Chemistry, 254, 8697-8706.

DETECTION OF A QUINOLE OXIDASE ACTIVITY IN CRUDE EXTRACTS OF THERMOPLASMA ACIDOPHILUM

Peter Gärtner

Max-Planck-Institut für Biophysik
Abt. Molekulare Membranbiologie
Heinrich-Hoffmann-Str. 7
6000 Frankfurt/Main 71
Federal Republik of Germany

INTRODUCTION

Thermoplasma acidophilum is a cell-wall free archaebacterium growing optimally at pH 2 and 58°C. The organism is an obligate aerobe. Living in a highly acidic environment with a neutral interior, T. acidophilum has to maintain a large proton gradient across the membrane. T. acidophilum contains a menaquinone methylated in pos. 5 and/or 8 (Collins, 1985).

By spectroscopic investigations cytochromes of b- and d-type, but not of c- or a-type could be detected. Therefore, the existence of a quinole oxidase was postulated which oxidizes menaquinole directly with molecular oxygen. As menaquinoles are rapidly oxidized nonenzymatically in presence of air, ubiquinole-10 was used as a more convenient substrate for the enzymatic assay.

Keywords: Archaebacteria, Thermoplasma,
quinole oxidase, Cyt-o, HOQNO

METHODS

T. acidophilum membranes were prepared by sonication of washed cells at a pH of 5.5 and solubilized Tris/HCl pH 7.5 with 2% Lubrol PX at a protein concentration of 20mg/ml. Ubiquinone was reduced according to the method of Rieske (1967).

Enzyme test protocol: Membranes or solubilisates were diluted to a volume of 1ml in 100mM Na-citrate pH 5.4 and preincubated at 50°C for 30 sec. Enzyme reaction was started by addition of 20ul 5mM $Q_{10}H_2$ suspension in 1.5% Lubrol PX. The reaction was stopped after 10 min by cooling the tube on ice and addition of 2 ml heptane/isopropanol 3+2 followed by vigorous shaking. The absorption of the organic phase was determined with the same sample without substrate as a reference. The absorptions at 270 and 294.7 nm, the maximum and minimum of the oxidized-reduced difference spectrum were used to calculate the percentage of oxidized and the total content of ubiquinone-10.

RESULTS

Using this protocol we were able to detect a $Q_{10}H_2$ Oxidase activity in membranes and solubilisates of T. acidophilum. The specific activity of Lubrol solubilisate was approximately 5uU/mg Protein

O_2-dependence: No QH_2 Oxidation was achieved when the solution was bubbled with nitrogen prior to addition of substrate. This was only true however, when the Lubrol used was free of peroxides. Otherwise QH_2 could be oxidized enzymatically in the absence of oxygen.

Detergent dependence: Enzyme activity has an optimum at 1-2 mg/ml Lubrol PX depending on whether membranes or solubilisate are tested.

Enzyme affinity for ubiquinole: Fig. 1 shows a plot of the reaction rate versus substrate concentration. The k_M-value derived from Hanes Plot (Fig. 2) is approximately 18uM.

Temperature and pH dependence: The enzyme showed a maximal activity at pH 5.4 and 50° C.

Inhibitors: Antimycin A, dibromothymoquinone, abietic acid, and Zn^{2+} ions have no inhibitory effect on the enzyme activity in concentrations up to 100 uM. 2-heptyl-4-hydroxyquinoline N-ocide shows inhibitons with an I_{50} of 150 nM. CO and KCN show inhibitory effects at concentrations above 100 uM.

DISCUSSION

Ubiquinole-10 is oxidized by an enzyme which presumably uses a menaquinone as a natural substrate. Indeed different menaquinones were oxidized by the same solubilisate at rates 4-5 times higher. These enzyme activities were measured with a different method. (Gärtner, unpublished data) which was not practicable for routine measurements. Ubiquinoles without side chain, such as duroquinol were also oxidized. However, the advantage of long chain derivatives is that they can be extracted after incubation, thus avoiding most UV absorbing material in the cuvette. The k_M-value of 18 uM lies in the range of other quinole oxidases. The anaerobic oxidation of QH_2 in presence of peroxide containing Lubrol indicates the use of peroxides as an alternative electron acceptor. The same phenomenon was seen when LDAO was used as detergent instead of Lubrol. The only substance inhibiting the enzyme activity in concentrations some orders of magnitude below substrate concentration is HOQNO. This substance has proved to inhibit ubiquinone oxidase of E. coli, which is, like that of T. acidophilum a Cyt-o type oxidase (Matsushita et al., 1987).

REFERENCES

Collins
FEMS Microbiology Letters 28 (1985), 21-23

Rieske,
Methods in Enzymology 10 (1967), 239

Matsushita et al.,
Methods in Enzymology 126 (1987), 113-122

ACKNOWLEDGEMENTS

Thanks are due to Prof. Hartmut Michel for helpful discussions and financial support.

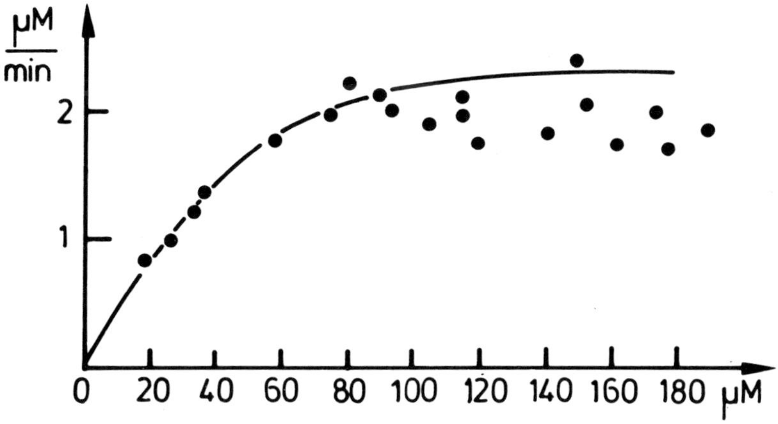

Fig.1 Enzyme activity as a function of ubiquinole -10 concentration

Protein concentration: 0.4mg/ml
Total Lubrol concentration:1mg/ml

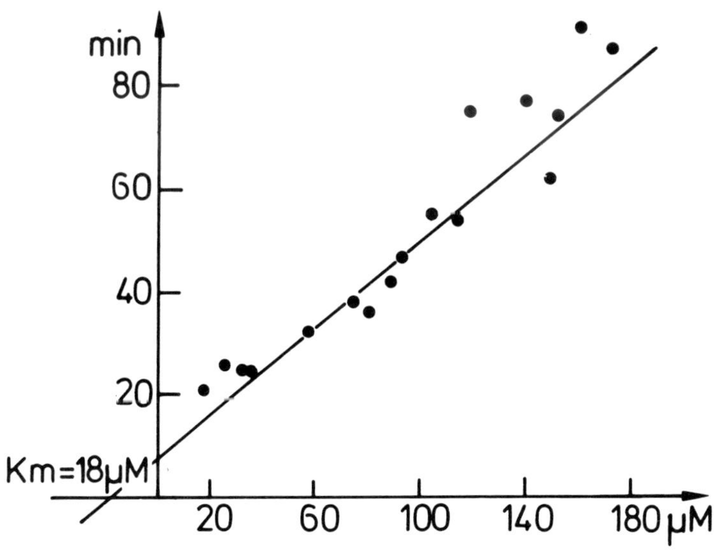

Fig.2 Hanes plot of enzyme activity

Abscissa: Substrate concentration
Ordinate : Substrate concentration/activity

THE MEMBRANE PROTEINS THAT INTERACT WITH QUINONES HAVE COMMON STRUCTURAL FEATURES.

M. Degli Esposti and M. Crimi

Department of Biology, University of Biologna, Via Irnerio 42, 40126 Bologna, Italy.

Keywords : cytochrome *b*; sequence analysis; quinone- binding; structural predictions.

Cytochrome *b* of quinol:acceptor reductases is the best characterized protein that interacts with quinones. The structure of this dihaem membrane haemoprotein can be deduced with reasonable accuracy by combining the results of mutagenesis (Di Rago & Colson, 1988) and of sequence analysis (Crofts *et al.*, 1987; Degli Esposti *et al.*, 1989). In this report we focus on the mapping of the conserved residues in order to deduce important structural features related to the function of cytochrome *b*. Fig. 1 shows a tentative 3D-model for cytochrome *b* of ubiquinol:cytochrome *c* reductase, adapted from that proposed previously (Degli Esposti *et al.*, 1989) in order to group together the mutation sites that lead to resistance towards either center "i" or center "o" inhibitors (Crofts *et al.*, 1987). The original nomenclature of the nine hydrophobic helices is maintained, even if helix IV does not span the membrane (Degli Esposti *et al.*, 1989).

Ten sequences more than those examined by Hauska *et al.* (1988) have been compared and this reveals only 34 chemically-conserved residues, of which 25 are identical. The location of these residues can be predicted on the basis of the structural analogy with the known features of the quinone-reacting subunits of bacterial reaction centers (Komya *et al.*, 1988), as shown in Fig. 1.

Clearly, most of the conserved residues of cytochrome *b* are clustered at the positive side of the membrane (Fig. 1). The mapping of the conserved residues in this domain of the protein overlaps that of the mutations leading to resistance towards center "o" inhibitors, thereby suggesting that such residues are involved in the reactions of center "o". However, some of the same conserved amino acids may be also required for protein contacts with the "Rieske" cluster and cytochrome *c*1 at the positive side of the membrane. Conversely, the few conserved residues in the other regions of cytochrome *b* are likely to be primarily involved in functional aspects.

Chemical conservation of the residues is required not only for functional interactions with quinone and quinol, but also for the binding of the haems. The four conserved histidines are the most likely axial ligands of the two haem irons. The positively-charged residues at positions 99 and 202 (H in *b* and R in chloroplast *b*6) are essential for binding the propionyls of the *b*H haem at the negative side of the membrane . Similarly, the positively-charged residues at positions 79, 178, 283 and 288 may bind the *b*L haem at the positive side of the membrane. Presumably, some of these basic residues function also as hydrogen bonding acceptors for ubiquinol at center "o".

At present, it is much easier to suggest which are the important residues in the interaction with quinones at center "i" than those at center "o", since too many hydrophilic residues are conserved in the protein domain of the latter (Fig. 1). D229 is the only acidic residue that is conserved at the negative side of the membrane and, therefore, should be important in the protonation of the quinone at center "i". The other conserved residues that could be involved in the quinone protonation are also necessary for the binding of the propionyls or the iron of the *b*H haem. S87 seems to be the best hydrogen bonding donor for the binding of the quinone in the middle of the membrane, a region that may functionally correspond to center "i". Q43 and possibly also P187 are located in the same region and may be involved in the stabilization of the semiquinone radical of center "i". Surprisingly, no mutation leading to resistance towards center "i" inhibitors maps close to these residues.

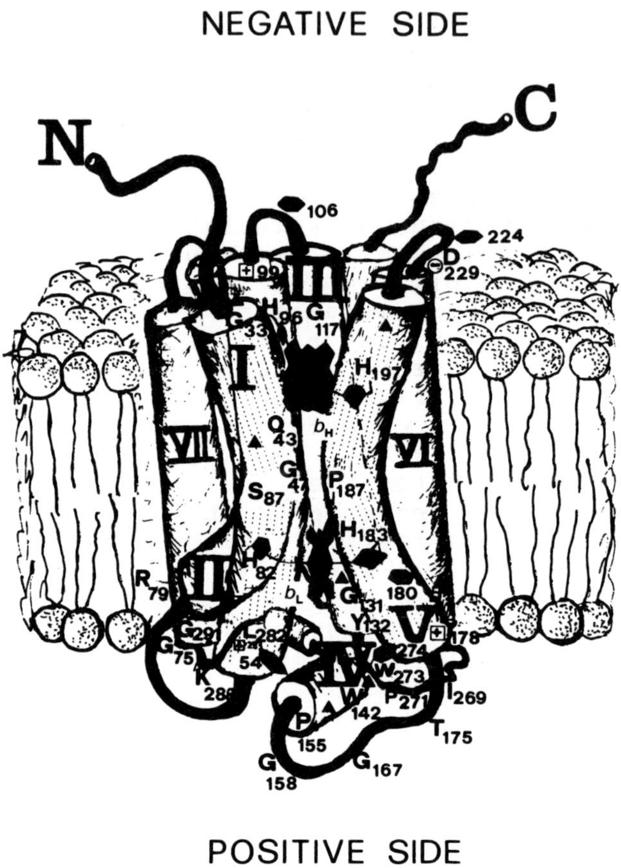

Fig. 1. <u>Mapping of the conserved residues in cyt *b*</u> The sequences of 21 mitochondrial *b*, 5 bacterial *b* and 4 chloroplast *b*6 are compared, and the 34 chemically-conserved residues observed are located in a 3D-model for mitochondrial cytochrome *b* (Degli Esposti *et al.*, 1989). See the article of Crofts *et al.* in this book for a somewhat different model, in which, however, the location of the residues would be similar. The black exagons identify conserved aromatic amino acids (F,Y,W or H). The black triangles locate conserved residues of mitochondrial and bacterial cytochrome *b* that are not present in chloroplast cytochrome *b*6.

The putative structures of center "i" and center "o" significantly differ also in the content of conserved aromatic residues, which is much larger around the *b*L haem (Fig. 1). Hence, the electron transfer reactions at center "o" might be critically dependent on the involvement of aromatic rings , possibly for tunnelling the electrons into the "Rieske" cluster. On the other hand, the structural determinants of the reactions at center "i" in cytochrome *b* look rather similar to those in the bacterial reaction center and also in other *b* cytochromes interacting with quinones, such as *b*556 of succinate:ubiquinone reductase in E.coli. (M.Degli Esposti, submitted).

REFERENCES

Crofts, A.R., Robinson, H.H., Andrews, K., Van Doren, S. and Berry, E. (1987) Catalytic sites for reduction and oxidation of quinones. In "Cytochrome Systems" , Papa, S., Chance, B. and Ernster, L. eds., pp. 617-624, Plenum Press, New York.

Degli Esposti, M., Ghelli, A., Luchetti, R., Crimi, M. and Lenaz, G. (1989) New approaches to the prediction of the folding of membrane proteins with redox function. Ital.J.Biochem. 38, 1-22.

Di Rago, J.P. and Colson, A.M. (1988) Molecular basis for resistance to antimycin and diuron, Q-cycle inhibitors acting at the Qi site in the mitochondrial ubiquinol:cytochrome *c* reductase in Saccharomyces cerevisiae. J.Biol.Chem. 263, 12564-12570.

Hauska G., Nitschke, W. and Herrmann, R.G. (1988) Amino acid identities in the redox center carrying polypeptides of cytochrome *bc*1/*b*6/ complexes. J.Bioenerg.Biomembr. 20, 211-228.

Komya,H., Yeates, T.O., Rees, D.C., Allen, J.P. and Feher,G. (1988) Structure of the reaction center from Rhodobacter sphaeroides R26: 2-4 symmetry relations and sequence comparisons between different species. Proc.Natl.Acad. Sci.USA. 85, 9012-9016.

REGULATION OF ELECTRON TRANSPORT ACTIVITY IN PLANT MITOCHONDRIA BY THE REDOX POISE OF THE QUINONE POOL

A.L. Moore[1], D.A. Day[2], I.B. Dry[3] and J.T. Wiskich[4]

[1]Department of Biochemistry, University of Sussex, Falmer, Brighton, BN1 9QG, UK
[2]Department of Botany, Australian National University, Canberra, A.C.T. 2601 Australia
[3]Department of Botany, University of Adelaide, Adelaide S.A. 5001, Australia

INTRODUCTION

Most plant mitochondria possess a branched respiratory pathway with routes of substrate oxidation that by-pass Complexes 1, 111 & 1V. In addition to a cyanide insensitive alternative oxidase, which branches from the main respiratory chain at the level of ubiquinone, there is a by-pass around Complex 1 that is insensitive to rotenone. As with other systems ubiquinone acts as a mobile carrier linking the dehydrogenases with either the bc complex or the alternative oxidase (Moore and Rich, 1985).

In the present paper we have investigated the relationship between the redox state of the quinone pool and respiratory flux under ADP limited conditions both in pea leaf and soybean cotyledon mitochondria in an attempt to determine if changes in the redox state of that pool regulate the partitioning of electron flux in a branched respiratory chain.

Keywords: Plant mitochondria, alternative oxidase, Q-pool

METHODS

Mitochondria were isolated and purified from pea leaf and soybean cotyledons as previously described by Moore et al. (1988) and Day et al. (1988). Oxygen consumption, steady-state membrane potentials and changes in the redox poise of ubiquinone were monitored simultaneously in a 2.0 ml cuvette housing an oxygen, $TPMP^+$-sensitive, glassy carbon and platinum electrodes (Moore et al., 1988).

RESULTS

Table 1 shows that the addition of glycine to pea leaf mitochondria oxidising malate under state 4 conditions leads to a dramatic increase in the oxygen consumption rate, mebrane potential and steady-state reduction levels of the ubiquinone pool. Such increases are observed even in the presence of rotenone suggesting that the Site 1 by-pass is operative under these conditions. It should be noted from Table 1 that although rotenone reduces all three parameters the subsequent addition of a second substrate stimulates the state 4 oxygen uptake rate, increases membrane potential and reduces the quinone pool to levels observed in the absence of this inhibitor. When rotenone is added during the simultaneous oxidation of both substrates it has relatively little effect on any parameter suggesting that under these conditions the rotenone insensitive by-pass is maximally engaged and that this pathway can act as a major route for the reduction of the Q-pool.

Table 1. Dual substrate oxidation under state 4 conditions

Condition		O_2 (nmols/min/mg)	Q_r(%)	$\Delta\Psi$(mV)
malate	state 4	43	19	182
	+ glycine	88	83	192
	+ rotenone	87	72	191
malate	state 4	60	27	182
	+ rotenone	44	20	174
	+ glycine	90	83	188

During an investigation into the modulation of the quinone pool both under state 3 and 4 conditions (Moore et al., 1988) it was found that, in agreement with the model of Kroger and Klingenberg (1973) the level of reduction of the quinone pool varies linearly with electron flux to the pool. Howevever, there was some evidence of a deviation from linearity in some tissues under state 4 conditions which we have suggested may be possibly due to the engagement of alternative pathway activity (Dry et al., 1989). In order to test this suggestion we investigated the relationship between the redox state of the Q-pool and respiratory flux via the cytochrome and alternative pathways with soybean cotyledon mitochondria which are strongly cyanide resistant (Day et al., 1988). In order to assess the potential influence of the alternative pathway activity on the linearity of the relationship in these mitochondria malonate titration experiments were performed during the oxidation of succinate both under state 3 and 4 conditions and in the presence of myxothiazol (Fig. 1).

Figure 1. Dependence of total respiratory flux on quinone redox state in soybean cotyledon mitochondria.

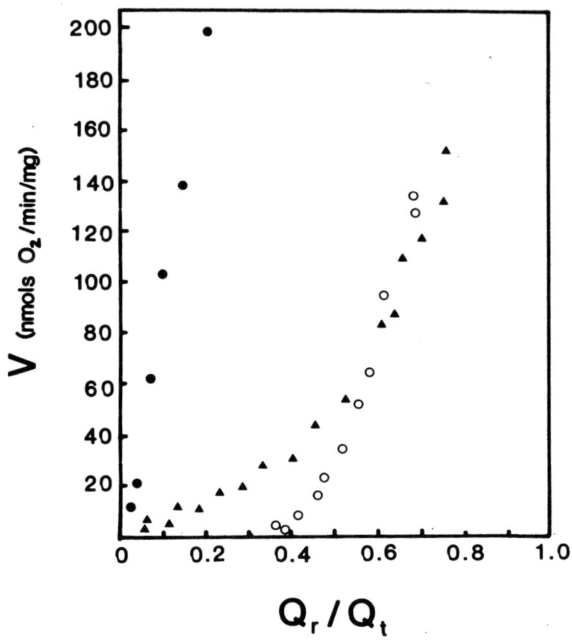

Assays were performed with 0.5 uM Q-1 as described by Moore et al. (1988). Symbols : (●) state 3 conditions, (▲) state 4 conditions, (○) in the presence of 2 uM myxothiazol.

It can be seen from Fig.1 that a linear relationship between the respiratory rate and the level of reduction of the Q-pool only exists under state 3 conditions. Both in state 4 or in the presence of myxothiazol the relationship is markedly non-linear. Furthermore the data indicate that the alternative pathway is not engaged to any significant extent until the level of reduction of the Q-pool reaches 35-40%. Below this level, rates of oxygen uptake via the alternative pathway were negligible and hence difficult to accurately determine. Similar results were obtained when malate was used as an electron donor thereby demonstrating that the non-linear reaction kinetics are truly a function

of interaction of the Q-pool with the alternative pathway and are independent of the mode of Q-pool reduction. An important conclusion to be drawn from this result is that a significant proportion of the uninhibited state 4 rate, in these mitochondria, was due to electron flux via the alternative pathway.

CONCLUSIONS

The results presented here demonstrate the marked influence of the alternative pathway activity on the relationship between Q_r levels and electron flux under restricted conditions. Furthermore they demonstrate that the engagement of alternative pathway activity is determined by the steady state level of Q-pool reduction as regulated by the input of electrons from substrate dehydrogenases and the reoxidative capacity of the cytochrome chain. The marked non-linearity of the relationship will limit the extent of energetically wasteful respiration in cyanide resistant mitochondria during oxidative phosphorylation. The finding that the rotenone insensitive by-pass is engaged even under state 4 conditions and can therefore act as a major pathway of electron flow to the Q-pool is important as the degree to which it can reduce the Q-pool will determine the potential cyanide resistance of NAD-linked substrate oxidation.

REFERENCES

Day, D.A., Moore, A.L., Dry, I.B., Wiskich, J.T. and Azcon-Bieto, J., 1988, Regulation of nonphosphorylating electron transport pathways in soybean cotyledon mitochondria and its implications for fat metabolism. Plant Physiology, 86, 1199-1204

Dry, I.B., Moore, A.L., Day, D.A. and Wiskich, J.T., 1989, Regulation of alternative pathway activity in plant mitochondria. Non-linear relationship between electron flux and redox poise of the quinone pool. Archives of Biochemistry and Biophysics. In press.

Kroger, A. and Klingenberg, M., 1973 The kinetics of the redox reactions of ubiquinone related to the electron transport activity in the respiratory chain. European Journal of Biochemistry, 34, 358-368

Moore, A.L. and Rich, R., 1985, Organisation of the respiratory chain and oxidative phosphorylation. In Encyclopedia of Plant Physiology, Vol. 18, (Springer-Verlag), pp. 134-172

Moore, A.L., Dry, I.B. and Wiskich, J.T., 1988 Measurement of the redox state of the ubiquinone pool in plant mitochondria. FEBS Letters, 235, 76-80.

INDISPENSABILITY OF RHODOQUINONE IN THE NADH-FUMARATE REDUCTASE SYSTEM OF *ASCARIS* MUSCLE MITOCHONDRIA

Kiyoshi Kita, Shinzaburo Takamiya, Rieko Furushima, *Hiroshi Suzuki, *Takayuki Ozawa and Hiroshi Oya

Department of Parasitology, Juntendo University, School of Medicine, Tokyo, JAPAN, *Department of Biomedical Chemistry, Faculty of Medicine, University of Nagoya, Nagoya, JAPAN

INTRODUCTION

The physiological significance of the NADH-fumarate reductase system has been well established in the energy metabolism of the parasitic nematode, *Ascaris suum*. This organism resides in the host's small intestine where oxygen tension is fairly limited. It is generally accepted that complex I(NADH-ubiquinone oxidoreductase) is the only one site for ATP formation in *Ascaris* adult mitochondria, coupled to electron transfer from NADH to fumarate. This may because *Ascaris* adult mitochondria contains a small amount of complex III(ubiquinol-cytochrome c oxidoreductase) and even smaller amount of cytochrome oxidase. However, no substantial information on the electron transfer from NADH to fumarate and the participating components in the chain has been documented. So, we have isolated complex I-III(Takamiya *et al*, 1984) and complex II(Takamiya *et al*, 1986) as succinate-ubiquinone oxidoreductase from *Ascaris* adult mitochondria and characterized them. Complex I-III catalizes the reduction of 1.68 μmolcytochrome c/min/mg protein with NADH, and this activity was sensitive to piericidin A and 2-heptyl-4-hydroxyquinoline-N-oxide. Subunit composition of *Ascaris* complex I-III is quite similar to that of bovine heart. Complex II is one of the major components in *Ascaris* adult mitochondria (Kita *et al*, 1988a), and shows high fumarate reductase activity in addition to succinate-ubiquinone oxidoreductase activity(Kita *et al*, 1988b). Thus, *Ascaris* mitochondria has two enzyme complexes, which are the candidates for the components in the electron transport system from NADH to fumarate. This paper reports the reconstitution of heterogous NADH- fumarate reductase activity from isolated bovine heart complex I and *Ascaris* complex II, and indispensability of rhodoquinone in this system.

Keywords; rhodoquinone, *Ascaris* mitochondria, fumarate reductase, complex II, anaerobic respiration, cytochrome b558.

METHODS

Reconstitution of NADH-fumarate reductase activity

A mixture(50 μl) of acetone-washed soybean phospholipids (asolectin, 0.45 mg), various amount of rhodoquinone or ubiquinone-10, bovine heart complex I(90 μg), *Ascaris* complex II(14 μg) and 30 mM phosphate buffer(pH 7.5) was incubated for 5 min in the cuvette at 25°C. Then 30 mM phosphate buffer(pH 7.5)was added to produce the final volume of 3 ml for the assay.

Other methods

protein chemical analysis, antibody preparation, immunochemical analysis and the determination of concentration of quinones were carried out as descrived previously(Kita *et al*, 1988b).

RESULTS

Reconstitution of NADH-fumarate reductase activity and effect of rhodoquinone.

NADH-fumarate reductase activity was reconstituted with bovine heart complex I, *Ascaris* complex II and phospholipid. We used complex I purified from bovine heart mitochondria because it is difficult to obtain *Ascaris* complex I with high NADH-ubiquinone oxidoreductase activity and high purity. Rhodoquinone was essential for this reconstitution, whereas ubiquinone showed little effect on it(Fig. 1). The dose response of rhodoquinone was linear up to 35 nmol per reaction mixture and muximum specific activity was 430 nmol /min/mg of complex II. This reconstitution provides direct evidence that complex I, complex II and rhodoquinone are the only components required for electron transport from NADH to fumarate.

Subunit structure of cytochrome b558 in Ascaris complex II and homology among the species

Complex II is composed of four polypeptides and this molecular organization of subunits is general feature of complex II from mitochondria of bovine heart(Hatefi, 1978), *Ascaris*(Takamiya, *et al*, 1986) and *E. coli*(Kita, *et al*, 1989). The requirement of two small hydrophobic subunits in electron transport in complex II has been suggested from capability in converting succinate dehydrogenase to succi-

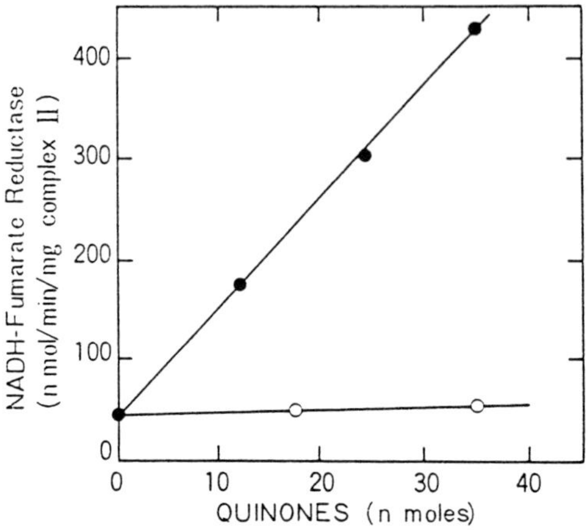

Fig. 1 Reconstitution of NADH-fumarate reductase.

nate-ubiquinone oxidoreductase, and the essential role of these subunits in quinol oxidation is also shown in the bacterial fumarate reductase system. So, properties of these hydrophobic subunits(cytochrome b558) in *Ascaris* complex II were analyzed. Antibodies against *Ascaris* cytochrome b558(15 KDa and 13.5 KDa) did not react with mammalian counterparts and those against mammalian cytochrome b in complex II also did not react with *Ascaris* counterparts. These hydrophobic subunits are seemed to have species specific epitopes. A little homology in N-terminal sequence was found between *Ascaris* smallest subunit(13.5 KDa) and *E. coli sdh D*(Fig. 2), whereas no homology was found between *Ascaris* third subunit(15 KDa) and *E. coli sdh C* and *frd C*(Kita, *et al*, 1988b).

E. coli (*frd D*)	M I N P N P K R S D E P V F W G L F G A G G M W S A I I A P V
E. coli (*sdh D*)	M V S N A S A L G R N G V H D F I L V R A T A I V L T L Y I I Y M V
Ascaris (13.5K)	G A T S A A V T G A A P P Q F D P I A A W K G F W P L H S
B. subt.(*sdh C*)	T R I A A Q M G A E V N - F D M A N I L S S P L
	120

Fig. 2 N-terminal sequence of smallest subunit of Comlex II.

DISCUSSION

From our previous observations and present results described here, we propose a scheme for the electron transport pathway of the NADH-fumarate reductase system in *Ascaris* adult muscle mitochondria. Similar scheme has been

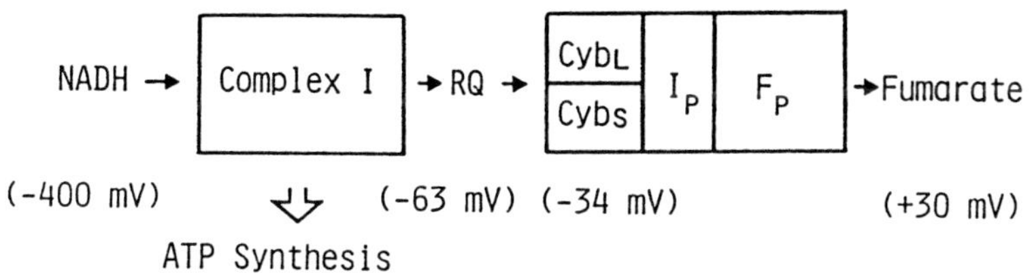

Fig. 3 Electron flow from NADH to fumarate in *Ascaris* mitochondria.

suggested (Köhler, 1985, Takamiya *et al*, 1986) and our biochemical data made clear its organization and indispensability of rhodoquinone. In rhodoquinone, the methoxy group of ubiquinone is substituted with an amino group, and Em'(-63 mV) is more negative than that of ubiquinone(+110 mV). In the case of bacteria, reducing equivalent from dehydrogenase are transfered to fumarate reductase via low potential naphthoquinone(menaquinone; -74 mV). From evolutional standpoint, Hiraishi (1988) suggested the fumarate-reducing process developed from a soluble and relatively simple system into the membrane-bound, menaquinone-mediated respiratory system and that, after oxygen appeared in the biosphere, most aerobic bacteria came to have not only succinate dehydrogenase evolving by modification of fumarate reductase but the higher potential quinone, ubiquinone, in the respiratory chain with concomitant loss of menaquinone. Association of b cytochrome with the fumarate reductase system may have occured in a later stage of the evolution of this enzyme and rhodoquinone may be new quinone compared with ubiquinone, because rhodoquinone is synthesized from ubiquinone. It is fascinating idea that rhodoquinone-associated fumarate reductase system, such as the case in *Ascaris*, is evolved from ubiquinone-associated succinate dehydrogenase system.

REFERENCES

Hatefi, Y. 1978, *Methods in Enzymology*, 53, 21-35

Hiraishi, A. 1988, *Archives of Microbiology*, 150, 56-60

Kita, K., Takamiya, S., Furushima, R., Ma, Y. and Oya, H. 1988, *Comparative Biochemistry and Physiology*, 89, 31-34

Kita, K., Takamiya, S., Furushima, R., Ma, Y., Suzuki, H., Ozawa, T. and Oya, H. 1988b, *Biochimica et Biophysica Acta*, 935, 130-140

Köhler, P. 1985, *Molecular and Biochemical Parasitology* 17, 1-18

Takamiya, S., Furushima, R. and Oya, H. 1986, *Biochimica et Biophysica Acta*, 848, 99-107

COMPARATIVE INVESTIGATIONS ON ANIMAL MITOCHONDRIA UBIQUINONE CONTENT. A QUALITATIVE AND QUANTITATIVE STUDY.

M. Battino*, E. Bertoli*, G. Formiggini°, A. Gorini^, R.F. Villa^ and G. Lenaz§.

*Istituto di Biochimica, Via Monte D'Ago - Ancona
°Dipartimento di Biochimica, Via Irnerio, 48 - Bologna
^Istituto di Farmacologia, Piazza Botta, 11 - Pavia
§Dipartimento di Biologia, Via Irnerio, 42 - Bologna
ITALY

INTRODUCTION

The role of Coenzyme Q (Ubiquinone) in electron transfer and its antioxidant properties have been extensively studied, reviewed and discussed and its importance in these functions has been well established.

Recent works (Battino et al., 1989; Lenaz and Fato, 1986) indicate that the inner mitochondrial concentration of Coenzyme Q is able to affect electron transfer being in the range of the K_m for the enzyme complexes in the Ubiquinone region.

A deficiency of the Q-pool molecules could have dramatic importance either for for the normal rate of the electron flow (i.e. the CoQ-pool becomes the limiting factor in the respiratory chain) or for its antioxidant properties and these situations are of great importance also for clinical implications.

On the other hand the great amount of data published in the past 20 years about the Ubiquinone levels in tissues (Ramasarma, 1985) are not always easily comparable because of different experimental conditions (i.e. many different methods and systems for extracting and evaluating ubiquinone concentration have been used) and of improving techniques in measuring the Ubiquinone levels in the last years.

We have undertaken a systematic study for detecting ubiquinone content by a previous standardized HPLC analysis performed after the same extraction procedure using different organs from different animals in order to obtain a comparable view of the real Ubiquinone concentrations not in the whole tissues or organs but only in mitochondria where Ubiquinone probably plays its main metabolic role.

Keywords: HPLC, Mitochondria

METHODS

Rat brain mitochondria were prepared according to Villa et al., 1989; rat and chicken liver mitochondria according to Fleischer et al., 1979; fish liver mitochondria according to Murphy and Houston, 1974; chicken, beef and rat heart mitochondria according to Blair, 1967; hamster brown adipose tissue mitochondria according to Cannon and Lindberg, 1979.

HPLC Ubiquinone determinations were performed using the method of Tsai et al., 1985 after an extraction procedure indicated by Kröger, 1978.

RESULTS AND CONCLUSIONS

Interesting preliminary results were displayed especially from rat that exhibited a wide distribution pattern of the homologs and their relative concentrations in the different organs.

As it is shown in Table I there are clear differences both in the total amount of Ubiquinone and in the CoQ_9/CoQ_{10} ratio which is of 2:1 in brain, 10:1 in heart and 16:1 in liver. A consistent amount of CoQ_7 and CoQ_8 is present only in liver.

Table I.
Ubiquinone levels in rat tissues by HPLC determinations (data are expressed in nmoles/mg mitochondrial protein)

	CoQ_7	CoQ_8	CoQ_9	CoQ_{10}
Brain				
Cortex			0.900	0.450
Hippocampus			0.800	0.430
Striatum			0.950	0.440
Heart		0.068	2.954	0.241
Liver	0.457	0.492	1.324	0.085

The homologs levels in the other mitochondria preparations are summarized in Table II.

Table II.
Ubiquinone levels in animal tissues by HPLC determinations.
(data are expressed in nmoles/mg mitochondrial protein)

		CoQ_9	CoQ_{10}
Beef	(HEART)		2.100
Hamster	(BAT)*	0.542	0.210
Chicken	(HEART)		2.712
	(LIVER)		2.041
Trout	(LIVER)		0.814

----
*BAT = Brown Adipose Tissue

The presence of consistent amount of Coenzyme Q_{10} in rat brain in addition to Coenzyme Q_9, the known major Ubiquinone component in rat tissues is not easily understandable and it could be ascribed both to particular metabolic processes and to dietary contribution.

The explanations of this unexpected situation might be of deep interest for its clinical and pharmacological involvements.

ACKNOWLEDGMENTS

Ubiquinone homologs were a kind gift from Eisai Co., Tokyo, Japan. This work was partially supported by grants from the Ministero della Pubblica Istruzione and the C.N.R., Rome, Italy.

REFERENCES

Battino M., Fato R., Parenti-Castelli G. and Lenaz G., 1989, Coenzyme Q can control the efficiency of oxidative phosphorylation. International Journal of Tissue Reactions, in press.

Blair P.U., 1967, The large scale preparation and properties of heart mitochondria from slaughterhouse material. Methods in Enzymology, 10, 78-81.

Cannon B. and Lindberg O., 1979, Mitochondria from brown adipose tissue: isolation and properties. Methods in Enzymology, 55, 65-72.

Fleischer S., Mc Intyre J.O. and Vidal J.C., 1979, Large csale preparation of rat liver mitochondria in high yield. Methods in Enzymology, 55, 32-39.

Kröger A., 1978, Determination of contents and redox states of ubiquinone and menaquinone. Methods in Enzymology, 53, 579-591.

Lenaz G. and Fato R., 1986, Is ubiquinone diffusion rate-limiting for electron transfer? Journal of Bioenergetics and Biomembranes, 18, 369-401.

Murphy P.G. and Houston A.H., 1974, Environmental temperature and the body fluid system of the fresh-water teleost-V. Comparative Biochemistry and Physiology 47B, 563-570.

Ramasarma T., 1985, Natural occurrence and distribution of Coenzyme Q. In Coenzyme Q edited by G. Lenaz (London: John Wiley & Sons), 67-82.

Tsai A.L., Kauten R. and Palmer G., 1985, Redox changes in in coenzyme Q in the millisecond time range: an approach using rapid quenching and High-Performance-Liquid-Chromatography. Analytical Biochemistry, 151, 131-136.

Villa R.F., Gorini A., Geroldi D., Lo Faro A. and Dall'Orbo C., 1989, Enzyme activities in perikaryal and synaptic mitochondrial fractions from rat hippocampus during development. Mechanisms of Ageing and Development, in press.

3. Biomedical and Clinical Applications of Ubiquinone

CHAIRMAN'S INTRODUCTION - UBIQUINONE, A DRUG IN SEARCH OF A DISEASE?

Harold Baum

Department of Biochemistry, King's College London, Campden Hill Road, London W8 7AH, UK

INTRODUCTION

Ubiquinone (Q) is a co-enzyme whose known function is central to energy metabolism. It is almost unique among bio-catalysts in being a lipid, which can be orally administered and absorbed unchanged and which is apparently non-toxic. It is synthesised in mammals (Ramasarma, 1985a) to a sufficient extent that no frank Q-deficiency syndrome has been described, so it is not a vitamin. Nevertheless, the availability of synthetic Q in adequate amounts for pharamaceutical administration has inevitably raised the question of whether there are circumstances in which the administration of exogenous Q might be therapeutically beneficial.

Keywords: Administration, Antioxidant, Biological effects, Membrane effects, Pharmokinetics, Redox pathways.

POSSIBLE ENHANCEMENT OF RESPIRATION BY Q ADMINISTRATION

The kinetics of Q in respiration have yet to be fully elucidated (Lenaz, 1988) and it may well be that there are circumstances where the Q content of mitochondria becomes functionally limiting, particularly in cases of increased respiratory demand. Some such possible circumstances are the following:

Changes in kinetic characteristics of respiratory complexes

As an example of the extent to which changes might occur in the kinetics of individual spans of the respiratory chain, we demonstrated some time ago (Baum and Pollak, 1977) that starvation brought about an apparent three-fold increase in the "kinetic competence" of Complex III in rat liver mitochondria. In other words, as deduced from an analysis of antimycin titration curves, it seemed that less than one third of the amount of functional Complex III was necessary in mitochondria from starved rats to sustain the same overall respiratory rate as observed for fed controls. The basis of this effect is unknown, and it might well be a secondary reflection of an altered lipid composition of the membrane. Clearly, whatever the basis, if such changes really do occur *in vivo* then the Q content required for maximal respiratory rate might vary very significantly.

Changes in membrane lipids

Any significant changes in the lipid:respiratory complex ratio of the inner membrane might be expected to affect the dependence of respiratory rate on the diffusion kinetics of Q. Similar effects might arise from any changes in lipid composition that affected the viscosity of the lipid domain of the membrane.

Decrease in mitochondrial Q content

It may well be that in some disease states, (or in ageing), there is a significant decrease in the Q content of the mitochondrial membrane to the extent that the amount of co-enzyme itself limits respiration. Such changes might arise from decreased synthesis, as seen in protein starvation in the rat (Joshi et al, 1963), increased metabolic turnover or oxidative destruction, or diversion of Q to other cellular locations and other roles.

Aspects of the above possibilities are explained in detail elsewhere in this volume. Also reviewed, in some of the contributions in this section, are the outcomes of attempts to treat with this safe, natural product a variety of diseases, particularly those affecting tissues with a high demand for respiratory energy, such as heart.

OTHER POSSIBLE BIOLOGICAL EFFECTS OF Q

Any success of treatment with Q, and any other prospects for Q therapy may not simply relate to its function in mitochondrial respiration. It must be remembered that Q is by no means exclusively located to the inner mitochondrial membrane (Ramasarma, 1985b), and there is a wide range of other functions and effects that one might speculate for it, each of which would have therapeutic (or indeed toxicological) implications. These include the following:

Membrane fluidity

Q has been reported to alter the fluidity of biological membranes (Lenaz and Degli Esposti, 1985). Such an effect could influence cell motility and also the behaviour of membrane receptors and enzymes, and hence, in turn, such key functions as stimulus-response coupling.

Membrane stability

Direct or indirect effects of Q on membrane stability, (the latter, for example, through an antioxidant role, see below), could influence membrane fusion, and hence uptake or secretion processes, cell division, or phospholipase action, and hence stimulus-response coupling or various lytic processes (Kobayashi et al, 1980).

Protein interactions

In principle, Q could interact specifically or non-specifically with hydrophobic regions of intra- or extra-cellular proteins, with possible effects on kinetics or stability. In this connection, we are presently investigating the extent to which Q will form complexes with elastin, and the effect of such complexation on the susceptibility to elastolysis. Also of interest is the fact that Q forms (non-reducible) complexes with albumin (Wrigglesworth, J.M., personal communication). Might the formation of such complexes affect not only the pharmokinetics of Q (see below) but also the normal functioning of the albumin pool?

Non-mitochondrial redox pathways

It has been suggested that Q might be involved in the redox pathway leading to acidification of Golgi-derived vesicles (Crane et al, 1984). There is also the possibility of its participation in other pathways, such as trans plasma membrane electron transport, a process which is now emerging as of possible importance in controlling the cell cycle (Crane et al, 1985). A role might also be speculated for Q in the oxidative metabolism of some xenobiotics.

Metabolic effects

Q is synthesised and degraded in the cell and the pathways involved share intermediates and co-enzymes with other pathways, notably those of steroid and tryrosine metabolism (Ramasarma, 1985a). Administration of Q might thus affect those pathways.

Some of these various possibilities are considered in more detail in other contributions to this volume.

ANTIOXIDANT FUNCTION OF Q

One possible non-respiratory role of Q deserves special mention and that is its potential function as an anti-oxidant. In man more than 2^{20} molecules of oxygen are reduced per g. tissue per day. This reduction usually takes place in a micro-environment where these are transition metal centres and/or complex aromatic species, and as a potential source of the direct or indirect generation of free oxygen radicals, these pathways may well be of more significance than environmental ionizing radiation. Ageing, cancer and a variety of specific degenerative diseases have been attributed to the toxic effects of these radicals. Might Q have a natural role in endogenous protection against oygen radicals? Might exogenously administered Q enhance this protection generally, or in specific cases such as anoxia and reperfusion injury, or exposure to radiation? For example, a component of the complex cascade of reperfusion injury is the failure of mitochondria to retain a calcium load, and we have shown that exogenous Q *in vitro* can be protective in this regard (Baum and Booth, 1984).

The whole question of antioxidant function of Q is dealt with in detail in a subsequent contribution. However it is worth pointing out that, quite apart from potential protective effects of exogenous Q against oxidative tissue damage, auto-oxidation might increase the demand for Q to sustain respiration, either because of resultant changes in membrane lipids, or through irreversible modification of endogenous Q. Additionally, the products of such modification might themselves be competitive inhibitors of respiration.

PHARMACOKINETICS OF Q

A final set of questions in considering the therapeutic potential of Q relates to its pharmacokinetics. How might it be administered, how is it transported from the site(s) of uptake, in which tissues does it concentrate, and how is it distributed between and within the cellular compartments of those tissues?

These problems are illustrated by the findings of Reahal and Wrigglesworth (this meeting). They show that, in the rat, orally administered Q_{10} was taken up almost exclusively in liver, where it was found to be mainly in the reduced form. On the other hand, Q_{10} administered intra-peritoneally to rats was detected in both liver (predominantly reduced) and in spleen (mainly oxidised). In these experiments no incorporation of Q_{10} was observed in heart or kidney following either route of admininstr-ation. A further complication is that the Q_{10} taken up in either case was mainly detected in the cytoplasmic fractions of the respective tissues, although whether this represented free droplets or complexes with carrier proteins is not known. Other work on this difficult and important problem is reported in the following section of this volume and the pharmacokinetic parameters of Q can now be seen to be being systematically addressed and elucidated.

CONCLUSION

It is to be hoped that a synthesis of the contributions to this volume will direct us to a more rational approach as to how Q might work as a drug, which conditions might benefit from treatment with it, and how best it might be administered in each case.

REFERENCES

Baum, H. and Pollack, J.K., 1977, Iron centres and rate-limiting spans in the respiratory chains of mito-chondria from adult and fetal rats. In _Iron Metabolism_, Ciba Foundation Symposium 51 (Amsterdam: Elsevier/ Excerpta Medica/North Holland Inc), pp 79-90.

Baum, H. and Booth, R.F.G., 1984, Transmembrane calcium fluxes and cell death. In _Membrane Processes, Molecular Biology and Medical Applications_, edited by Gh. Benga, H. Baum and F.A. Kummerow (New York: Springer-Verlag Inc.), pp. 167-176.

Crane, F.L., Sun, I.L., Barr, R. and Morre, D.J., 1984. In _Biomedical and Clinical Aspects of Coenzyme Q_, Vol. IV, edited by K. Folkers and Y. Yamamura (Amsterdam: Elsevier), pp. 77-86.

Crane, F.L., Sun, I.L., Clark, M.G., Grebing, C. and Low, H., 1985, Transplasma-membrane redox systems in growth and development. _Biochimica et Biophysica Acta_, 811, 233-264.

Joshi, V.C., Jayaraman, J. and Ramasarma, T., 1963, Tissue concentrations of coenzyme Q, ubichromenol and tocopherol in relation to protein status in the rat. Biochemical Journal, 88, 25-31.

Kobayashi, M., Shimomura, Y., Suzuki, H., Takana, M., Baum, H. and Ozawa, T., 1980, Structural effects of ubiquinones on the mitochondrial inner membrane. Journal of Applied Biochemistry, 2, 270-279.

Lenaz, G., 1988, Role of mobility of redox components in the inner mitochondrial membrane. Journal of Membrane Biology, 104, 193.

Lenaz, G. and Degli Esposti, M., 1985, Physical properties of ubiquinones in model systems and membranes. In Coenzyme Q, edited by G. Lenaz (Chichester: John Wiley & Sons Ltd.), pp. 83-105.

Ramasarma, T., 1985a, Metabolism of coenzyme Q. In Coenzyme Q, edited by G Lenaz (Chichester: John Wiley & Sons Ltd.), pp. 131-142.

Ramasarma, T., 1985b, Natural occurrence and distribution of coenzyme Q. In Coenzyme Q, edited by G. Lenaz (Chichester: John Wiley & Sons Ltd.), pp. 67-81.

THE ANTIOXIDANT ROLE OF COENZYME Q

Robert E. Beyer* and Lars Ernster**

*Department of Biology, The University of Michigan, Ann Arbor, Michigan 48109-1048, USA
and
**Department of Biochemistry, Arrhenius Laboratories, University of Stockholm, S-106 91 Stockholm, Sweden

INTRODUCTION

Over the last three decades it has become generally recognized that coenzyme Q (CoQ, ubiquinone) is an intrinsic component of the mitochondrial and bacterial electron transport systems involved in energy conservation for ATP synthesis. Table 1, quoted from a presentation at an earlier CoQ symposium, is a historical survey of some key discoveries that form the basis for our current perception of the function of CoQ in mitochondrial electron transport. Further details of these and subsequent developments are found in various symposium volumes and monographs (e.g., Wolstenholme & O'Connor, 1961; Folkers & Yamamura, 1977, 1981, 1984, 1986; Yamamura et al., 1980; Trumpower, 1982; Ernster, 1984b; Lenaz, 1985; Lee, 1987) as well as numerous review articles and original reports, including those presented at this Meeting. Considerable progress has been made in recent years especially concerning the identification of separate CoQ-binding proteins as components of the NADH-CoQ reductase, succinate-CoQ reductase and $CoQH_2$-cytochrome *c* reductase complexes (Complexes I, II and III, respectively) of the mitochondrial respiratory chain and the elucidation of the pathways of electron transfer within and between these complexes. Whether other enzymes that feed electrons into the respiratory chain via CoQ, such as α-glycerophosphate dehydrogenase, choline dehydrogenase or ETF dehydrogenase, also contain CoQ-binding proteins, remains to be investigated. Another open question is whether the proton translocation during the operation of Complex III takes place directly by way of CoQ, in accordance with the mechanisms known as the "protonmotive Q cycle" (Mitchell, 1975b), or whether it involves additional components, e.g. the *b* cytochromes (Wikström & Krab, 1980; Papa, 1982 and this Meeting). Yet

Table 1. Some key discoveries concerning the role of coenzyme Q (CoQ) in the mitochondrial respiratory chain.

(From Ernster, 1984)

Year	Discovery	Reference
1957	Evidence for the involvement of CoQ as a redox carrier in the mitochondrial respiratory chain.	Crane et al., 1957
1959-66	Restoration of succinate and NADH oxidation in CoQ-depleted mitochondria by excess CoQ.	Lester & Fleischer, 1959 Szarkowska, 1966
1967	Proposal of participation of CoQ semiquinone in electron transport through Complex III of the respiratory chain.	Baum et al., 1967
1969-74	Restoration of succinate and NADH oxidation in CoQ-depleted submitochondrial particles by quantitative reincorporation of CoQ.	Ernster et al., 1969 Norling et al., 1974
1970-73	Kinetic evidence for the participation of CoQ in mitochondrial electron transport. The "Q pool" concept.	Kröger & Klingenberg, 1970, 1973
1970	Evidence for the occurrence of CoQ semiquinone during mitochondrial electron transport.	Bäckström et al., 1970
1972	Proposal of a role of CoQ semiquinone in electron transport through Complex III, accounting for the oxidation-dependent reduction of cytochrome <u>b</u>.	Wikström & Berden, 1972
1975	Postulation of a cyclic electron-transport pathway involving CoQ semiquinone through Complex III of the respiratory chain ("Q cycle").	Mitchell, 1975a,b
1975-80	Evidence for stabilized forms of CoQ semiquinone in mitochondria.	Ruzicka et al., 1975 Case et al., 1976 Salerno & Ohnishi, 1980
1977-80	Demonstration of CoQ-binding proteins in mitochondria.	Yu et al., 1977, 1978 Yu & Yu, 1980
1979-80	Evidence for a Q cycle, based on studies of the iron-sulfur protein component of Complex III.	Trumpower & Edwards, 1979 Slater & de Vries, 1980

another unresolved problem relates to the site and mechanism of superoxide-radical generation that has been shown to accompany mitochondrial respiration (cf. Chance et al., 1979) and, in particular, the possible involvement of CoQ in this process. This subject has been critically reviewed elsewhere (Ozawa, 1985; Nohl et al., 1986; Beyer et al., 1986, 1987; Beyer, 1989) and will not be treated herein.

The present report deals with another functional aspect of CoQ that has acquired growing attention in recent years, namely, its role as an antioxidant in biological membranes. This subject, too, has been reviewed on several occasions (Ozawa, 1985; Beyer et al., 1986, 1987; Beyer, 1989). The sections below are a brief summary and an update of the evidence supporting this concept at the cellular, subcellular and molecular levels, with a discussion of some of its physiological and pathological implications. The paper will be concluded by considering some possible future directions of this field of research.

Keywords: Coenzyme Q, Ubiquinone, Antioxidant, Review

EVIDENCE FOR AN ANTIOXIDANT FUNCTION OF COENZYME Q

Subcellular systems

The earliest indications that CoQ may function as an antioxidant were reported by Lea and Kwietny (1962) and by Mellors and Tappel (1966a), who showed that both the quinone and quinol forms of CoQ possessed about 20% of the antioxidant activity of α-tocopherol against lipid peroxidation. However, these experiments were performed at relatively high oxygen pressure, which favors the oxidation of quinols to quinones. Mellors and Tappel (1966b) subsequently tested CoQ_6 in both the oxidized and reduced forms under normoxic conditions and found that the reduced form was about four times more effective than the oxidized form in reducing the stable free radical diphenyl-p-picrylhydrazyl, and was almost as reactive as α-tocopherol. They also showed that reduced CoQ_6 was considerably more potent than the oxidized form as an inhibitor of light-induced peroxidation of arachidonic acid emulsions by haemoglobin, and that, in this system, the quinol was as effective as α-toxopherol. Similar results were obtained with mitochondrial phopsholipids. Mellors and Tappel (1966b) were also pioneers in showing an inhibition of mitochondrial lipid peroxidation by CoQ reduced through the respiratory chain using β-hydroxybutyrate as substrate in the presence of antimycin (Fig. 1). A functional relationship between CoQ and α-tocopherol with respect

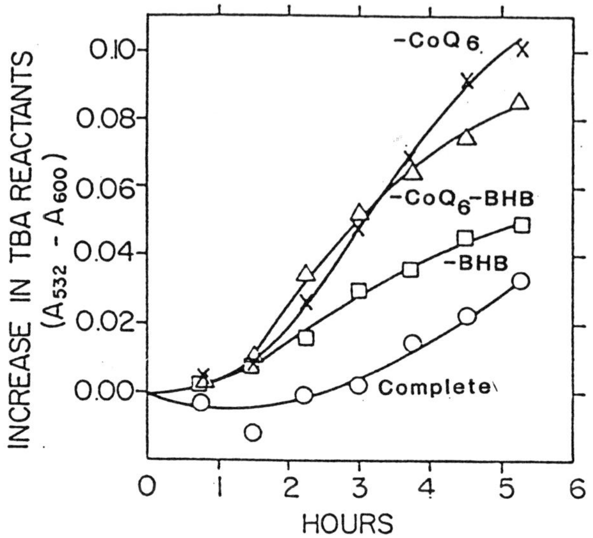

Fig. 1. Inhibition of photo-oxidation of rat liver mitochondria by CoQ_6H_2 formed by enzymic reduction of exogenous CoQ_6. The complete mixture contained 2 mg CoQ_6 in 0.02 ml ethanol, 25 mM sodium β-hydroxybutyrate (BHB), 20 μg antimycin in 0.02 ml ethanol, and mitochondria (68 mg protein) in Tris-KCl, pH 7.5), in a final volume of 20 ml. (From Mellors and Tappel, 1966b, with permission)

to the stabilization of mitochondrial membranes has been suggested by Fitch and Folkers (1967).

Takeshige et al. (1980) extended these observations with submitochondrial particles from beef heart, which were depleted of CoQ by pentane extraction and in which CoQ was reincorporated to varying extents using the method described by Ernster and associates (Ernster et al., 1969, 1978; Norling et al., 1974). Lipid peroxidation was initiated by the addition of ADP-Fe^{3+} and either NADH or NADPH. Reincorporation of CoQ into the depleted particles resulted in an inhibition of lipid peroxidation to a degree that was dependent on the extent of the replenishment of CoQ. Moreover, when NADH oxidation was inhibited by rotenone, the inhibition of lipid peroxidation required the addition of succinate and was dependent on the concentration of the latter (Fig. 2A). Similar results were obtained when, instead of inhibiting NADH oxidation with rotenone, the level of reduced CoQ was diminished by using limiting concentrations of NADH or rep-

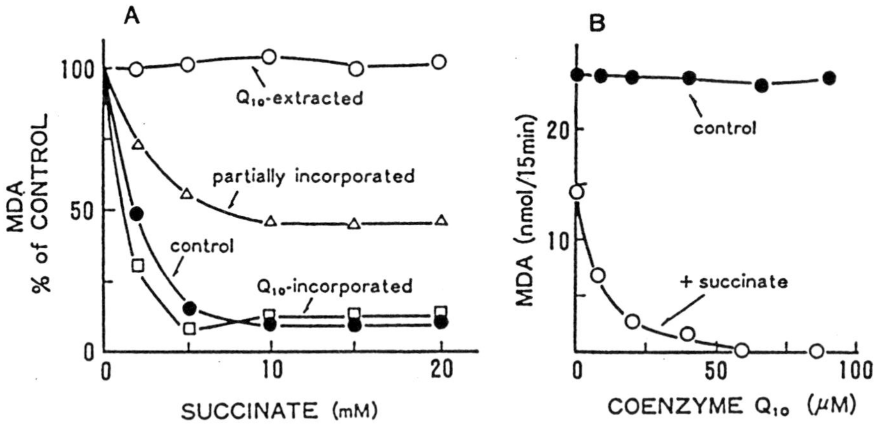

Fig. 2. Effects of CoQ_{10} extraction and reicorporation and of succinate on NADH- and ADP-Fe^{3+}-induced lipid peroxidation in the presence of rotenone (A) in beef-heart submitochondrial particles, and (B) in a mixture of submitochondrial particles and mitochondrial phospholipid liposomes. (From Takeshige et al., 1980, with permission)

lacing NADH by NADPH. When the particles were incubated together with liposomes containing mitochondrial phospholipids, peroxidation of the latter was also inhibited by succinate in a CoQ-dependent manner (Fig. 2B). Apparently, the CoQ reduced by succinate in the particles was able to act as an antioxidant in the liposomes, the latter probably undergoing fusion with the submitochondrial particles.

Additional evidence for an antioxidant function of reduced CoQ was reported by Booth et al. (1982), using ascorbate + Fe^{3+} to induce lipid peroxidation in egg-yolk phosphatidyl choline liposomes, and by Landi et al. (1984, 1985) in sonic lipid vesicles containing various CoQ homologs.

Solaini et al. (1987), employing beef heart mitochondria partially depleted of coenzyme Q by pentane extraction, have reported that lipid peroxidation induced by an adriamycin-iron complex proceeds at a higher rate than in mitochondria into which coenzyme Q has been reincorporated. Moreover, in mitochondria depleted of coenzyme Q, both NADH and succinate oxidase activities were inactivated more rapidly than in mitochondria in which coenzyme Q had been replenished. This suggests a protective role for coenzyme Q in free radical-associated inactivation of electron transfer chain function as well as membrane lipid peroxidation.

A protective effect of reduced CoQ against lipid peroxidation could also be demonstrated with rat liver microsomes

Table 2. Inhibition of NADPH- and ADP-Fe^{3+}-initiated lipid peroxidation in rat liver microsomes by reduced CoQ_{10}. (From Beyer et al., 1987)

Additions	Conc. (μM)	Oxygen consumption (natoms/min/mg protein)	MDA* formed (nmoles)
Solvent**		224	5.87
CoQ_{10}	150	320	5.71
$CoQ_{10}H_2$	150	59	1.19
$CoQ_{10}H_2$	300	31	0.26
CoQ_{10} + $CoQ_{10}H_2$	150 + 75	97	2.01

*malondialdehyde **ethanol + cardiolipin

(Beyer et al., 1987; Beyer, 1988, 1989). In these experiments, lipid peroxidation was initiated by the addition of NADPH and ADP-Fe^{3+} and was followed by measuring oxygen consumption and MDA formation (cf. Hochstein et al., 1964; Ernster & Nordenbrand, 1967). The result of an experiment in which CoQ_{10} was added to the oxygen electrode chamber in the fully reduced state is shown in Table 2. As with liposomes, mitochondria, and submitochondrial particles, reduced CoQ is an effective antioxidant in microsomes. CoQ in the oxidized form had no antioxidant effect.

Cell cultures

Takahashi et al. (1988) used myocardial cells in culture to test the hypothesis that coenzyme Q is able to prevent myocardial cell uptake of doxorubicin. They have observed that 1) coenzyme Q had no effect on the uptake of doxorubicin by the cells, 2) the intact doxorubicin molecule was taken up by the cells, and 3) treatment of the cells with CoQ prior to doxorubicin administration protected the beating of the cells from doxorubicin toxicity. These observations lend further support for an antioxidant role of CoQ, adding to the vast literature (see e.g. Okuma & Ota, 1986) on the efficacy of CoQ as an agent protecting cells from free radical damage during the one-electron reduction of antitumor quinones (cf. Powis, 1989 for review).

Table 3. Distribution of CoQ in subcellular fractions from rat liver.

The values are the means ± S.E. of 7 experiments.

(From Kalén et al., 1987)

Fraction	CoQ_9 (µg/mg protein)
Homogenate	0.79 ± 0.08
Golgi vesicles	2.62 ± 0.15
Lysosomes	1.86 ± 0.18
Mitochondria	1.40 ± 0.16
Inner mitochondrial membrane	1.86 ± 0.13
Microsomes	0.15 ± 0.02
Peroxisomes	0.29 ± 0.04
Plasma membrane	0.74 ± 0.07
Supernatant	0.02 ± 0.004

Intracellular distribution and biosynthesis of CoQ

In view of its role in the respiratory chain, CoQ in eukaryotes was initially assumed to be located and synthesized exclusively in the mitochondria (for reviews, see Ramasarma, 1985a, b). However, studies by Sastry et al. (1961) and Jayaraman and Ramasarma (1963) have revealed that CoQ also occurs in the nucleus and the microsomes of rat liver. Subsequently Crane and Morré (1977) reported evidence for the occurrence of substantial amounts of CoQ in the Golgi apparatus of various tissues.

Recent work in Dallner's laboratory has shown that CoQ in rat liver occurs in most cellular membranes (Kalén et al., 1987; see Table 3). By studying the time course of incorporation of both ^{14}C- or ^{3}H-tyrosine and ^{3}H-mevalonate into rat liver CoQ in vivo or in tissue slices it was also found that the microsomal CoQ was labeled more rapidly than the mitochondrial CoQ, with kinetics consistent with a precursor-product relationship. This conclusion was further substantiated by demonstrating (Kalén et al., 1989) that nonaprenyl-4-hydroxybenzoate transferase, a key enzyme in the biosynthesis of CoQ_9, is concentrated in the endoplasmic reticulum Golgi system of rat liver. Evidence was also presented for a discharge of newly synthesized CoQ with lipoproteins to rat liver perfusate and to the bile (Elmberger et al., 1989). In the perfusate CoQ was primarily associated with the very low density lipoprotein fraction, while in rat blood plasma it was evenly distributed among the very low density, low density and high density lipoproteins. These results strongly indicate that (a) CoQ is synthesized in the endoplasmic reticulum and transported from there to other cellular membranes; and (b) there exists a transport mechanism for the secretion of CoQ into the blood stream and the bile.

Dallner and associates have recently also investigated the effect of various inducers of membrane proliferation on the intracellular distribution and content of CoQ in rat liver (Kalén et al., 1989b). Phenobarbital, 3-methylcholanthrene and N-nitrosodiethylamine, which are known inducers of cytochrome P-450 and endoplasmic reticulum proliferation, enhanced both the rate of CoQ synthesis and the CoQ content of microsomes. Clofibrate and di(2-ethylhexyl)phthalate, which are known as selective peroxisomal inducers but also cause an increase in the number of mitochondria, gave rise to an increase in mitochondrial and microsomal CoQ contents. It should be noted that all three organelles, endoplasmic reticulum, mitochondria and perioxisomes, contain enzyme systems that can reduce molecular oxygen and are thus potential sites of oxygen-radical generation. Changes in the intracellular distribution and content of CoQ in rat liver under the influence of various agents had also been reported by Jayaraman and Ramasarma (1963). These data, too, are consistent with a relationship between oxidative metabolism and CoQ biosynthesis.

Physiological and pathological factors influencing tissue CoQ levels in vivo

A number of reports have been published on intact animals which, while not as incisive as experiments at the subcellular or cellular levels, are nevertheless consistent with an antioxidant function of CoQ. Studies on cold-acclimated rats (Beyer et al., 1962a), rats treated with thyroid hormone (Pedersen et al., 1963), and other conditions also affecting the overall rate of energy metabolism of intact rats (Beyer et al., 1962b), revealed alterations of CoQ concentrations in highly aerobic tissues, the direction of change being the same for metabolic rate, and thus free radical production, and for tissue CoQ content. Significantly, the increase in tissue CoQ content following thyroid hormone treatment occurred _after_ the increase in metabolic rate, suggesting that the increase in CoQ may have been an adaptation to, rather than a cause of, increased oxidative activity (Pedersen et al., 1963). In this context it may be of interest to note that no increase in the CoQ content of skeletal muscle was found in "Luft's syndrome" (Luft et al., 1962), a state of severe hypermetabolism of non-thyroid origin, with a defect of mitochondrial respiratory control. Morphologically, the syndrome is characterized, among other features, by a vast accumulation of mitochondria in the perinuclear region of the skeletal muscle cells and a lack of opaque bodies in the mitochondrial matrix. Interestingly, similar morphological changes have recently been reported by Yagi (1987) in the endothelial cells of the thoracic aorta of rabbit upon intravenous injection of linoleic acid hydroperoxide.

High in vivo oxidative activity such as that which accompanies exhaustive exercise results in a two- to three-fold increase in free-radical formation and lipid peroxidation in muscle and liver (Davies et al., 1982; Salminen & Vihko, 1983a). Adaptation to endurance training reduces the susceptibility of tissue to the damaging effects of free radicals and lipid peroxidation (Salminen & Vihko, 1983b). Beyer et al. (1984a) have shown that cardiac and other red muscles of endurance-trained rats contain levels of CoQ which are elevated to a greater extent than other components of the mitochondrial electron transport chain. Boveris et al. (1969) have also reported an example of regulation of CoQ concentrations independent of cytochrome c and protein in mitochondria from diabetic animals.

Further evidence consistent with an antioxidant role for CoQ has been derived from analyses of CoQ concentrations in tissues from rats over their life span. Beyer et al. (1985) have reported that in elderly animals CoQ levels drop drastically in heart, kidney, and highly aerobic (red) muscles (Table 4). These observations on ageing animals are consistent with the "free radical theory of ageing" (Harman, 1983) and with a report on an inverse correlation between longevity and peroxide-producing potential of mammalian tissues (Cutler, 1985). Kalén et al. (1989c) have recently reported a similar age-related decrease of CoQ levels in human organs (Table 5). Their studies indicate an intricate interplay among the three major biosynthetic products of mevalonate metabolism, i.e., CoQ, dolichol and cholesterol. Parallel to the decrease in CoQ upon ageing there is a drastic increase in dolichol - in certain organs more than 100-fold - while the cholesterol levels are relatively unchanged.

Beyer (1988) investigated the combined effects of endurance training and ageing on the CoQ content of rat heart and the susceptibility of rat heart submitochondrial particles to lipid peroxidation. The CoQ content decreased 30% between 3 and 25 months of age in sedentary animals while a

Table 4. Age-related changes in the CoQ content of various tissues of Sprague-Dawley rats.
(From Beyer et al., 1985)

Tissue	Age (months)		
	2	18	25
Heart	1134	2036	825
Kidney	501	883	688
Gastrocnemius	210	340	233
DVL*	365	522	408
Soleus	364	338	240

*Deep (red) portion of the vastus lateralis muscle.
CoQ contents are expressed in nanomoles/g wet weight tissue.

Table 5. Age-related changes in the CoQ content of human organs. (From Kalén et al., 1989c)
CoQ contents are expressed in µg/mg wet-weight tissue.

Organ	Age group				
	1-3 days	0.7-2 years	19-21 years	39-43 years	77-81 years
Lung	2.2	6.4	6.0	6.5	3.1
Heart	36.7	78.5	110.0	75.0	47.2
Spleen	20.7	30.2	32.8	28.6	13.1
Liver	12.9	45.1	61.2	58.3	50.8
Kidney	17.4	53.4	98.0	71.1	64.0
Pancreas	9.2	38.2	21.0	19.3	6.5
Adrenal	17.5	57.9	16.1	12.2	8.5

6-8 samples were analyzed for each organ. For S.E.M. see original paper.

regimen of endurance running 5 days each week for 5 months resulted in an increase of the CoQ content in heart submitochondrial particles by 41% compared to sedentary elderly animals. In addition, small differences were observed in the rate of NADH-supported lipid peroxidation catalyzed by submitochondrial particles isolated from hearts of the three groups of animals. These differences were in the direction of changes in overall respiratory activity reported previously (Beyer et al., 1984b; Starnes et al., 1981, 1983) under these experimental circumstances.

Effect of CoQ administration on experimentally induced oxidative damage

A number of reports have appeared describing protective effects of CoQ administration against oxidative tissue injury. Sugiyama et al. (1980a) found a lowering of the level of lipid peroxides in the liver and heart of animals after CoQ treatment. The same group (Sugiyama et al., 1980b) also reported that administration of CoQ to animals subjected to cardiac ischemia resulted in improved function of mitochondria isolated from the affected area when compared to controls. In studies of the protection from ischemic myocardial injury by intravenous administration of CoQ, Katagiri et al. (1986) have shown that a portion of the administered CoQ is bound to sarcoplasmic reticular membranes. In animals so treated prior to coronary artery occlusion, degradation of sarcoplasmic reticular membrane components was reduced compared to tissue from animals not treated with CoQ. Shibata et al. (1986) have reported that in experimental myocardial ischemia, 12-hydroxyeicosatetraenoic acid, a metabolite of arachidonic acid resulting from the action of lipoxygenase associated with injury of sarcoplasmic reticulum as well as mitochondria, shows less of an increase when previously treated with CoQ. Kawasaki et al. (1986) have demonstrated that administration of CoQ one hour prior to the production of liver ischemia protects this organ from oxidative cellular damage during reperfusion. Of particular interest is the finding that the large decrease of glutathione, α-tocopherol and CoQ (both oxidized and reduced) was prevented by prior administration of CoQ, suggesting a role for CoQ in maintaining the efficacy of general intracellular antioxidant function. The same authors have demonstrated a protective role of CoQ in experimental endotoxemia persuant to administration of lipopolysaccharide prepared from *E. coli*. The administered CoQ was shown to maintain the levels of reduced CoQ and αtocopherol in the endotoxemic mouse liver.

Beyer and Starnes (1985) have reported that administration of CoQ partially reversed the severe decline of respira-

Table 6. MDA* in 24-hour urine samples and in liver fractions from rats treated with CCl_4, ethanol (EtOH) and CoQ. (From Beyer, 1988)

Treatment	Urine	Liver fractions		
		Whole homogenate	Mito-chondria	Micro-somes
Control	0.92	86	8.2	5.4
CoQ	0.62	13	3.7	3.2
CCl_4	4.88	766	28.3	10.7
CCl_4 + CoQ	1.46	465	11.1	6.5
EtOH	2.42	188	19.8	8.4
EtOH + CoQ	0.81	15	4.6	3.4

*MDA (malondialdehyde and other "TBA-reacting material" is expressed in microgram/24 hrs in the case of urine, and in nanomoles MDA/mg protein in the liver fractions. 5-7 animals in each group. For statistical comparison of groups, see the original paper.

tory metabolism of cardiac tissue that occurs in animals at advanced age. This effect was most striking with the oxidation of long-chain fatty acids, the preferred substrate of cardiac muscle. Quinn et al. (1980) have reported that CoQ administration protects liver cells against damage resulting from exposure to carbon tetrachloride. Bertelli et al. (1986) have confirmed these observations. Employing carbon tetrachloride and ethanol to produce profound and moderate free-radical damage, respectively, Beyer (1988) has reported that CoQ administration results in an inhibition of carbon tetrachloride- and ethanol-stimulated lipid peroxidation using urinary malondialdehyde as an index (Table 6). In the same study, CoQ administration was shown to lower malondialdehyde levels in liver homogenate, mitochondrial and microsomal fractions from both carbon tetrachloride and ethanol treated rats (cf. Table 6). As it is well established that both carbon tetrachloride (Brattin et al., 1985) and ethanol (Thurman, 1973; Uysal et al., 1985) cause free-radical damage, these data are consistent with an antioxidant function for CoQ in vivo.

In an interesting study of the influence of in vitro CoQ antagonists on doxorubicin toxicity in vivo, Tàbora et al.

(1986) found a toxic effect of such antagonists *per se* which was prevented completely by coenzyme pretreatment with CoQ, but only partly by pretreatment with α-tocopherol. CoQ antagonists enhanced doxorubicin toxicity in a synergistic manner and this effect was not diminished by CoQ treatment.

Clinical studies

In addition to a vast literature on successful applications of CoQ in a variety of clinical situations (cf. Folkers & Yamamura, 1977, 1981, 1984, 1986; Yamamura et al., 1980; Trumpower, 1982; Lenaz, 1985), recent findings have further stimulated the medical community's interest in CoQ therapy, especially in the area of congestive heart failure and certain neurologic disorders. Karlsson et al. (1986) have reported that patients under treatment with adriamycin have diminished CoQ levels compared to normal controls in heart and skeletal muscle as well as in venous blood, CoQ being essentially depleted in skeletal muscle of several patients. In addition, another study of CoQ and congestive heart failure (Mortensson et al., 1986) reports that upon withdrawal of long-term CoQ therapy, clinical decline, relapse, and death in two subjects occurred within two weeks. Several patients who relapsed after CoQ withdrawal regained previous functional status after retreatment with CoQ.

Kearns-Sayre syndrome is a mitochondrial encephalomyopathy characterized by the triad of progressive external ophthalmoparesis, heart conduction block and pigmentary degeneration of the retina which start in childhood. Such patients show reduced levels of CoQ in serum and mitochodnria isolated from skeletal muscle and respond favorably to CoQ therapy (Takahashi et al., 1986; Ogasahara et al., 1985, 1986; Bresolin et al., 1988). Zierz et al. (1989) have confirmed these observations in patients with Kearns-Sayre syndrome and ophthalmoplegia plus, strengthening the possibility that discrete pathology may result from reduced tissue levels of CoQ and enforcing the rationale for the treatment of such clinical entities with CoQ.

In studies with isolated muscle mitochondria from patients with Kearns-Sayre syndrome Lee et al. (1987) have concluded that the electron-transfer system of these mitochondria is impaired and that the site of the defect is on the path of electron transfer between cytochrome *c* and cytochrome oxidase. Interestingly, a similar impairment of electron transfer between cytochrome *c* and cytochrome oxidase has recently been shown to occur upon a partial loss of cardiolipin - an essential component of the mitochondrial cytochrome *c*-cytochrome oxidase system - due to lipid peroxidation occurring in skeletal-muscle upon exposing rats to ischemia and reperfusion in vivo (Soussi et al., 1989a) or by chemically generated superoxide radicals in vitro (Soussi et al., 1990). The opposite effect, i.e., an enhanced interaction between cytochrome c and cytochrome oxidase, was found by the same authors (Soussi et al., 1989b) to ensue upon endurance training of rats, a treatment known to lead to an increased CoQ level in skeletal muscle mitochondria (Beyer et al., 1984a; Beyer & Starnes, 1985).

As pointed out at a previous CoQ meeting (Ernster, 1984), there seems to be a striking parallel between those pathologic conditions in which CoQ has been reported to have a therapeutic effect and the growing number of human disease processes that are attributed to oxygen free-radical damage (cf. Bulkley, 1983). Awareness of this relationship may provide a promising rational basis for continued research regarding the application of CoQ in clinical medicine.

Molecular and cellular mechanisms

Quinones have a highly complex redox chemistry. As pointed out in a recent review by Brunmark and Cadenas (1989), "every quinone has a chemistry of its own", its redox properties being dependent on its substituents and its chemical environment. Consequently, the precise molecular and cellular mechanisms undelying the antioxidant function of reduced CoQ in various biological systems remain to be elucidated. The present section is based on, and is an update of, a previous discussion of this topic (Beyer et al., 1987).

Reduced CoQ ($CoQH_2$), like hydroquinones in general, is oxidized by the superoxide radical ($O_2^{-\cdot}$), giving rise to the semiquinone ($CoQ^{-\cdot}$) and hydrogen peroxide:

$$CoQH_2 + O_2^{-\cdot} \rightarrow CoQ^{-\cdot} + H_2O_2 \quad (1)$$

Semiquinones, including that of CoQ, are readily autoxidized to the quinone with the formation of the superoxide radical:

$$CoQ^{-\cdot} + O_2 \rightarrow CoQ + O_2^{-\cdot} \quad (2)$$

However, the semiquinone of CoQ may be stabilized by binding to various "Q proteins" (cf. King, 1985 and refs. in Table I) and reduced in that form by various quinone reductases to $CoQH_2$. $CoQ^{-\cdot}$ may also be removed by the disproportionation reaction:

$$2CoQ^{-\cdot} + 2H^+ \rightarrow CoQ + CoQH_2 \quad (3)$$

the equilibrium of which is shifted to the right by the removal of $CoQH_2$ through reaction 1.

In addition, as recently shown by Cadenas and associates (Cadenas et al., 1988; Cadenas, 1989; Brunmark & Cadenas, 1989; Öllinger et al., 1990) superoxide dismutase promotes the reduction of certain semiquinones by $O_2^{-\bullet}$, acting as a "superoxide-semiquinone oxidoreductase", with the formation of the hydroquinone and O_2. Whether $CoQ^{-\bullet}$ belongs in this category of semiquinones has not yet been tested and remains to be established. If this were the case, the reaction could be written as:

$$CoQ^{-\bullet} + O_2^{-\bullet} + 2H^+ \rightarrow CoQH_2 + O_2 \qquad (4)$$

The sum of this reaction and reaction (1) would be identical with the conventional reaction catalyzed by superoxide dismutase:

$$2O_2^{-\bullet} + 2H^+ \rightarrow O_2 + H_2O_2 \qquad (5)$$

where the role of $CoQH_2$ would be to react with $O_2^{-\bullet}$ generated in the lipid phase of the membrane which may not be readily accessible to superoxide dismutase.

As discussed earlier (Beyer et al., 1987), $CoQH_2$ may also react with perferryl radicals (e.g. ADP-Fe^{3+}-$O_2^{-\bullet}$), which are believed to be the species responsible for the initiation membrane lipid peroxidation (Hochstein & Ernster, 1963), or even with a lipid free radical ($L^\bullet$) or lipid peroxide free radical ($LOO^\bullet$), in a manner similar to its reaction with free $O_2^{-\bullet}$ (reaction 1).

The possible role of the mitochondrial "CoQ pool" in maintaining the antioxidant function of CoQ has also been discussed previously (Beyer et al., 1987) and will not be elaborated on further here. Likewise, we have discussed the possible involvement of the two-electron quinone reductase DT diaphorase (cf. Ernster et al., 1987) in the reduction of CoQ in both mitochondria and microsomes. Crane & Morré (1977) have provided evidence for quinone-reducing enzyme activity in the Golgi apparatus. In general it appears that there exist enzymic pathways for quinone reduction in the various types of biological membranes where CoQ is known to occur. A detailed exploration of these pathways is of crucial interest for the understanding of the molecular and cellular mechanisms involved in the antioxidant role of reduced CoQ.

FUTURE DIRECTION FOR CoQ RESEARCH

Obviously, a number of aspects of CoQ function remain open to investigation. As pointed out above, we need more information about enzymes involved in CoQ reduction in various biological membranes, enzymes which are crucial not only for its function in electron transport but also for its role as an antioxidant. The occurrence of CoQ in different cellular locations also raises the problem of how CoQ is transported from its site of synthesis - most likely the endoplasmic reticulum - to its final destinations in the cell, the mitochondria, the Golgi apparatus, and the plasma membrane through which it apparently can be secreted into the blood and the bile. What are the mechanisms responsible for the sorting and transport of CoQ within the cell and for its transport in and uptake from the extracellular space?

Another question of great importance concerns the regulation of CoQ biosynthesis and biodegradation in response to altered physiological and pathological conditions, especially in relation to those of other polyisoprenoid compounds, primarily dolichol and cholesterol. Important advances in this field have been made in recent years, and further rapid progress is to be expected.

The question of the means by which CoQ administration results in the myriad of clinical and experimental benefits reported in the literature remains essentially unanswered. Whether such treatment restores CoQ to a level required to maintain an adequate rate of electron transfer and energy conservation for ATP synthesis or, alternatively, as an antioxidant to protect tissue at jeopardy from peroxidative damage, or both, will require further incisive research. Nevertheless, it would appear that sufficient evidence exists to allow the assignment of an antioxidant role to CoQ in addition to its other well-established cellular functions.

ACKNOWLEDGEMENTS

We thank our colleagues Dr. Kathleen D. Timberlake, Ms. Terri Tanaka, Ms. Ann Nielsen and Ms. Kerstin Nordenbrand for their most valuable cooperation in the preparation of this paper. Financial support from the Elsa U. Pardee Foundation, The University of Michigan Office of the Vice President for Research, the LSA College Faculty Travel Office, the Swedish Cancer Foundation and the Swedish Natural Science Research Council is gratefully acknowledged. In 1985-86 Dr. Beyer was a Visiting Professor at the Arrhenius Laboratory, University of Stockholm and a Senior Fellow of the Fogarty International Center for Advanced Study in the Health Sciences, National Institutes of Health, USA.

REFERENCES

Bäckström, D., Norling, B., Ehrenberg, A., Ernster, L., 1970, Electron spin resonance measurement on ubiquione-depleted and ubiquinone-replenished submitochondrial particles. Biochim. Biophys. Acta, 197, 108-111.

Baum, H., Rieske, J.S., Silman, M.I., Lipton, S.H., 1967, On the mechanism of electron transfer in complex III of the electron transfer chain. Proc. Nat. Acad. Sci. U.S.A., 57, 798-805.

Bertelli, A., GIovannini, L., Bertelli, A.A.E., Maltinti, G., Scalori, V., Romano, M.R., 1986, Tissue concentrations of coenzyme Q in liver of rats intoxicated by carbon tetrachloride. Int. J. Tiss. Reac., 8, 343-346.

Beyer, R.E., 1988, Inhibition by coenzyme Q of ethanol- and carbontetrachloride-stimulated lipid peroxidation in vivo and catalyzed by microsomal and mitochondrial systems. Free Rad. Biol. Med., 5, 297-303.

Beyer, R.E., 1989, The role of coenzyme Q in free radical formation and antioxidation, in CRC Handbook of Free Radicals and Antioxidants in Biomedicine, Volume III, Miquel, J., Quintanilha, A.T., and Weber, H., eds., CRC Press, Boca Raton, pp. 45-62.

Beyer, R.E., Starnes, J.W., 1985, Coenzyme Q and myocarcial function in aging and exercise. In Pathology of Cardiovascular Injury, Stone, H.L. and Weglicki, W.B., eds., Martinus Nijhoff, Boston, pp. 489-511.

Beyer, R.E., Noble, W.M., Hirschfield, T.J., 1962a, Coenzyme Q (ubiquinone) levels of tissues of rats during acclimation to cold. Can. J. Biochem. Physiol., 40, 511-518.

Beyer, R.E., Noble, W.M., Hirschfeld, T.J., 1962b, Alterations of rat-tissue coenzyme Q (ubiquinone) levels by various treatments. Biochim. Biophys. Acta, 57, 376-379.

Beyer, R.E., Morales-Corral, P.G., Ramp, B.J., Kreitman, K.R., Falzon, M.J., Rhee, S.Y.S., Kuhn, T.W., Stein, M., Rosenwasser, M.J., Cartwright, K.J., 1984a, Elevation of tissue coenzyme Q (ubiquinone) and cytochrome c concentrations by endurance exercise in the rat. Arch. Biochem. Biophys., 234, 323-329.

Beyer, R.E., Starnes, J.W., Edington, D.W., Lipton, R.W., Compton, R.T., III, Kwasman, M.A., 1984b, Exercise-induced reversal of age-related declines of oxidative reactions, mitochondrial yield, and flavins in skeletal muscle of the rat. Mech. Ageing Dev., 24, 309-323.

Beyer, R.E., Burnett, B.-A., Cartwright, K.J., Edington, D.W., Falzon, M.J., Kreitman, K.R., Kuhn, T.W., Ramp, B.J., Rhee, S.Y.S., Rosenwasser, M.J., Stein, M., An, L.C.-I., 1985, Tissue coenzyme Q (ubiquinone) and protein concentrations over the life span of the laboratory rat. Mech. Ageing Dev., 32, 267-281.

Beyer, R.E., Nordenbrand, K., Ernster, L., 1986, The role of coenzyme Q as a mitochondrial antioxidant: a short review, in Biomedical and Clinical Aspects of Coenzyme Q, Vol. 5, Folkers, K. and Yamamura, Y., eds., Elsevier, Amsterdam, pp. 17-24.

Beyer, R.E., Nordenbrand, K., Ernster, L., 1987, The function of coenzyme Q in free radical production and as an antioxidant: A review, Chemica Scripta, 27, 145-153.

Booth, R.F.G., Galanopoulos, D.G., Quinn, P.J., 1982, Protection by ubiquinone and ubiquinol against lipid peroxidation in egg yolk phosphatidylcholine liposomes. Biochem. Int., 5, 151-156.

Boveris, A.A., Ramos, M.C.P., Stoppani, A.O.M., Foglia, V.G., 1969, Phosphorylation, oxidation and ubiquinone content in diabetic mitochondria. Proc. Soc. Exp. Biol. Med., 132, 170-174.

Brattin, W.J., Glende, E.A., Jr., Recknagel, R.O., 1985, Pathological mechanisms in carbon tetrachloride hepatotoxicity. J. Free Radical Biol. Med., 1, 27-38.

Bresolin, N., Bet, L., Binda, A., Moggio, M., Comi, G., Nador, F., Ferrante, C., Carenzi, A., Scarlato, G., 1988, Clinical and biochemical correlations in mitochondrial myopathies treated with coenzyme Q_{10}. Neurology, 38, 892-899.

Brunmark, A., Cadenas, E., 1989, Redox and addition chemistry of quinoid compounds and its biological implications. Free Rad. Biol. Med., 7, 435-477.

Bulkley, G.B., 1983, The role of oxygen free radicals in human disease processes. Surgery, 94, 407-411.

Cadenas, E., 1989, Biochemistry of oxygen toxicity. Annu. Rev. Biochem., 58, 79-110.

Cadenas, E., Mira, D., Brunmark, A., Lind, C., Segura-Aguilar, J., Ernster, L., 1988, Effect of superoxide dismutase on the autoxidation of various hydroquinones. - A possible role of superoxide dismutase as superoxide semiquinone oxidoreducase. Free Rad. Biol. Med., 5, 71-79.

Case, G.D., Ohnishi, T., Leigh, J.S., 1976, Intramitochondrial positions of ubiquinone and iron-sulphur centers determined by dipolar interactions with paramagnetic ions. Biochem. J., 160, 785-795.

Chance, B., Sies, H., Boveris, A., 1979, Hydroperoxide metabolism in mammalian organs. Physiological Reviews, 59, 527-605.

Crane, F.L., Hatefi, Y., Lester, R.L., Widmer, C., 1957, Isolation of a quinone from beef heart mitochondria. Biochem. Biophys. Acta, 24, 220-221.

Crane, F.L., Morré, D.J., 1977, Evidence for coenzyme Q function in Golgi membranes, in Biomedical and Clinical Aspects of Coenzyme Q, Vol. 1, Folkers, K. and Yamamura, Y., eds., Elsevier, Amsterdam, pp. 3-14.

Cutler, R.G., 1985, Peroxide-producing potential of tissues: Inverse correlation with longevity of mammalian species. Proc. Natl. Acad. Sci. U.S.A., 82, 4798-4802.

Davies, K.J.A., Quintanilha, A.J., Brooks, G.A., Packer, L., 1982, Free radicals and tissue damage produced by exercise. Biochem. Biophys. Res. Commun., 107, 1198-1205.

Elmberger, P.G., Kalén, A., Brunk, U., Dallner, G., 1989, Discharge and newly-synthesized dolichol, dolichyl phosphate and ubiquinone with lipoproteins to rat liver perfusate and to the bile. Lipids, in press.

Ernster, L., 1984a, Ubiquinone: Redox coenzyme, hydrogen carrier, antioxidant, in Biomedical and Clinical Aspects of Coenzyme Q, vol. 4, Folkers, K., and Yamamura, Y., eds., Elsevier, Amsterdam, pp. 3-13.

Ernster, L., ed., 1984b, Bioenergetics, Elsevier, Amsterdam.

Ernster, L., Nordenbrand, K., 1967, Microsomal lipid peroxidation. Methods

Enzymol., 10, 574-580.

Ernster, L., Lee, I.-Y., Norling, B., Persson, B., 1969, Studies with ubiquinone-depleted submitochondrial particles. Essentiality of ubiquinone for the interaction of succinate dehydrogenase, NADH dehydrogenase, and cytochrome b. Eur. J. Biochem., 9, 299-310.

Ernster, L., Glaser, E., Norling, B., 1978, Extraction and reincorporation of ubiquinone in submitochondrial particles. Methods Enzymol., 53, 573-579.

Ernster, L., Estabrook, R.W., Hochstein, P., Orrenius, S., eds., 1987, DT-Diaphorase - A quinone reductase with special functions in cell metabolism and detoxication. Chimica Scripta, 27A.

Fitch, C.D., Folkers, K., 1967, Coenzyme Q and the stability of biological membranes. Biochem. Biophys. Res. Commun., 26, 128-131.

Folkers, K., Yamamura, Y., eds., 1977, Biomedical and Clinical Aspects of Coenzyme Q, Vol. 1, Elsevier, Amsterdam.

Folkers, K., Yamamura, Y., eds., 1981, Biomedical and Clinical Aspects of Coenzyme Q, Vol. 3, Elsevier, Amsterdam.

Folkers, K., Yamamura, Y., eds., 1984, Biomedical and Clinical Aspects of Coenzyme Q, Vol. 4, Elsevier, Amsterdam.

Folkers, K., Yammaura, Y., eds., 1986, Biomedical ann Clinical Aspects of Coenzyme Q, Vol. 5, Elsevier, Amsterdam.

Harman, D., 1983, The free-radical theory of ageing: Consequences of mitochondrial ageing. Age, 6, 86-92.

Hochstein, P., Ernster, L., 1963, ADP-activated lipid peroxidation coupled to the TPNH oxidase system of microsomes. Biochem. Biophys. Res. Commun., 12, 388-394.

Hochstein, P., Nordenbrand, K., Ernster, L., 1964, Evidence for the involvement of iron in the ADP-activated peroxidation of lipids in microsomes and mitochondria. Biochem. Biophys. Res. Commun., 14, 323-328.

Jayarman, J., Ramasarma, T., 1963, Intracellular distribution of coenzyme Q in rat liver. Arch. Biochem. Biophys., 103, 258-266.

Kalén, A., Appelkvist, E.L., Chojnacki, T., Dallner, G., 1989a, Nonaprenyl-4-hydroxybenzoate transferase, an enzyme involved in ubiquinone biosynthesis, in the endoplasmic reticulum - Golgi system of rat liver. J. Biol. Chem., in press.

Kalén, A., Appelkvist, E.L., Dallner, G., 1989b, The effects of inducers of the endoplasmic reticulum, peroxisomes and mitochondria on the amounts and synthesis of ubiquinone in rat liver subcellular membranes. Chem.-Biol. Interactions, in press.

Kalén, A., Appelkvist, E.L., Dallner, G., 1989c, Age-related changes in the lipid compositions of rat and human tissue. Lipids, 24, 579-584.

Kalén, A., Norling, B., Appelkvist, E.L., Dallner, G., 1987, Ubiquinone biosynthesis by the microsomal fraction from rat liver. Biochim. Biophys. Acta, 925, 70-78.

Karlsson, J., Folkers, K., Åström, H., Jansson, E., Pernow, B., Holmgren, A., Mellstedt, H., Diamant, B., 1986, Effect of adriamycin on heart and skeletal muscle coenzyme Q (CoQ_{10}) in man, in Biomedical and Clinical Aspects of Coenzyme Q, Vol. 5, Folkers, K. and Yamamura, Y., eds., Elsevier, Amsterdam, pp. 241-245.

Katagiri, T., Konno, N., Yanagishita, T., Tanno, F., Geshi, E., Kitsu, T., Akiyama, K., Niitani, H., 1986, Protection of ischemic myocardial injury by Coenzyme Q_{10} - mechanism of action, in Biomedical and Clinical Aspects of Coenzyme Q, Vol. 5, Folkers, K. and Yamamura, Y., eds., Elsevier, Amsterdam, pp. 167-177.

Kawasaki, T., Sugino, K., Maraubayashi, S., 1986, The role of cellular antioxidants and of administered coenzyme Q_{10} in oxidative cellular damage - experimental liver ischemia and endotoxemia, in Biomedical and Clinical Aspects of Coenzyme Q, Vol. 5, Folkers, K. and Yamamura, Y., eds., Elsevier, Amsterdam, pp. 145-155.

King, T.E., 1985, Ubiquinone proteins, in Coenzyme Q, G. Lenaz, ed., Wiley, New York, pp. 391-408.

Kröger, A., Klingenberg, M., 1970, Quinones and nicotinamide nucleotides associated with electron transfer. Vitamins and Hormones, 28, 533-559.

Kröger, A., Klingenberg, M., 1973, The kinetics of the redox reactions of ubiquinone related to the electron-transport activity in the respiratory chain. Eur. J. Biochem., 34, 358-368.

Landi, L., Cabrini, L., Sechi, A.M., Pasquali, P., 1984, Antioxidative effect of ubiquinones on mitochondrial membranes. Biochem. J., 222, 463-466.

Landi, L., Carbrini, L., Tadolini, B., Sechi, A.M., Pasquali, P., 1985, Incorporation of ubiquinone into lipid vesicles and inhibition of lipid peroxidation. Ital. J. Biochem., 34, 356-363.

Lea, C.H., Kwietny, A., 1962, The antioxidant properties of ubiquinone and related compounds. Chem. Ind. (London), 24, 1245-1246.

Lee, C.P., ed., 1987, Current Top. Bioenerg., Vol. 15, Academic Press, New York.

Lee, C.P., Martens, M.E., Peterson, P.L., Hatfield, J.S., 1987, Cytochrome oxidase and Neuromuscular Diseases, In Cytochrome Systems, Papa, S., Chance, B. and Ernster, L., eds., Plenum Press, New York, pp. 407-414.

Lenaz, G., ed., 1985, Coenzyme Q: Biochemistry, Bioenergetics and Clinical Applications of Ubiquinone, Wiley, New York.

Lester, R.C., Fleischer, S., 1959, The specific restoration of succinoxidase activity by coenzyme Q compounds in acetone-extracted mitochondria. Arch. Biochem. Biophys., 80, 470-472.

Luft, R., Ikkos, D., Palmieri, G., Ernster, L., Afzelius, B., 1962, A case of severe hypermetabolism of nonthyroid origin with a defect in the maintenance of mitochondrial respiratory control: A correlated clinical, biochemical and morphological study. J. Clin. Invest., 41, 1776-1804.

Mellors, A., Tappel, A.L., 1966a, Quinones and quinols as inhibitors of lipid peroxidation. Lipids, 1, 282-284.

Mellors, A., Tappel, A.L., 1966b, The inhibition of mitochondrial peroxidation by ubiquinone and ubiquinol. J. Biol. Chem., 241, 4353-4356.

Mitchell, P., 1975a, Protonmotive redox mechanisms of cytochrome b-c_1 complex in the respiratory chain: protonmotive ubiquinone cycle. FEBS Lett., 45, 1-6.

Mitchell, P., 1975b, The protonmotive Q cycle: a general formulation. FEBS

Lett., 59, 137-139.

Mortensen, S.A., Bouchelouche, P., Muratsu, K., Folkers, K., 1986, Clinical decline and relapse of cardiac patients on coenzyme Q_{10} withdrawal, in Biomedical and Clinical Aspects of Coenzyme Q, Vol. 5, Folkers, K. and Yamamura, Y., eds., pp. 281-290.

Nohl, H., Jordan, W., Youngman, R.J., 1986, Quinones in Biology: Functions in electron transfer and oxygen activation. Adv. Free. Rad. Biol. Med., 2, 211-279.

Norling, B., Glazek, E., Nelson, B.D., Ernster, L., 1974, Studies with ubiquinone-depleted submitochondrial particles. Quantitative incorporation of small amounts of ubiquinone and its effects on the NADH and succinate oxidase activities. Eur. J. Biochem., 47, 475-482.

Ogasahara, S., Yorifuji, S., Nishikawa, Y., Takahashi, M., Wada, K., Hazama, T., Nakamura, S., Kono, N., Tarui, S., 1985, Improvement of abnormal pyruvate metabolism and cardiac conduction defect with coenzyme Q_{10} in Kearns-Sayre syndrome. Neurology, 35, 372-377.

Ogasahara, S., Nishikawa, Y., Yorifuji, S., Soga, F., Nakamura, Y., Takahashi, M., Hashimoto, S., Kono, N., Tauri, S., 1986, Treatment of Kearns-Sayre syndrome with coenzyme Q_{10}. Neurology, 36, 45-53.

Okuma, K., Ota, K., 1986, The effect of coenzyme Q_{10} on ECG changed induced by doxorubicin (adriamycin), in Biomedical and Clinical Aspects of Coenzyme Q, Vol. 5, Folkers, K. and Yamamura, Y., eds., Elsevier, Amsterdam., pp. 247-256.

Öllinger, K., Buffinton, G.D., Ernster, L., Cadenas, E., 1990, Effect of superoxide dismutase on the autoxidation of substituted hydro- and semi-naphthoquinones. Chem.-Biol. Interactions, in press.

Ozawa, T., 1985, Formation of oxygen radicals in the electron transfer chain and antioxidant properties of coenzyme Q, in Coenzyme Q: Biochemistry, Bioenergetics and Clinical Applications of Ubiquionne, Lenaz, G., ed., Wiley, New York, pp. 441-456.

Papa, S., 1982, Molecular mechanism of proton translocation by the cytochrome system and the ATPase of mitochondria. Role of proteins. J. Bioenerg. Biomembr., 14, 69-86.

Pedersen, S., Tata, J.R., Ernster, L., 1963, Ubiquinone (coenzyme Q) and the regulation of basal metabolic rate by thyroid hormones. Biochim. Biophys. Acta, 69, 407-409.

Powis, G., 1989, Free radical formation by antitumor quinones. Free Rad. Biol. Med., 6, 63-101.

Quinn, P.J., Baum, H., Harris, E.J., Franklin, C.S., Trivedi, P., 1980, The protective role of coenzyme Q against mercurial and carbon tetrachloride toxicity, in Biomedical and Clinical Aspects of Coenzyme Q, Vol. 2, Yamamura, Y., Folkers, K. and Ito, Y., eds., Elsevier, Amsterdam., pp. 435-446.

Ramasarma, T., 1985a, Natural occurence and distribution of coenzyme Q, in Coenzyme Q, Lenaz, G., ed., Wiley, New York, pp. 67-81.

Ramasarma, T., 1985b, Metabolism of coenzyme Q, in Coenzyme Q, Lenaz, G., ed., Wiley, New York, pp. 131-142.

Ruzicka, F.J., Beinert, H., Schepler, K.L., Dunham, W.R., Sands, R.H., 1975, Interaction of ubisemiquinone with a paramagnetic component in heart tissue. Proc. Natl. Acad. Sci. U.S.A., 72, 2886-2890.

Salerno, J.C., Ohnishi, T., 1980, Studies on the stabilized ubisemiquione species in the succinate-cytochrome c reductase segment of the intact mitochondrial membrane system. Biochem. J., 192, 769-781.

Salminen, A., Vihko, V., 1983b, Endurance training reduces the susceptibility of mouse skeletal muscle to lipid peroxidation in vitro. Acta Physiol. Scand., 117, 109-113.

Salminen, A., Vihko, V., 1983a, Lipid peroxidation in exercise myopathy. Exp. Mol. Pathol., 38, 380-388.

Sastry, P.S., Jayaraman, J., Ramasarma, T., 1961, Distribution of coenzyme Q in rat liver cell fractions. Nature, 189, 577.

Slater, E.C., DeVries, S., 1980, Identification of the BAL-labile factor. Nature, 288, 717-718.

Shibata, N., Akagami, H., Asada, M., Sanma, H., 1986, Increase of 12-hydroxyeicosatetraenoic acid (12-HETE) in experimental ischemic myocardium and an effect of coenzyme Q_{10} on the phenomenon, in Biomedical and Clinical Aspects of Coenzyme Q, Vol. 5, Folkers, K. and Yamamura, Y., eds., Elsevier, Amsterdam, pp. 179-190.

Solaini, G., Landi, L., Pasquali, P., Rossi, C.A., 1987, Protective effect of endogenous coenzyme Q on both lipid peroxidation and respiratory chain inactivation induced by an adriamycin-iron complex. Biophys. Biophys. Res. Commun., 147, 572-580.

Soussi, B., Bylund-Fellenius, A.-C., Scherstén, T., Ångström, J., 1990, ^{1}H NFR evluation of the ferricytochrome c-cardiolipin interaction: effect of superoxide radicals. Biochem. J., 265, in press.

Soussi, B., Idström, J.-P., Scherstén, T., Bylund-Fellenius, A.-C., 1989a, Cytochrome c oxidase and cardiolipin alterations in response to skeletal muscle ischemia and reperfusion. Acta Physiol. Scand., in press.

Soussi, B., Idström, J.-P., Scherstén, T., Bylund-Fellenius, A.-C., 1989b, Kinetic parameters of cytochrome c oxidase in rat skeletal muscle: effect of endurance training. Acta Physiol. Scand., 135, 373-379.

Starnes, J.W., Beyer, R.E., Edington, D.W., 1981, Effects of age and cardiac work in vitro on mitochondrial oxidative phosphorylation and (^{3}H)-leucine incorporation. J. Gerontol., 36, 130-135.

Starnes, J.W., Beyer, R.E., Edington, D.W., 1983, Myocardial adaptation to endurance exercise in aged rats. Am. J. Physiol., 245, H560-H566.

Sugiyama, S., Kitazawa, M., Ozawa, T., Suzuki, K., Izawa, Y., 1980a, Antioxidative effect of coenzyme Q_{10}. Experentia, 36, 1002-1003.

Sugiyama, S., Ozawa, T., Kato, T., Suzuki, W., 1980b, Recovery time-course of ventricular vulnerability after coronary reperfusion in relation to mitochondrial function in ischemic myocardium. Am. Heart J., 100, 829-837.

Szarkowska, L., 1966, The restoration of DPNH oxidase activity by coenzyme Q (ubiquinone). Arch. Biochem. Biophys., 113, 519-522.

Tàbora, O., Lewandowski, E., Combs, A.B., 1986, Influence of in vitro ubiquinone antagonists on doxorubicin toxicity in vivo. J. Toxicol. Environ. Health, 18, 232-240.

Takahashi, K., Mayumi, T., Kishi, T., 1988, Influence of coenzyme Q_{10} on doxorubicin uptake and metabolism by mouse myocardial cells in culture. Chem. Pharm. Bull., 36, 1514-1518.

Takahashi, M., Ogasahara, S., Nishikawa, Y., Yorifuji, S., Nakamura, Y., Soga, F., Ueno, S., Tarui, S., 1986, Therapeutic effect of coenzyme Q_{10} in patients having the Kearns-Sayre syndrome, in Biomedical and Clinical Aspects of Coenzyme Q, Vol. 5, Folkers, K. and Yamamura, Y., eds., Elsevier, Amsterdam, pp. 345-352.

Takeshige, K., Takayanagi, K., Minakami, S., 1980, Reduced coenzyme Q_{10} as an antioxidant of lipid peroxidation in bovine heart mitochondria, in Biomedical and Clinical Aspects of Coenzyme Q, Vol. 2, Yamamura, Y., Folkers, K. and Ito, Y., eds., Elsevier/North Holland, Amsterdam, pp. 15-26.

Thurman, R.G., 1973, Induction of hepatic microsomal NADPH-dependent production of hydrogen peroxide by chronic prior treatment with ethanol, Mol. Pharmacol., 9, 670-675.

Trumpower, B.L., ed., 1982, Functions of Quinones in Energy Conserving systems, Academic Press, New York.

Trumpower, B.L., Edwards, C.A., 1979, Purification of a reconstitutively active iron-sulphur-protein (oxidation factor) from succinate-cytochrome c reductase complex of bovine heart mitochondria. J. Biol. Chem., 254, 8697-8707.

Uysal, M., Aykac, G., Kocak-Toker, N., Silvas, A., Yalçin, S., Oz, H., 1985, Lipid peroxidation in liver, plasma, and erythrocytes of rats chronically treated with ethanol. Biochem. Med., 34, 370-372.

Wikström, M., Berden, J.A., 1972, Oxidoreduction of cytochrome b in the presence of antimycin. Biochim. Biophys. Acta, 283, 403-420.

Wikström, M., Krab, K., 1980, Current Top. Bioenerg., 10, 51-101.

Wolstenholme, G.E.W., O'Connor, C.M., 1961, eds., Ciba Foundation Symposium on Quinones in Electron Transport, Churchill, London.

Yagi, K., 1987, Membrane damage provoked by lipid peroxides, in Advances in Membrane Biochemistry and Bioenergetics, Kim, C.H., Tedeschi, H., Diwan, J.-J., Salerno, J.C., eds., Plenum, New York, pp. 553-560.

Yamamura, Y., Folkers, K., Ito, Y.L., eds., 1980, Biomedical and Clinical Aspects of Coenzyme Q, Vol, 2, Elsevier, Amsterdam.

Yu, C.A., Yu, L., 1980, Isolation and properties of a mitochondrial protein that converts succinic dehydrogenase into succinate-ubiquinone oxidoreductase. Biochemistry, 19, 3579-3585.

Yu, C.A., Yu, L., King, T., 1977, The existence of an ubiquinone binding protein in the reconstitutively active cytochrome b-c_1 complex. Biochem. Biophys. Res. Commun., 78, 259-265.

Yu, C.A., Nagaoka, S., Yu, L., King, T.E., 1978, Evidence for the existence of a ubiquinone protein and its radical in the cytochrome b and c_1 region in the mitochondrial electron transport chain. Biochem. Biophys. Res. Commun., 82, 1070-1078.

Zierz, S., Jahns, G., Jerusalem, F., 1989, Coenzyme Q in serum and muscle of 5 patients with Kearns-Sayre syndrome and 12 patients with ophtalmoplegia plus. J. Neurol., 236, 97-101.

Multiple Role of Coenzyme Q in Mammalian Cells

Takayuki Ozawa, Satoru Sugiyama, Hiroshi Suzuki and Masashi Tanaka

Department of Biomedical Chemistry, Faculty of Medicine, University of Nagoya, Nagoya 466, Japan

INTRODUCTION

The occurrence of coenzyme Q (CoQ) in mitochondria has long been known and their role as electron carriers in the mitochondrial respiratory chain has been well established. It has, however, been reported that CoQ is distributed (apparently as intrinsic membrane lipid) among various other organella such as Golgi apparatus, microsomes, uncleus, and plasma membrane. The relatively high molar content of CoQ, compared with that of other respiratory carriers, in mitochondria also suggests possible functions other than serving as a simple respiratory carrier. For instance, regulatory effects of CoQ on those respiratory enzymes for which they also act as cosubstrates have been suggested.

Key words; anti-phospholipase, CoQ-binding protein, cDNA, DNA mutation, by-pass formation

ANTI-PHOSPHOLIPASE ACTION

Beside the antioxidant role of CoQ (Sugiyama et al., 1980), we found, in collaboration with Dr. Baum, that it has membrane stabilizing action and anti-phospholipase action (Kobayashi et al., 1980). Accordingly, CoQ was demonstrated to be useful to prevent ischemia and reperfusion-induced ventricular arrhythmias (Nagai et al., 1985). *In vitro* addition of CoQ to lung microsome effectively prevented phospholipase A_2 attack to microsomal membrane phospholipids, as shown in Fig. 1.

cDNA OF CoQ-BINDING PROTEIN

Concerning electron transport activity of CoQ, it has been proposed that a certain amount of quinone exists in a bound form with protein, because stable ubisemiquinone

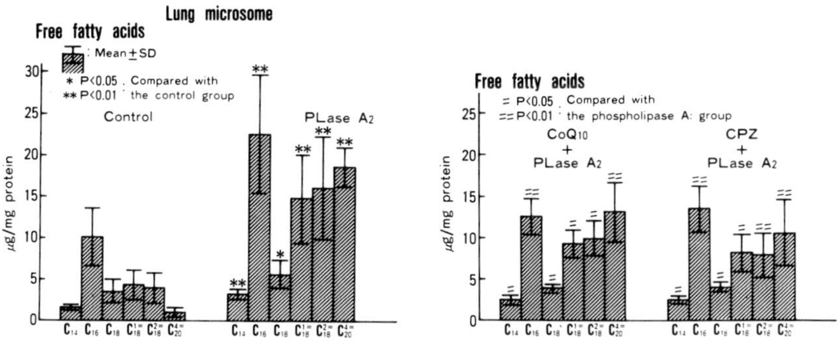

Fig. 1. The effects of phospholipase A_2 (PLase A_2), CoQ_{10}, and chlorpromazine (CPZ) on lung microsomal free fatty acids (FFA) in vitro. The addition of PLase A_2 significantly elevated the FFA levels. The simultaneous addition of either CoQ_{10} or CPZ with PLase A_2 significantly inhibited the elevation of each of those FFA.

could be detected in the electron transport chain. Recently, we succeeded to clone and to sequence a cDNA for human mitochondrial CoQ-binding protein of Complex III (QPc) (Suzuki et al., 1988), as shown in Fig. 2. From its amino acid sequence deduced from nucleotide sequence, a computer aided 3D model of QPc is presented in Fig. 3. We also isolated three genomic DNA genes which hybridize to the cDNA for QPc (Suzuki et al, in press). It was found that two of them are psuedogenes and one is functional. QPc was found to exist one molecule in Complex III (Suzuki and Ozawa, 1984a) and to form 1:1 complex with CoQ (Suzuki and Ozawa, 1984b). Thus, a stoichiometric amount of CoQ plays the role of an electron carrier in the electron transfer chain. However, there is a certain amount of pooled form CoQ in mitochondrial membrane. A part of it may play a role in Q-cycle as proposed by Dr. Mitchell.

HUMAN MITOCHONDRIAL DNA MUTATION

Recently, mitochondrial myopathy which is caused by mutation of mitochondrial DNA has attracted much interest. With Southern blot analysis on muscle mitochondrial DNA from a mother and daughter both with mitochondrial myopathy, we first presented evidence (Ozawa et al., 1988) that the mutant mitochondrial DNA is responsible for the maternal inheritance of the disease, as shown in Fig. 4. The results suggested that the lesions of mitochondrial DNA expanded with aging. We presented a hypothesis (Linnane et al., 1989) that mitochondrial DNA mutations are important

```
CGGAGAAGGCAACGCTTCTCTTTCTGGTCAAA ATG GCT GGT AAG CAG GCC GTT TCA GCA TCA GGC AAG
                                 Met Ala Gly Lys Gln Ala Val Ser Ala Ser Gly Lys
                                      -   -  Arg Pro  -   -   -   -   -  Ser Arg
                                  1                                   10

TGG CTG GAT GGT ATT CGA AAA TGG TAT TAC AAT GCT GCA GGA TTC AAT AAA CTG GGG TTA
Trp Leu Asp Gly Ile Arg Lys Trp Tyr Tyr Asn Ala Ala Gly Phe Asn Lys Leu Gly Leu
 -   -  Glu  -   -   -   -   -   -   -   -   -   -   -   -   -   -   -   -   -
                             20                                      30

ATG CGA GAT GAT ACA ATA TAC GAG GAT GAA GAT GTA AAA GAA GCC ATA AGA AGA CTT CCT
Met Arg Asp Asp Thr Ile Tyr Glu Asp Glu Asp Val Lys Glu Ala Ile Arg Arg Leu Pro
 -   -   -   -   -   -  His  -  Asn Asp  -   -   -   -   -   -   -   .   -   -
                             40                                      50

GAG AAC CTT TAT AAT GAC AGG ATG TTT CGC ATT AAG AGG GCA CTG GAC CTG AAC TTG AAG
Glu Asn Leu Tyr Asn Asp Arg Met Phe Arg Ile Lys Arg Ala Leu Asp Leu Asn Leu Lys
 -   -   -   -  Asp  -   -  Val  -   -   -   -   -   -   -   -   -  Ser Met Arg
                             60                                      70

CAT CAG ATC TTG CCT AAA GAG CAG TGG ACC AAA TAT GAA GAG GAA AAT TTC TAC CTT GAA
His Gln Ile Leu Pro Lys Glu Gln Trp Thr Lys Tyr Glu Glu Glu Asn Phe Tyr Leu Glu
Gln  -   -   -   -   -   -   -   -   -   -   -   -   -  Asp Lys Ser  -   -   -
                             80                                      90

CCG TAT CTG AAA GAG GTT ATT CGG GAA AGA AAA GAA AGA GAA GAA TGG GCA AAG AAG TAA
Pro Tyr Leu Lus Glu Val Ile Arg Glu Arg Lys Glu Arg Glu Glu Trp Ala Lys Lys ***
 -   -   -   -   -   -   -   -   -   -   -   -   -   -   -   -   -   -   -
                            100                                     110

TCATGTAGTTGAAGTCTGTGGATGCAGCTGTTATGAAGATGGTTAAACTTGAAACAAACAATTTTAAGAATTATTTGGT
CTGAAGATGTTTTACTTTAAATAAATGTCTATTGTAATGGCTGGAGTTTTTGAATTCCAAACCTTATACTG
```

Fig. 2. Nucleotide sequence of human QPc cDNA and its deduced amino acid sequence.

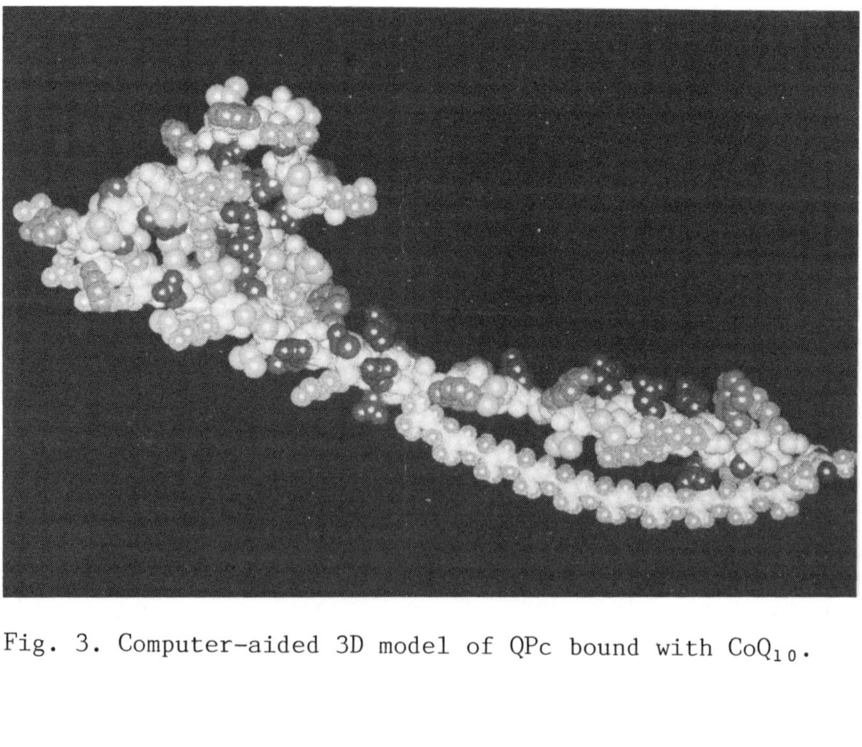

Fig. 3. Computer-aided 3D model of QPc bound with CoQ_{10}.

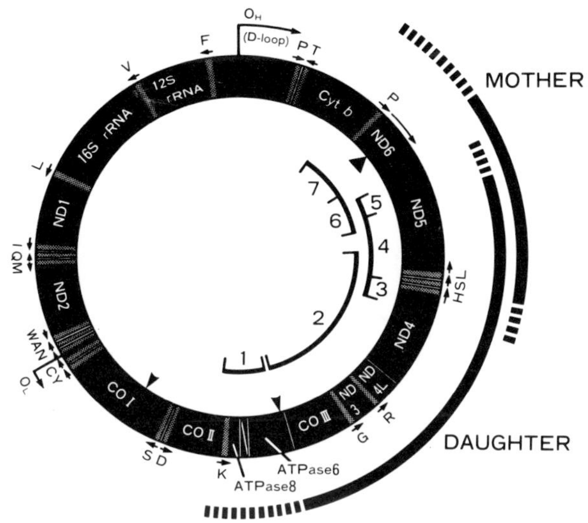

Fig. 4. Map of the human mitochondrial genome showing the locations of coding regions and the probes used. The extent of the deleted regions in mutant mitochondrial DNA in the mother and the daughter is indicated by the solid line. Interrupted lines at the ends of deletion define upper and lower limits. Probes, the same are indicated inside the circle by numbers. The restriction sites of PstI and BamI are indicated by arrowheads and a triangle, respectively.

contributors to aging and degenerative diseases.

CoQ AS A FAIL-SAFE MECHANISM

In the case of mitochondrial myopathy, especially cytochrome oxidase-defficient patients, it was demonstrated that CoQ administration could dramatically deduce lactic acid level in serum induced by excise (Jinnai et al., 1989), as shown in Fig. 5. Using KCN-treated bovine submitochondrial particles, as a model of the disease, externally added CoQ could accept electrons from the reduced respiratory chain to form hydrogen peroxide converting to water and oxygen by catalase (Ozawa et al., 1987), as shown in Fig. 6. Thus, a part of excess free CoQ could form a by-pass of the electron transfer chain to oxygen playing an important role in the fail-safe mechanism of our bio-engine (Ozawa et al., 1987), as schematically presented in Fig. 7.

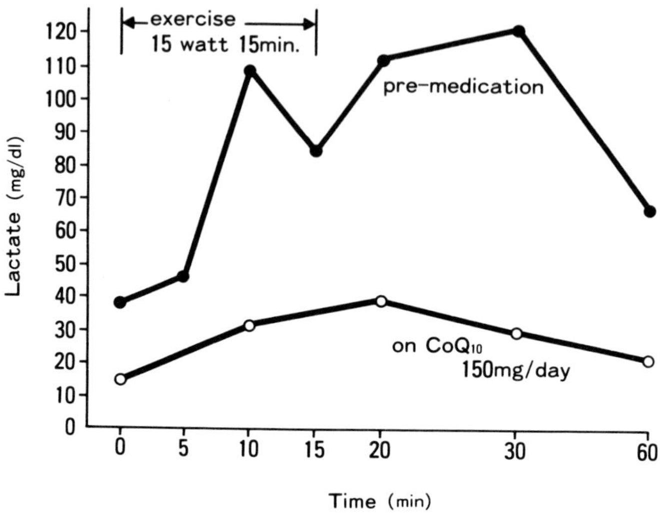

Fig. 5. Aerobic exercise tests by using an ergometer before medication and on CoQ_{10}. On CoQ_{10}, lactate level in the blood before exercise is lower and the increment of them by the exercise is less than those of pre-medication.

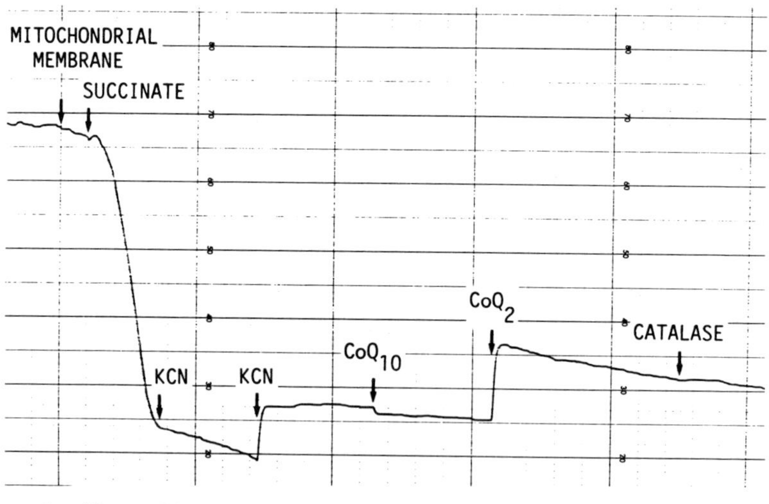

Fig. 6. The effects of external added CoQ on KCN-blocked respiration of bovine heart submitochondrial particles. Respiration of submitochondrial particles (2 mg/ml) was completely blocked by 3 mM KCN, then 145 μM CoQ_{10}, 50 μM CoQ_2 and catalase (20 μg/ml) were added to the mixture as indicated in the figure.

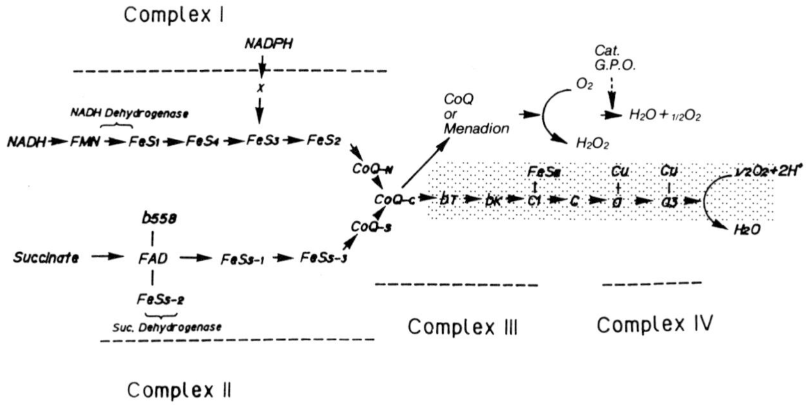

Fig. 7. Schematic representation of the components in the mitochondrial electron-transfer chain and the formation of a by-pass on the administration of CoQ or menadione in the case of a deficiency of Complex III or IV.

REFERENCES

Jinnai K, Yamada H, Kanda F, Masui Y, Tanaka M, Ozawa T, Fujita T. Eur J Neurol (in press)

Kobayashi M, Shimomura Y, Suzuki M, Tanaka M, Baum H, Ozawa T, 1980, J Appl Biochem 2, 270-279.

Linnane AW, Maruzuki S, Ozawa T, Tanaka M, 1989, Lancet i:642-645.

Nagai S, Miyazaki Y, Ogawa K, Satake T, Sugiyama S, Ozawa T, 1985, J Mol Cell Cardiol 17, 873-884.

Ozawa T, Tanaka M, Suzuki H, Nishikimi M, 1987, Brain Develop 9, 76-81.

Ozawa T, Yoneda M, Tanaka M, Ohno K, Sato W, Suzuki H, Nishikimi M, Yamamoto M, Nonaka I, Horai S, 1988, Biochem Biophys Res Commun 154, 1240-1247.

Sugiyama S, Kitazawa M, Ozawa T, Suzuki K, Izawa Y, 1980, Experientia 36,1002.

Suzuki H, Ozawa T, 1984a, Biochem Int 9, 563-568.

Suzuki H, Ozawa T, 1984b, Biochem Biophys Res Commun 124, 889-895.

Suzuki H, Hosokawa Y, Toda H, Nishikimi M, Ozawa T, 1988, Biochem Biophys Res Commun 156, 987-994.

Suzuki H, Hosakawa Y, Toda H, Nishikimi M, Ozawa T, Biochem Biophys Res Commun (in press)

FACTORS AFFECTING BLOOD AND TISSUE LEVELS OF CoQ_{10}: IN VITRO AND IN VIVO STUDIES

G.P.Littarru, S.Lippa*, A.Oradei*, V.Aureli*, F.Serino**

Institute of Biochemistry, Faculty of Medicine, University of Ancona, * Institute of Human Physiology and ** Institute of General Surgery, Catholic University, Faculty of Medicine, Roma, Italy.

INTRODUCTION

Coenzyme Q was initially isolated from beef heart mitochondria and from human heart tissue: its role in the mitochondrial respiratory chain has been thereafter deeply inquired. As emphasized elsewhere in this book new roles have been proposed, also on the basis of CoQ location, which is not exclusive of the inner mitochondrial membrane.

Coenzyme Q content in mammal tissue depends both on endogenous biosynthesis and on exogenous intake. The biosynthetic pathways leading to the formation of the quinone ring, starting from tyrosine, and of the polyisoprenoid side chain, starting from acetyl CoA, are known (Ramasarma 1985). Up to a certain point this route is common to the cholesterol biosynthetic pathway. The CoQ content of many common types of food are also known (Kamei et al., 1986).

The essentiality of exogenous intake was outlined by the very low blood CoQ_{10} levels found in patients after one week of total parenteral nutrition (Kishi et al., 1986).

On a theoretical basis a CoQ deficiency could arise also from an insufficient biosynthesis, such as it may be determined by the use of the new agents aimed to pharmacologically block cholesterol biosynthesis. We might have a deficiency also in terms of higher metabolic needs not fulfilled by the biosynthetic capabilities of the cell in a certain moment, or higher catabolic destruction of CoQ due to increased rate of peroxidative processes. It is worthwhile to inquire if these situations are characterized by abnormal blood levels of CoQ.

KEY WORDS: CoQ_{10} plasma concentration, hyperthyroidism, physical training, kidney transplant.

MATERIALS AND METHODS

0.5 ml of plasma (or tissue homogenate), supplemented with 2 ml of ethanol: isopropanol (95:5), 0.5 ml of 0.1 M sodium dodecylsulphate and 0.5 µg of CoQ_8 as internal standard, were extracted twice with 4 ml of n-hexane. Combined extracts were brought to dryness under N_2, then purified through a silica gel glass column (70-230 mesh), 120 x 5 mm, using a 2-step elution. First washing was done with 1 ml of 2.5 % ethanol in n-hexane, then CoQ_{10} was eluted with 4 ml of hexane: dichloromethane: ethanol (60 : 10 : 30). Eluate was evaporated to dryness under N_2 (at 45-50°C) and redissolved in 100 µl of ethanol. 20 µl were injected into the HPLC apparatus, whose conditions were as follows. Column: ultrasphere ODS, 250 x 4.6 mm; mobile phase: ethanol-methanol (70-30), detector: U.V. 280 nm.

RESULTS AND DISCUSSION

Plasma CoQ_{10} levels in patients affected by different thyroid syndromes are reported in Fig.1 (Mancini et al., 1987). An inverse relationship between blood T3, or T4, and CoQ_{10} concentration is evident.

Fig. 1. Correlation between plasma concentration of CoQ_{10}, T_3 and T_4 in a group of patients affected by thyroid disorders

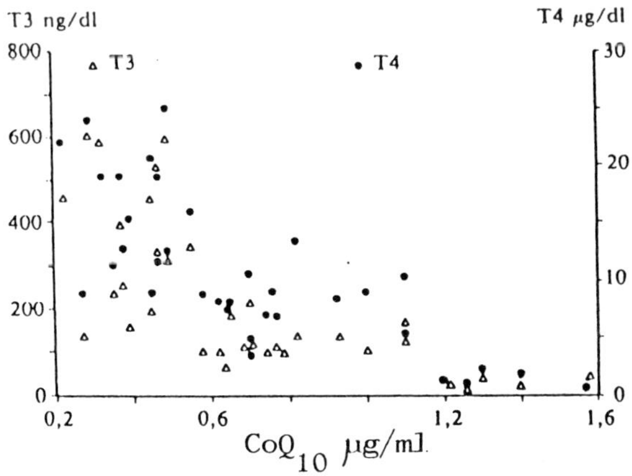

Fig. 2 shows an overall view of plasma CoQ_{10} levels in patients affected by hyperthyroidism (Basedow syndrome or toxic adenoma), in euthyroid patients (black dot = simple

goitre; empty circles = hyperthyroid patients during chronic antithyroid treatment) and in hypothyroid subjects.

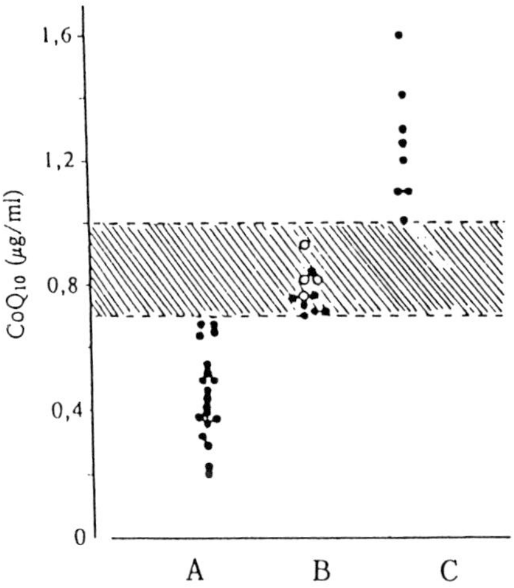

Fig. 2. Plasma CoQ_{10} in

A: hyperthyroidism

B: simple goitre or cronically treated hyperthyroidism

C: hypothyroidism

Our data confirm early observations by Ogura et al. (1980) furthermore they show a definitely better correlation between high CoQ_{10} plasma levels and hypothyroidism.

It is interesting to notice that in rats experimentally induced hyperthyroidism also led to significantly lower CoQ_9 plasma levels (Ikeda et al. 1984).

The content of ubiquinone was increased by 75% in the mitochondria from thyroxine-treated rats, the percentage of reduced ubiquinone being also higher (Horrum et al., 1986). Succinate respiration was increased in State III and State IV. Since thyroxine treatment produced an increase in the concentration of cytochromes b, c, a and a3, but not C_1, it can be reasonably hypothesized that the increase in Ubiquinone reduction could be due to enhanced electron flow from the sites of substrate oxidation. Decreased plasma levels of CoQ_{10} in humans might correspond to increased tissue levels of CoQ_{10}, or to an accelerated turnover rate of tissue CoQ_{10} due to increased metabolic demand.

Another situation also leading to elevation of tissue coenzyme Q and cytocrome C in the rat is endurance training (Beyer et al., 1984). We checked CoQ_{10} plasma concentration in a group of young cyclists (Guerra et al.,

1987) and found levels which were significantly lower than the values of a sedentary age-matched goup. Furthermore, among the athletes themselves significantly lower plasma CoQ_{10} concentrations were found in the peak training season, compared to the winter months when the cyclists had practically stopped their aerobic training (see Tab.1).

Table 1. CoQ_{10} plasma concentration (μg/ml) and training.

A Sedentary control group	B Athletes resting-season	C Athletes training period
0.79 ± 0.21	0.68 ± 0.14	0.53 ± 0.16

A vs B : N.S.
A vs C : $p < 0.01$
B vs C : $p < 0.005$

Karlsson (1987) found an inverse correlation between plasma CoQ_{10} levels and aerobic capacity, as determined by VO_2 max, in a group of endurance runners. CoQ_{10} availability in endurance training may become inadeguate.

Increased CoQ catabolic rate might also be present, due to the well known increase in free radical formation and lipid peroxidation which occurs in muscle and liver upon exhaustive exercise. We might therefore conclude, also in this case, that lowered blood CoQ_{10} levels in man might reflect higher metabolic demand, especially from heart and skeletal muscle, leading to increased CoQ_{10} extraction from the blood compartment and/or to a lesser extent of release from tissues.

One situation characterized by increased peroxidative level with ensueing overdistruction of different classes of lipids, and other chemical components of the cell, is ischemia. It was previously shown by our group that plasma CoQ_{10} levels in human patients diminished during the first minutes following kidney or liver transplant. One possible interpretation of this finding is that there is an increased uptake of plasma CoQ_{10} by the newly transplanted organ, which had undergone a certain degree of CoQ_{10} destruction during the ischemic period of organ preservation before transplant. To verify this hypothesis we checked the levels of CoQ_{10} in serial biopsies from pig kidneys subjected to "cold" or "warm" ischemia.

In the cold ischemia experiment kidneys were excised from the animal and kept in ice-cold preserving medium. In the warm ischemia renal artery was clamped and the organ was kept in situ. As shown in Fig.3 there was a considerable decrease of tissue CoQ_{10} in both models of ischemia. In the warm ischemia experiment renal artery was declamped after 60 min and blood readmitted: immediately after declamping there was a little but consistent increase in kidney CoQ_{10} content. In our experiment the CoQ_{10} uptake by the previously ischemic organ was apparently high enough to overcome CoQ_{10} peroxidative destruction which might ensue following riperfusion.

The time course of peroxidative destruction of CoQ in a simplified system of linoleic acid micellae is described elsewhere in this book. In those experiment we clearly demonstrated that CoQ_{10} is capable of conferring antioxidant protection to the system, being itself destroyed while providing this protective effect.

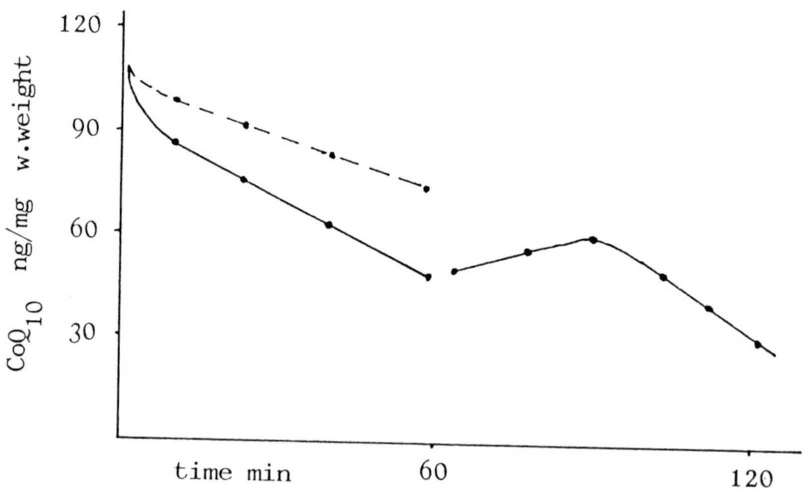

Fig. 3. Levels of CoQ_{10} in serial biopsies from pig kidney subjected to cold ischemia (-------) or warm ischemia followed by riperfusion (———).

REFERENCES

Beyer,R.E., Morales Corral,P., Ramp Bobbi,J. Kreitman K.R. Falzon,M.J., Rhee,S.Y.S., Kuhun,T.W., Stein,M., Rosenwasser,M.J., Cartwright,K.J. 1984, Elevation of tissue Coenzyme Q (ubiquinone) and cytochrome C concentrations by endurance exercise in the rats. Archives of Biochemistry and Biophysics, 234, 2, 323, 331.

Guerra,G., Ballardini,E., Lippa,S., Oradei,A., Littarru, G.P. 1987, Effetti della somministrazione di ubidecarenone sul consumo massimo di ossigeno e sulla performance fisica in un gruppo di giovani ciclisti. Medicina dello Sport, 46, 6, 359-366.

Horrum,M.A., Tobin,R.B., Ecklund,R.E. 1986. Thyroxine induced changes in rat liver mitochondrial ubiquinone. Biochemical Biophysical Research Communication 138,381.

Ikeda,S., Hamada,N., Morii,H., Inaba,M., Yamakawa,J. 1984, Serum and tissue coenzyme Q_9 in rats with thyroid dysfunction, Hormones and Metabolism Research, 16,585-594.

Kamei,M., Fujiita,T., Kanke,T., Sasaki,K., Oshiba,K., Otami,S., Matsui,T., Morisawa,S. 1986, The distribution and content of Ubiquinone in foods. International Journal of Vitamins and Nutrition Research., 56, 57-68.

Karlsson,J. 1987, Heart and skeletal muscle ubiquinone or CoQ_{10} as a protective agent against radical formation in man. In: Advances in Myochemistry I, edited by G. Benzi, pp.305-312.

Kishi,T., Okamoto,T., Kishi,H., Okada,A. 1986, Serum levels of Coenzyme Q_{10} in patients receiving total parenteral nutrition and relationship to serum lipids. In: Biomedical and clinical aspects of Coenzyme Q Vol.5, edited by K. Folkers and Y. Yamamura (Amesterdam: Elsevier) pp.113-120.

Mancini,A., De Marinis,L., Calabrò,F., Oradei,A., Lippa,S. Littarru,G.P., Barbarino,A., Folli,G. 1987, La determinazione del Coenzima Q_{10} plasmatico nella fisiopatologia diagnostica delle patologie tiroidee. Rivista Italiana di Biologia e Medicina, 7, 44-50.

Ogura,S., Morii,H., Ohno,M., Ueno,T., Kitabatake,S., Hamada,N., Ito,K. 1980, Serum coenzyme Q_{10} levels in thyroid disorders. Hormones and Metabolism Research, 12, 537-545.

CLINICAL PERSPECTIVES IN THE TREATMENT OF CARDIOVASCULAR DISEASES WITH COENZYME Q10

Mortensen S.A., MD, PhD, Heidt P., MD, Sehested J., MD, PhD.

Rigshospitalet (University Hospital)
Medical Department B, DK-2100 Copenhagen Ø, Denmark.
Deutsches Herzzentrum Berlin
Augustenburger Platz 1, 1000 Berlin 65, West Germany.

INTRODUCTION

Cardiomyopathy is a persistent challenge to the investigator but at the same time a frustration to the clinician especially in the cases with poor systolic function. When congestive heart failure is clinically overt, treatment is extremely difficult. It may improve symptoms for some time, but the therapies are not causal. Recent trials with ACE-inhibitors have offered the patients relief of symptoms and also better survival rates when compared to the classical antifailure principles. But the disease itself has not been cured and the general outlook is poor.

Till this date, heart transplantation has been the only avenue to success regarding survival in patients with severe heart failure. However, serious problems remain with respect to the large number of potential recipients waiting, and dying, due to the limited donor availability. Therefore it is necessary to continue the efforts clinically and investigatively, to improve the classification and thereby the treatment of patients with dilated myocardial disease.

Endomyocardial biopsy is a rather new technique enabling the obtainment of fresh cardiac muscle from patients with a variety of cardiac diseases. The method allows us to investigate not only the pathoanatomical changes but also it has provided a means of studying selected metabolic intermediates in the myocardium.

Theoretically, a poor utilization efficiency of oxygen may be due to exhaustion of the myocardial stores of bioenergetics. Deficiency of essential substances such as Coenzyme Q10 (CoQ10) (Folkers et al., 1977) may be involved initially or a depletion may occur during the progression of myocardial failure leading to a vicious cycle. CoQ10 has probably multiple roles in the human heart cell (Lenaz and Castelli, 1984) acting as a membrane stabilizer and free radical scavenger. As a key-substance in the oxidative phosphorylation and the synthesis of ATP, CoQ10 is indispensable for the function of the myocyte. There are several potential applications of CoQ10 within the field of cardiovascular disease to be summarized in the following four headings.

Keywords: cardiac failure, treatment, prophylaxis, coenzyme Q10, cardiac transplantation.

FAILING HEART

A biochemical rationale for using CoQ10 as a therapy in human heart disease has been established long ago (Folkers K. et al. 1977). Lately, however, this concept has been further strengthened via investigations of endomyocardial biopsies from patients with cardiomyopathy (Mortensen et al., 1984). Endomyocardial levels of CoQ10 determined by HPLC were found significantly lower in groups with more advanced heart failure (NYHA classes III & IV) compared with those patients in the milder stages of heart failure (classes I & II). Furthermore, the tissue CoQ10 deficiency was significantly restored by oral therapy (Mortensen et al., 1985).

Our preliminary clinical results from CoQ10 therapy in 12 patients with heart failure were impressive, why we dared to use the rather great words, that CoQ10 was "a major advance in therapy". This statement was based on the following: its clinical effectiveness, its safety and as it was a novel and alternative principle in the treatment of heart failure.

In a subsequently larger series of 40 patients with heart failure of various origin nearly two thirds revealed clinical improvement. It was expressed in percentages as high as 69% in patients with dilated cardiomyopathy and comparatively lower (43%) in patients with chronic failure due to coronary artery disease.

The improved cardiac function in patients treated with CoQ10 (Langsjoen et al., 1984; Judy et al. 1984; Mortensen et al., 1985) supports our hypothesis that heart failure is a condition characterized by mitochondrial dysfunction and energy starvation, and that the high ADP/ATP ratio - due to an increased demand on the respiratory chain - is corrected by CoQ10 supplementation.

ISCHEMIC HEART

Whereas CoQ10 seems an obvious choice in the treatment of chronic heart failure the substance has been considered rather experimental as anti-anginal treatment. A possible therapeutic benefit from oral treatment with CoQ10 may, however, be obtained in patients with myocardial ischemia, both in stable angina and supposedly also in subacute or acute phases of the disease.

Available clinical experiences have been limited to stable angina pectoris as in studies by Kamikawa et al. (1985). In a double-blind trial these authors found a significant rise in exercise time as well as a prolonged time until onset of ischemic ST-response during CoQ10 therapy. The frequency of anginal attacks and the nitroglycerin consumption tended to decrease with CoQ10. Interestingly, a correlation was found between changes in blood CoQ10 levels and the improvements in exercise duration.

Additional clinical studies seem warranted in ischemic heart disease, especially in myocardial infarction as CoQ10, theoretically, may salvage tissue in the border-zone of the infarct.

THREATENED HEART

Clinical use of CoQ10 as a prophylactic agent is gradually emerging in patients subjected to treatment with anthracyclines as well as in various forms of cardiac surgery.

Non-invasive monitoring of myocardial function during doxorubicin treatment (+/- CoQ10) has been carried out via measurements of the systolic time intervals and impedance cardiography. In a preliminary study of lung cancer patients (Judy et al., 1984) CoQ10 seemed to exert a prophylactic effect in patients with normal cardiac function (at baseline). Thus, a significantly larger doxorubicin-dose (mean 900 mg/square meter) was tolerated without any signs of reduced myocardial contractility in this group.

Effective protection in open heart surgery is important to improve the operative result and the postoperative myocardial performance. Improved myocardial preservation is achieved by a combination of hypothermia and cardioplegia, however, some degree of myocardial necrosis from the transient global ischemia is inevitable.

In cardiac valve replacement CoQ10 seemed to have a prophylactic effect against low cardiac output (Tanaka et al., 1982), and in coronary artery bypass grafting (Sunamori et al., 1984) it was noted that patients pretreated with CoQ10 required less inotropic support. CoQ10 may also have a role as a protecting agent during cardiological interventions such as percutaneous transluminal coronary angioplasty, which is a future research area.

TRANSPLANTED HEART

Several factors may influence the CoQ10 levels in the donor heart before and after transplantation such as: the duration of cold cardioplegia, the use of immunosuppressives, the occurrence rejection etc. The hypothetical finding of low levels of CoQ10 - especially in acute rejection - may provide a biochemical basis of a randomized, trial with oral CoQ10 versus placebo in human cardiac transplantation.

Therefore we are currently investigating transplant patients coming to the follow-up examinations with respect to the levels of CoQ10 in blood and endomyocardial biopsies (Heidt P., Mortensen S.A, Sehested J. et al., unpubl. data).

The preliminary result of a pilot study revealed a tissue CoQ10 level by HPLC which was found within the supposed normal range. There was no difference between the mean values of CoQ10 in a group with rejection (N=11) versus a group without rejection (N=10): 0.36 and 0.38 µg/mg dry weight, respectively. The blood levels were found highest in patients without rejection: mean 1.21 µg/ml, and it is noteworthy that the lowest blood-levels were found in cases having histological signs of moderate or severe rejection: mean 0.84 µg/ml.

Theoretically, the tissue stores of CoQ10 may be exhausted during repeated attacks of rejection Further results are needed to clarify the role of CoQ10 as a possible serological marker of acute cardiac rejection.

CONCLUSION

CoQ10 may definitely have a role as a part of a "medical bridge" to cardiac transplantation. CoQ10 may support the recipient heart and probably postpone the time of the operation in some cases. CoQ10 may preserve the donor heart during transplantation, and further research will clarify the possible use of CoQ10 as a prophylactic substance in the long-term follow up after cardiac transplantation.

REFERENCES

Folkers, K., Watanabe, T., Pinphanichakarn, V. et al., 1977, Biomedical and clinical research on coenzyme Q. In: Biomedical and Clinical Aspects of Coenzyme Q, Amsterdam, edited by K. Folkers and Y. Yamamura, (Elsevier/North-Holland Biomedical Press), vol. 1, pp. 299-313.

Judy, W.V., Hall, J.H., Toth, P.D., Folkers, K., 1984, Myocardial effects of co-enzyme Q10 in primary heart failure. In: Biomedical and Clinical Aspects of Coenzyme Q, Amsterdam, edited by. K. Folkers and Y. Yamamura, (Elsevier Science Publishers B.V.), vol. 4, pp. 353-367.

Judy, W.V., Hall, J.H., Dugan, W., Toth, P.D., Folkers, K., 1984, Coenzyme Q10 reduction of adriamycin cardiotoxicity. In: Biomedical and Clinical Aspects of Coenzyme Q, Amsterdam, edited by K. Folkers and Y. Yamamura, (Elsevier Science Publishers B.V.), vol. 4, pp. 231-241.

Kamikawa, T., Kobayashi, A., Yamashita, T., Hayashi, H., Yamazaki, N., 1985, Effects of coenzyme Q10 on exercise tolerance in chronic stable angina pectoris. The American Journal of Cardiology, 56, 247-251.

Langsjoen, P.H., Vadhanavikit, S., Folkers, K., 1984, Effective treatment with coenzyme Q10 of patients with myocardial disease, classes III and IV. In: Biomedical and Clinical Aspects of Coenzyme Q, Amsterdam, edited by K. Folkers and Y. Yamamura, (Elsevier Science Publishers B.V.), vol. 4, pp. 325-332.

Lenaz, G., Parenti Castelli, G., 1984, Multiple roles of ubiquinone in mammalian cells. Drugs Under Experimental and Clinical Research, 10 (7), 481-490.

Mortensen, S.A., Vadhanavikit, S., Folkers, K., 1984, Deficiency of coenzyme Q10 in myocardial failure. Drugs Under Experimental and Clinical Research, 10 (7), 497-502.

Mortensen, S.A., Vadhanavikit, S., Baandrup, U., Folkers, K., 1985, Long-term coenzyme Q10 therapy: A major advance in the management of resistant myocardial failure. Drugs Under Experimental and Clinical Research, 11 (8), 581-593.

Sunamori, M., Okamura, T., Amano, J., Suzuki, A., 1984, Clinical application of coenzyme Q to coronary artery bypass graft surgery. Biomedical and Clinical Aspects of Coenzyme Q, Amsterdam, edited by K. Folkers and Y. Yamamura, (Elsevier Science Publishers B.V.), vol. 4, pp. 333-342.

Tanaka, J., Tominaga, R., Yoshitoshi, M., et al., 1982, The prophylactic effect on low cardiac output following cardiac valve replacement. Annals of Thoracic Surgery, 33, 145-151.

ACKNOWLEDGEMENT

We are greately indebted to P. O. Edlund and B. Lindeke, ACO Research Department, Sweden, for their scientific interest and help with the HPLC analyses of blood and biopsies from the cardiac transplant patients.

EFFECT OF COENZYME Q_{10} ON LOAD-INDUCED CARDIAC DYSFUNCTION: DOUBLE BLIND STUDY AND INVESTIGATION OF DOSE-RESPONSE RELATIONSHIP

Teiichi ODA, M.D., F.C.C.P.

Department of Paediatrics, School of Medicine, Fukuoka University, Fukuoka 814-01, Japan

INTRODUCTION

There is a specific form of left ventricular (LV) dysfunction which can be disclosed only by an afterload increase. In this "latent" or "load-induced" cardiac dysfunction (LCD), being different from the congestive heart failure, normal or usually supernormal LV ejection fraction (EF) is observed *at rest*. However, by adding afterload, a significant decrease in EF emerges, as several authors have reported (Kanaya & Arakawa, 1979; Gottdiener *et al.*, 1981; Newman *et al.*, 1981; Oda, 1985). During childhood, LCD is detectable most frequently in symptomatic patients with mitral valve prolapse (MVP). Although many children with MVP are entirely healthy, some of them acquire symptoms as Jeresaty (1979) described, i.e., fatigue, chest pain, dyspnoea, palpitations, dizziness, syncope, TIA (transient ischemic attacks) and psychiatric manifestations. We found that MVP patients who have such symptoms, also have LCD almost always and v.v. Sudden deaths were also reported and experienced by us. The LCD in MVP patients usually has a self-limited course, with an average duration of 12 months (6 to 18 months). Coxsackie virus antibodies elevated in about 90% of them. Cardiac catheterization studies on 8 patients performed by us revealed that the left and right ventricular (RV) end-diastolic pressures (EDP) were high. The mean LVEDP was 19.1±5.0 (SD) and the RVEDP was 12.5±5.0 mmHg. Such high EDPs rose higher again, by 100 to 200%, after infusion of angiotensin II. Above data strongly suggest that LCD hearts may work usually along the right-uppermost part of Frank-Starling' curve by homeostatic augmentation of the preload. If some increase in afterload or some decrease in preload takes place, a sudden fall in the cardiac stroke work may occur due to bankruptcy of Frank-Starling' mechanism. An impaired myocardial contractility may be responsible for this phenomenon. Myocardial biopsies performed by us revealed a presence of mild degenerations in the myocytes and in their nuclei.

We have reported that coenzyme Q_{10} (CoQ) is effective on the LCD, in consequence of open randomized and single blind studies. Possible existence of certain relationships have been suggested between dose and response and between dose and serum level of CoQ (Oda, 1985, 1986).

In asymptomatic healthy children, regardless of the presence or absence of MVP, the EF always increases in consequence of various loading tests, i.e., ergometer, angiotensin II, handgrip (Oda, 1985). Among those tests, quantitative isometric handgrip (HG) seemed to be the most suitable one for routine clinical procedure.

Our objectives at present are:

(1) Further evaluation of efficacy of CoQ on LCD by performing a double blind study, and

(2) Further collection of cases for more reliable investigation of dose-response relationship of CoQ.

Keywords: Double blind study, Mitral valve prolapse, Latent cardiac dysfunction, Dose-response relationship, Serum CoQ level.

METHOD

Load application: Quantitative HG, during which the patient grips a small manometer cuff from 20 mmHg to 100 mmHg for maximally endurable minutes, was used.

Evaluation of EF: M-mode echocardiogaphy (echo) was used. In the double blind study, two-dimensional (2D) echo, apical right oblique position (30°) was also used. The stroke volume (SV) and the cardiac output (CO) were subsequently calculated.

Double blind study: Consecutive 40 MVP patients with LCD, aged 6 to 17 years, were divided randomly into 2 finally matched groups as follows:

(1) Placebo group (N=20), in which a medication of inactive placebo for 7 days, followed by CoQ (3.1 to 3.4 mg/kg/day b.i.d.) for another 7 days, and

(2) CoQ group, in which a medication of CoQ (3.1 to 3.4 mg/kg/day b.i.d.) throughout 14 days was given.

Before treatment (time 0), on 7th day (1W) and on 14th day (2W), HG, serum CoQ determination, hematological tests and liver function examinations were also made. Changes of clinical symptoms were also recorded. For most statistical analysis, Student *t* test, paired or unpaired, was used. For pertinent objects, Fisher's direct calculation of probability and Wilcoxon's signed rank test were applied. Dose-response relationship: Consecutive patients, total 600, were divided into 6 groups randomly according to doses of CoQ, as are shown in Table 7. Prospective follow-up study was made on each patient and weeks needed for continuous normalization in HG was recorded.

RESULTS

Results obtained by the double blind study are shown in Tables 1 to 6 (see text for abbreviations). The majority of patients regained normal HG response along with clinical improvement within 7 days of CoQ treatment. However, the treatment for 14 days broght better results than for 7 days. Changes in serum CoQ level coincided with above data. No side effects were observed in any aspect.

Table 7 gives an overall view of the dose-response relationship. An eminent difference was seen between doses of $\leq$2.9 and $\geq$3.0 mg/kg/day. A nearly rectilinear relationship existed between mean weeks needed for normalization in HG and mean log dose (Figure 1).

The EF obtained by 2D echo, which is said to have a close relationship with the EF obtained by angiocardiography, was lower than the EF by M-mode. However, the rates of change (ΔEF%) were almost the same (Tables 3, 4, 5). In patients with LCD, the left ventricular diastolic dimension (LVDd) decreased and the systolic dimension (LVDs) increased, in contrast with normalized patients.

Table 1. Number of patients who showed normal HG response.

Placebo Group			CoQ Group		
0	1W	2W	0	1W	2W
0	0	19	0	18	19

—p<0.001 by Fisher's test-----

Table 2. Change of clinical symptoms.

	Placebo Group		CoQ Group	
	1W	2W	1W	2W
No change	13		1	1
Improved (+)	5	1	2	
Improved(++)	2	7	6	
Disappeared		12	11	19

—p<0.05 by Wilcoxon's test---

Table 3. Change of mean EF in % by HG.

		Placebo Group			CoQ Group		
		0	1W	2W	0	1W	2W
M-mode	REST	77.6±3.1	77.7±3.2	67.2±3.8	76.6±2.7	67.2±3.2	63.4±4.5
	H G	63.8±3.5	62.6±3.2	75.3±3.9	62.5±4.4	74.6±3.4	77.9±2.8
2D	REST	62.6±3.4	62.3±3.5	50.4±4.9	63.7±3.0	52.6±5.1	48.6±4.9
	H G	47.5±3.0	46.2±3.2	60.5±3.2	47.3±3.5	60.5±4.2	65.3±2.6

Underlined: Abnormal Response

Table 4. Rate of increase of EF, LVDd, LVDs by HG (M-mode echo)

	Placebo Group			CoQ Group		
	0	1W	2W	0	1W	2W
ΔEF%	-14.0±4.2	-15.2±3.2	8.1±4.9	-14.1±3.6	7.1±4.0	14.5±3.6
		-p<0.001- (1W vs 2W)		*--p<0.001-* (0 vs 1W)	*-p<0.001-* (1W vs 2W)	
		---------p<0.001------------ (Placebo 1W vs CoQ 1W)				
			----------p<0.001------------- (Placebo 2W vs CoQ 2W)			
ΔLVDd%	-5.2±3.3	-3.8±2.4	5.3±4.2	-4.0±2.6	5.1±3.9	9.5±4.1
		-p<0.001- (1W vs 2W)		*--p<0.001-* (0 vs 1W)	*-p<0.005-* (1W vs 2W)	
		---------p<0.001------------ (Placebo 1W vs CoQ 1W)				
			----------p<0.005------------- (Placebo 2W vs CoQ 2W)			
ΔLVDs%	9.9±6.2	15.3±3.6	-5.3±3.6	12.8±4.9	-3.4±4.1	-7.7±5.5
		-p<0.001- (1W vs 2W)		*--p<0.001-* (0 vs 1W)	*-p<0.025-* (1W vs 2W)	
		---------p<0.001------------ (Placebo 1W vs CoQ 1W)				
			----------p<0.025------------- (Placebo 2W vs CoQ 2W)			

Table 5. Rate of increase of EF, SV, CO by HG (2D echo)

	Placebo Group			CoQ Group		
	0	1W	2W	0	1W	2W
ΔEF%	-14.7±3.2	-16.7±3.3	9.5±4.6	-16.5±3.5	7.0±5.5	16.7±4.1
		*-p<0.001-x (1W vs 2W)		* p<0.001-* (0 vs 1W)	*-p<0.001--* (1W vs 2W)	
		---------p<0.001------------ (Placebo 1W vs CoQ 1W)				
			----------p<0.001------------- (Placebo 2W vs CoQ 2W)			
ΔSV%	-33.2±7.2	-34.2±6.4	37.5±21.5	-36.7±7.1	28.8±22.5	66.0±22.9
		-p<0.001- (1W vs 2W)		*-p<0.001-* (0 vs 1W)	*-p<0.001--* (1W vs 2W)	
		---------p<0.001-- (Placebo 1W vs CoQ 1W)				
			----------p<0.001------------- (Placebo 2W vs CoQ 2W)			
ΔCO%	-24.7±8.7	-23.6±7.1	63.3±35.3	-21.5±9.8	51.5±34.0	99.3±39.4
		-p<0.001- (1W vs 2W)		*-p<0.001-* (0 vs 1W)	*-p<0.001--* (1W vs 2W)	
		---------p<0.001------------ (Placebo 1W vs CoQ 1W)				
			----------p<0.001------------- (Placebo 2W vs CoQ 2W)			

Table 6. Serum CoQ level (μg/ml).

0	1W	2W	0	1W	2W
0.48±0.09	0.53±0.10	1.96±0.10	0.57±0.11	2.05±0.65	2.34±0.52

—p<0.001— (1W vs. 2W, first group); *—p<0.001—**—p<0.025—* (0 vs.
————p<0.001———— (1W, first group vs.
————p<0.01———— (2W, first group vs.

Table 7. Number of MVP patients normalized in HG: Dose vs. time (weeks) needed.

Dose of CoQ (mg/kg/day)	Weeks 1	2	3	4	8	12	16	20	24	28	32	36	40	44	52	>60	Total
0.6 – 0.9						1	2	2				1	1		1	3	11
1.0 – 1.4				3	10	4	3	2	2	2	3	4	1	6		4	44
1.5 – 1.9		1	5	12	15	11	7	6	6	4	3	4	3	3		1	81
2.0 – 2.4	1	6	23	73	33	16	5	3	2	4	2						168
2.5 – 2.9	2	6	6	30	8	5	2	1									60
3.0 – 3.4	200	23	6	4	3												256
Total	203	36	40	122	69	37	19	14	10	10	8	9	5	9	1	8	600

r= −0.67 (p<0.001)

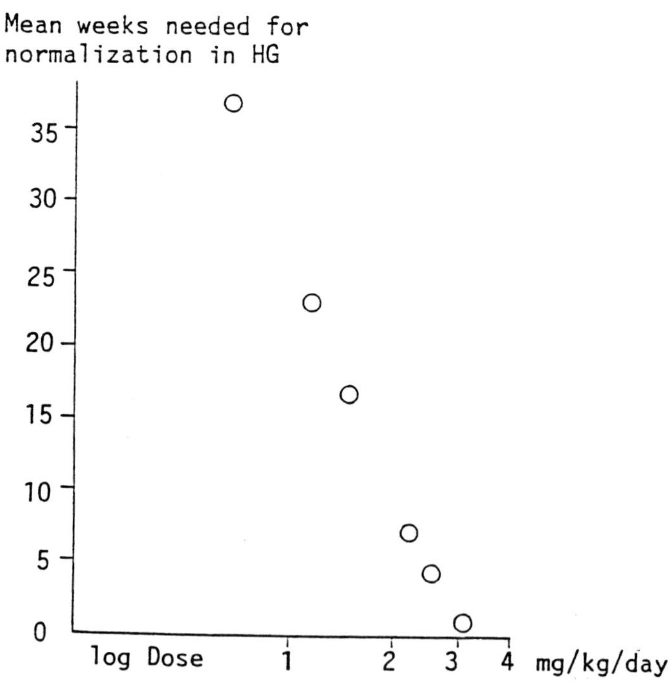

Figure 1. Dose vs. response of CoQ.

CONCLUSIONS

The double blind study gave a convincing conclusion that CoQ is effective on MVP patients with LCD. The CoQ treatment (3.1 to 3.4 mg/kg/day) broght normal cardiac function evaluated by handgrip test within 7 days in the majority, however, CoQ for 14 days gave better results.

A close relationship between the dose and the time needed for the appearance of response (dose-response relationship) existed. The maximally effective dose may be inferred to be 3.3 to 3.4 mg/kg/day by extrapolation. These results may give some information for the treatment of other diseases to which CoQ is applicable.

REFERENCES

Gottdiener, J.S.,Borer,J.S.,Bacharach,S.L., Green, M.V. and Epstein, S.E.,1981, Left ventricular function in mitral valve prolapse:Assessment with radionuclide cineangiography, American Journal of Cardiology, 47, 7-13.

Jeresaty, R.M.,1979, Mitral Valve Prolapse, (New York, Raven Press), pp.38-44.

Kanaya,H and Arakawa,K., 1979, Ultrasonodiagnosis of subclinical heart failure by increasing afterload with angiotensin II , Japanese Heart Journal, 20,407-417.

Newman, G.E.,Gibbons, R.J. and Jones, R.H.,1981, Cardiac function during rest and exercise in patients with mitral valve prolapse. Role of radionuclear angiocardiography, American Journal of Cardiology, 47, 14-17.

Oda, T., 1985, Effect of coenzyme Q_{10} on stress-induced cardiac dysfunction in paediatric patients with mitral valve prolapse:A study by stress echocardiography, Drugs under Experimental and Clinical Research,11,557-576.

Oda, T.,1986, Coenzyme Q_{10} therapy on the cardiac dysfunction in patients with mitral valve prolapse:Dose vs. effect & dose vs. serum level of coenzyme Q_{10} , In Biochemical and Clinical Aspects of Coenzyme Q, Vol.5, edited by K.Folkers and Y.Yamamura (Amsterdam, Elsevier), pp. 269-280.

Addendum and acknowledgement: These studies were performed on the basis of the informed consent of patients and their families. The author wishes to thank Professor Takeshi Kuroda of Fukuoka University Hospital for taking charge of the controller in the double blind study, and Dr. Kunihiro Hamamoto and Dr. Hisaharu Sagara of our Department for their critical collaboration in reading echocardiograms.

INFLUENCE OF UBIQUINONE ON SYSTOLIC AND DIASTOLIC CARDIAC FUNCTION IN THE FAILING AND ISCHEMIC MYOCARDIUM

W. V. Judy, Ph.D. and K. Folkers, Ph.D.*

Department of Medical Research, St. Vincent Hospital and Health Care Center, Indianapolis, Indiana and *The Institute for Medical Research, The University of Texas at Austin, Austin, Texas, U.S.A.

INTRODUCTION

The biochemistry of Ubiquinone (CoQ10) in mitochondrial electron transfer is well recognized by the scientific community as is its essential role in the respiratory chain for ATP production and therefore aerobic metabolism. A deficiency of CoQ10 or ATP at the cellular level as a causative factor for muscle dysfunction is not generally recognized by the clinical community. For example, heart failure has been classified as to the conditions which lead to such failure, rather than being attributed to a primary deficiency of energy substrate.

In the cellular metabolic machinery, CoQ10 is uniquely positioned and is the beginning of a new era for naturally occurring intracellular enzyme treatment modalities, all of which may be energy dependent.

The purpose of this report is to present data showing the effects of CoQ10 treatment on resting and exercise cardiac function in three groups and to present between group comparisons.

METHODS

Eighty-eight patients were treated with CoQ10. These patients were of three groups: Group I (N=44) were Class IV Congestive Heart Failure (CHF) patients; Group II (N=24) were Ischemic Heart Disease (IHD) patients who had stable resting and exertional angina, but normal resting cardiac functions; Group III (N=20) were normal (Normal) middle aged individuals with no known cardiac pathology or other active disease process.

In all groups, CoQ10 was given orally in doses of 100 mg. per day. In the CHF and IHD groups, CoQ10 was given supplemental to conventional therapy. Pretreatment (control) data was collected on two occasions over a thirty day period to show stability or a nonimproving condition.

Resting and exercise cardiac functions (cardiodynamics and cardiokinetics) were measured by the noninvasive bioelectrical impedance method. This method allows serial studies with a low between test variation in normals, and a good correlation for cardiac output and left ventricular ejection fraction compared to the respective clinical standards of thermodilution cardiac output and nuclear cardiology ejection fraction in clinical patients. The cardiodynamic parameters measured were Heart Rate (HR), Stroke Index (SI), Cardiac Index (CI) End Diastolic Volume Index (EDVI), End Systolic Volume Index (ESVI) and systolic blood pressure (SP). The cardiokinetic parameters derived were Left Ventricular Ejection Fraction (EF), Stroke Work Index (SWI), Systolic Elastance (SEea) and Diastolic Elastance (DEea). Resting data were collected in the supine position. Exercise data were collected during graded bicycle ergometry. Blood pressure was measured by the manual arm cuff method. Venous blood samples were taken before and during CoQ10 treatment for the determination of blood CoQ10 level. The HPLC method was used to measure blood CoQ10 content.

The data presented are group means ± SE. Statistical differences between control and treatment groups means, and differences between groups were derived from the student t- test. The data presented is for control and after one year of CoQ10 treatment. *The level of significant difference was selected at p§0.01 (§ means less than).

RESULTS

Blood CoQ10 levels for the various groups before (control) and after one (1) year CoQ10 treatment are shown in table 1..

Table 1. Blood CoQ10 levels

GROUP	CONTROL	CoQ10 (1 year)
CHF	0.52±0.14*	2.26±0.11*
IHD	0.88±0.13	2.36±0.16*
NORMAL	0.92±0.16	2.30±0.14*

Control levels were in the normal range for the IHD and normal groups, but were reduced (p§0.01) in the CHF group. CoQ10 treatment increased blood CoQ10 levels (p§0.01) to almost equivalent levels in each group.

Control resting cardiodynamics for each group are presented in table 2.

Table 2. Control resting cardiodynamics

VARIABLE	CHF	IHD	NORMAL
HR	93.4±2.3*	75.6±2.4	79.6±2.3
SI	23.2±1.6*	40.3±1.3	41.7±1.8
CI	2.1±0.06*	3.1±0.03	3.3±1.2
EDVI	113.0±9.7*	71.1±3.7	70.4±3.6
ESVI	91.2±6.2*	31.2±2.8	28.3±2.6
SP	118.6±5.3	133.6±6.6	128.6±5.9

For each measured variable there were no significant differences between the IHD and normal groups. In the CHF group, HR, EDVI and ESVI were elevated (p§0.01), whereas SI and CI were reduced (p§0.01).

Resting cardiokinetics for the different groups are presented in table 3.

Table 3. Control resting cardiokinetics

VARIABLE	CHF	IHD	NORMAL
EF	20.2±1.6*	59.2±2.1	60.3±2.4
SWI	37.6±1.1*	70.4±2.4	78.3±2.5
SEea	0.75±0.10*	2.6±1.0	2.6±0.07
DEea	0.37±0.04*	0.60±0.02	0.62±0.02

Again, no significant difference in the measured variables was found between the IHD and normal groups. However, each variable was reduced (p§0.01) in the CHF group.

The influence of CoQ10 on resting cardiodynamics in the three groups are presented in table 4.

Table 4. Post CoQ10 resting cardiodynamics

VARIABLE	CHF	IHD	NORMAL
HR	83.7±2.7	78.3±2.3	74.0±2.6
SI	38.8±4.2	43.0±1.6	44.0±1.5
CI	2.9±0.08*	3.3±0.03	3.3±0.04
EDVI	87.6±7.9	79.0±3.6	71.2±3.7
ESVI	47.3±5.6*	31.6±2.2	29.0±2.1
SP	130.5±7.4	126.6±6.4	126.0±7.0

After one year of CoQ10 treatment there was no improvenent in resting cardiodynamics in either the IHD or normal groups. In the CHF group, CoQ10 treatment significantly changed resting cardiodynamics for each variable. Heart Rate, EDVI and ESVI were all reduced (p§0.01), whereas SI,CI and SP were all significantly improved (p§0.01). Compared to the IHD and normal groups, CI and ESVI were still lower (p§0.01) in the CHF after CoQ10 treatment.

The influence of CoQ10 on resting cardiokinetics in the three different groups are presented in table 5.

Table 5. Post CoQ10 resting cardiokinetics

VARIABLE	CHF	IHD	NORMAL
EF	37.6±1.17*	59.2±2.1	61.4±2.5
SWI	62.0±1.8*	70.4±2.4*	79.5±2.1
SEea	1.5±0.17*	2.6±1.0	2.7±0.08
DEea	0.44±0.06*	0.6±0.02	0.60±0.03

After one year of CoQ10 treatment, resting cardiokinetics were not significantly changed in either the IHD or normal groups, and there were no meaningful differences between these groups. In the CHF group, resting cardiokinetics were increased (p§0.01). None of these contractile indices returned to normal but remained less (p§0.01) than those of the other groups.

Response to exercise was highly variable within and between groups. This variability was limited in the CHF and IHD groups relative to their heart failure and metabolic induced ischemia. The exercise workload and duration was significantly (p§0.01) greater in the normal groups as was expected. Group cardiodynamic responses for each measured variable at peak exercise are presented in table 6.

Table 6. Pre CoQ10 exercise cardiodynamics

VARIABLE	CHF	IHD	NORMAL
HR	118.3±3.2*	122±6.3*	156.5±4.7
SI	21.4±2.0*	39±2.0*	49.6±1.8
CI	2.4±0.05*	4.8±0.2*	7.7±0.4
EDVI	123.3±8.9*	92±4.8*	68.8±2.5
ESVI	102.1±7.1*	53±2.4*	19.6±0.8
SP	140.6±6.6*	150±6.8*	184.3±8.5

At peak exercise, the normal group was able to attain a higher (p§0.01) HR, SI, CI and SP at a lower (p§0.01) EDVI and ESVI. The peak exercise cardiodynamics was greater (p§0.01) in the IHD group for SI and CI, and less (p=0.01) for EDVI and ESVI than in the CHF group. The resting to exercise response was variable between groups. In the normal group, all variables increased except EDVI and ESVI. In the IHD, group all variables except SI increased during exercise. In this group the ischemic event was accompanied by an increase in EDVI and ESVI. In the CHF group SI and CI decreased during exercise, while EDVI, ESVI and SP increased (p§0.01). Obviously, in this group exercise induced a congestive failure mode as signified by the reduced cardiac pumping and increased ventricular volumes.

Pre CoQ10 exercise cardiokinetics at peak exercise for the respective groups are shown in table 7.

Table 7. Pre CoQ10 exercise cardiokinetics

VARIABLE	CHF	IHD	NORMAL
EF	17.3±1.0*	43.2±2.0*	73.0±3.2
SWI	37.3±1.7*	80.3±3.4*	125.2±6.6
SEea	0.6±0.04*	1.5±0.05*	5.4±0.3
DEea	0.4±0.02*	0.45±0.02	0.48±0.02

Exercise, in the normal group, increased (p§0.01) all indices of systolic contractile function, while DEea decreased (p§0.01). In the IHD group, EF, SEea and DEea decreased (p§0.01) during exercise, whereas SWI increased (p§0.01). All these changes were significantly different from resting levels. Similarly, in the CHF group, EF, SEea and DEea also decreased (p§0.01) during exercise as the heart was forced into volume overload failure.

Exercise cardiodynamics for each group after CoQ10 treatment are shown in table 8.

Table 8. Post CoQ10 exercise cardiodynamics

VARIABLE	CHF	IHD	NORMAL
HR	140.6±4.7*	156.0±8.3*	170.2±5.1
SI	37.6±4.6*	48.4±2.2*	56.3±2.6
CI	5.3±0.11*	7.5±0.4*	9.5±0.8
EDVI	88.6±6.9*	77.1±3.5*	72.2±3.7
ESVI	51.6±4.9*	31.4±1.3*	16.3±0.7
SP	168.6±6.6*	182.0±9.4	227.5±10.6

In the normal group the CoQ10 treatment exercise levels of HR, SI, CI and SP were greater (p§0.01) than those found in the CHF or IHD groups. Only CI in the normal group was changed (p§0.01) above the pre CoQ10 exercise level. In both the IHD and CHF groups HR, SI, CI and SP increased during exercise, whereas EDVI and ESVI decreased. All of these changes were different (p§0.01) from pre CoQ10 exercise levels except SI and ESVI in the IHD group.

Exercise cardiokinetics after CoQ10 treatment in the different groups are shown in table 9.

Table 9. Post CoQ10 exercise cardiokinetics

VARIABLE	CHF	IHD	NORMAL
EF	44.6±1.6*	62.6±3.4*	77.2±3.4
SWI	63.9±3.2*	113.4±6.6*	172.3±8.0
SEea	2.7±0.1*	4.9±0.21*	6.5±0.3
DEea	0.54±0.02	0.50±0.03	0.52±0.22

In all groups, each contractile indices increased with CoQ10 treatment. All variables were increased (p§0.01) above control exercise levels in the CHF group. In the IHD group all except DEea were significantly (p§0.01) increased. In the normal group only SWI and SEea were increased significantly. Although these contractile indices increased above pre CoQ10 levels in the CHF and IHD groups, they were still significantly less than those attained in the normal group.

CONCLUSIONS

The control, resting and exercise, between group differences found in the study were expected because of the patient selection criteria. The major findings were clear and quite significant. Coenzyme Q10 therapy will improve resting and exercise cardiac function in Class IV CHF patients, but will not return such to normal. In patients who have no CoQ10 deficiency (IHD and Normal groups), CoQ10 treatment has no significant effects on resting systolic or diastolic cardiodynamics or kinetics. In IHD patients with normal resting cardiac function, CoQ10 therapy will prevent the myocardial depression, delay the onset of myocardial ischemia and blunt the ischemic response associated with increased myocardial metabolic demand. In normal subjects, CoQ10 therapy is only beneficial in conditions of high metabolic demand where the cellular (cardiac and skeletal muscle) CoQ10 pools may be depleted by a high aerobic demand and ATP turnover rate.

The significantly reduced blood CoQ10 level and the associated significantly depressed myocardial function in the CHF group, suggest a possible cause and effect relationship between

CoQ10 and the degree of heart failure. The fact that raising blood CoQ10 to three times normal in this group did not improve systolic or diastolic cardiac function to normal show that, in Class IV failure, the metabolic and contractile machinery cannot be returned to normal. This does not mean that in lesser classifications of CHF more normal cardiac functions cannot be achieved. It may be that CoQ10 therapy will delay or prevent the progression of failure in the earlier.stages. Indeed, there are data which show improved resting cardiac functions, reduced degree of failure classification and improved long term survival through five years in Class II, III and IV CHF patients. The mechanism of CoQ10 action in the CHF patient may not only be related to correcting a deficiency state and improving the available ATP for bioenergetics. It may also include a return of the viable cardiac cells to a more normal physiological and biochemical state by its action as a free-radical scavenger and as a membrane stabilizer via increasing the available ATP to energize membrane ion exchange pumps and positive effects on calcium ATPase and calcium channel function.

The above mechanisms are also probable in the ischemic heart group. Under conditions of increased metabolic demand, CoQ10 prevents ATP depletion and reduces mitochondrial oxygen utilization. In this way CoQ10 increased the oxygen delivery across stenotic coronary vessels. These characteristics suggest that CoQ10 acts as an ischemic modifier and provides myocardial protection from metabolically or pathologically induced myocardial ischemia.

In the normal group the beneficial effects of CoQ10 on exercise cardiac functions is probably related to total body bioenergetics. Both myocardial and skeletal ATP synthesis and resynthesis is enhanced by CoQ10. During exercise a more stable ATP pool should improve aerobic metabolism thereby increasing oxygen consumption and reducing oxygen debt.

INHIBITION OF LIPID PEROXIDATION BY ADRENOCHROME

A. Bindoli, M.P. Rigobello and L. Galzigna

Dip. Chimica Biologica and Centro Studio Fisiologia Mitocondriale, via Marzolo 3, 35131, Padova, ITALY

Keywords: Adrenochrome, o-quinones, lipid peroxidation, free radicals

INTRODUCTION

Adrenochrome is an ortho-quinone derived from the autoxidation of adrenalin; it exhibits neurotoxic and cardiotoxic properties referable both to its interaction with sulfhydryl groups and to the production of oxygen reduced forms through a redox cycling process (Bindoli *et al.*,1989).

Several papers indicate that quinonoid compounds, particularly in their reduced form, can act as powerful antioxidants. Nevertheless, depending on the experimental conditions and on the specific properties of each considered quinone, they can either stimulate or inhibit lipid peroxidation; for instance anthracycline antibiotics, e.g. adriamycin stimulate lipid peroxidation (Goodman *et al.*, 1977), while menadione (Wills, 1972) and coenzyme Q (Beyer, 1988) act as powerful antioxidants. Adrenochrome was reported to exert its toxic effects through a peroxidative process (Gudbjarnason, *et al.*, 1978); in the present paper the effects of adrenochrome on different lipid peroxidation systems was examined and the effects are those of an antioxidant. Yet, this property does not exclude a toxic action through an oxidative stress-linked mechanism (depletion of GSH and production of oxygen reduced species).

METHODS

Microsomes (0.5 mg protein/ml) and liposomes (0.5 mg/ml) were incubated in 0.125 M KCl, 5 mM Tris, 15 mM Hepes (pH 7.4) at 30° C with the peroxidizing systems indicated under the appropriate figures and tables. Lipid peroxidation was measured as malondialdehyde (MDA) with the thiobarbituric acid method.

RESULTS

Microsomal lipid peroxidation is strongly stimulated by NADPH in the presence of iron ions, and depends on the concentrations of both NADPH and Fe^{2+}, while the presence of iron chelators appears also to play a fundamental role (Ursini *et al.*, 1989). In Table 1 the inhibitory effects of increasing concentrations of adrenochrome on the NADPH/Fe/-ADP-induced lipid peroxidation is reported. The extent of inhibition is directly proportional to the concentration of adrenochrome and is inversely related to that of added iron. The inhibitory action exerted by adrenochrome appears far more marked at relatively low concentrations of NADPH (0.1 mM) and Fe^{2+} (2 μM), where an IC_{50} of about 15 μM is apparent; at higher concentrations of both NADPH (2mM) and Fe^{2+}

(20 μM) an IC_{50} of about 150 μM was calculated. From the time course of the inhibition of the NADPH/Fe/ADP-induced lipid peroxidation by adrenochrome (Fig.1) it appears that relatively high concentrations of adrenochrome (200 μM) almost completely inhibited the lipid peroxidation induced by 0.1 mM NADPH/2 μM Fe^{2+}/ 20 μM ADP (Fig.1a) up to about 50 min (Fig. 1a), while in the presence of 2 mM NADPH/20 μM Fe^{2+}/ 0.2 mM ADP (Fig. 1b) the inhibition by increasing concentrations of adrenochrome is less evident and reaches a plateau.

Table 1.: Adrenochrome (ADR) inhibition of microsomal lipid peroxidation induced by NADPH/Fe^{2+}/ADP: **a**, **d**: 0.1 mM NADPH; **b**, **e**: 0.5 mM NADPH; **c**, **f**: 2mM NADPH. ADP:Fe^{2+} ratios were 10:1. Incubation time: 15 min. Values are % of the control (in brackets nmoles MDA/mg protein).

ADR (μM)	a	b	c	d	e	f
	2 μM Fe^{2+}			20μM Fe^{2+}		
0	100 (4.6)	100 (8.2)	100 (16.1)	100 (29.9)	100 (38.8)	100 (54.8)
5	62	90	96	92	96	94
10	55	75	87	66	80	90
25	30	39	75	48	70	77
50	18	36	66	46	59	73
100	16	25	42	34	48	60
200	14	18	23	27	38	35

Fig. 1. *Time course of the inhibition of microsomal lipid peroxidation induced by increasing concentrations of adrenochrome.*

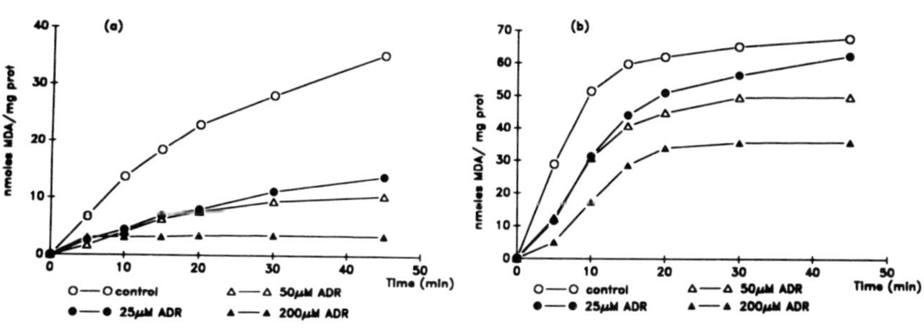

In a previous paper (Bindoli *et al*, 1988) it was shown that adrenochrome can be reduced by the mitochondrial and microsomal electron transport chain giving rise to the corresponding semiquinone that autoxidizes with the formation of the superoxide anion subsequently dismutating to H_2O_2. These reduced forms of oxygen might have a role in the induction of lipid peroxidation, particularly in the presence of metal catalysts. On the other hand the addition of superoxide dismutase and catalase does not substantially increase the inhibitory effect exerted by adrenochrome on lipid peroxidation (not shown) indicating that $O_2^{\cdot -}$ and H_2O_2 are not involved in the induction of lipid peroxidation. In addition, this indicates that the inhibition of lipid peroxidation by adrenochrome is not dependent on a depletion of oxygen from the reaction vessel due to the redox cycling of adrenochrome. Both adrenochrome and ferric ions are reduced by the flavoprotein of the microsomal electron transport chain, but they do not appear to compete in the reduction process. Consequently, the inhibition of lipid peroxidation by adrenochrome is not referable to an interference with the iron-reduction process.

Cumene hydroperoxide induces microsomal lipid peroxidation with a process dependent on cytochrome P-450 and independent of iron ions catalysis. In the absence of NADPH there is a scarce inhibition of lipid peroxidation by adrenochrome, while this inhibition, even though not large, is increased by addition of NADPH (data not shown), indicating that the reduced form of adrenochrome is more effective than the oxidized one in inhibiting lipid peroxidation. Adrenochrome, similarly to several other quinones can be reduced by the enzyme DT-diaphorase in a two-electrons process. In Fig. 2 the inhibition by adrenochrome of lipid peroxidation induced in liposomes by cumene hydroperoxide and hemin is reported. Adrenochrome *per se* is scarcely effective (20% inhibition), while the addition of DT-diaphorase increases its inhibitory effect to about 50%.

***Fig.** 2. Effect of DT-diaphorase on adrenochrome-induced inhibition of MDA formation in liposomes.*

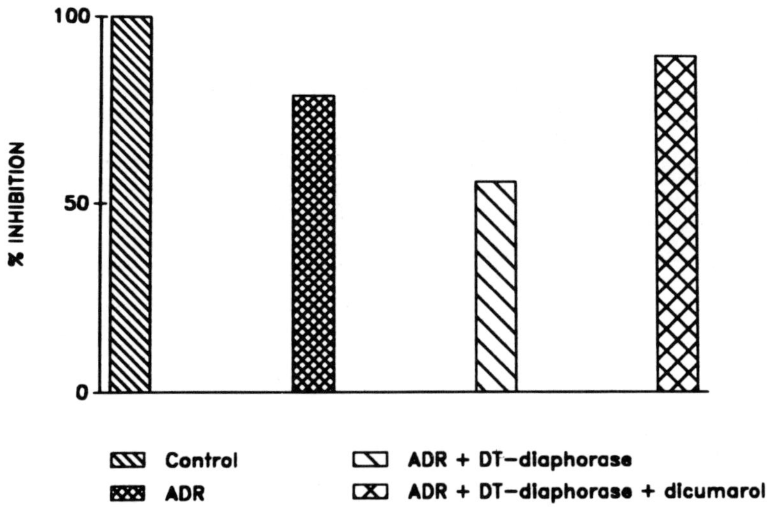

Liposomes were incubated for 15 minutes in the presence of 1 mM cumene hydroperoxide, 12µg/ml hemin and 0.5 mM NADPH. Other additions: 200 µM adrenochrome (ADR), 100 µM dicumarol and 1mg prot of DT-diaphorase obtained as dialysed cytosol after 105.000 x g centrifugation of a rat liver homogenate.

CONCLUSION

From the reported results it appears that the toxicity of adrenochrome is not referable to a peroxidative process. On the contrary, adrenochrome inhibits lipid peroxidation, induced with three different systems either dependent or independent of iron ions. Consequently, the toxicity of aminochromes should essentially be referred to arylation of nucleophilic groups and to redox cycling that brings about the formation of reactive semiquinones and oxygen reduction products. The reduced forms of adrenochrome (leucoadrenochrome or the corresponding semiquinone) are particularly efficient in eliciting the antioxidant effect; nevertheless a partial antioxidant effect is also exhibited by the oxidized form.

REFERENCES

Beyer, R.E., 1988, Inhibition by coenzyme Q of ethanol and carbon tetrachloride-stimulated lipid peroxidation in vivo and catalyzed by microsomal and mitochondrial systems. Free Radical Biology and Medicine, 5, 297-303.

Bindoli, A., Rigobello, M.P. and Galzigna, L., 1988, Production of reduced forms of oxygen by adrenochrome in the presence of ascorbate, microsomes and submitochondrial particles. In: Free Radicals: Chemistry, Pathology and Medicine, London, edited by C. Rice-Evans and T. Dormandy (London: Richelieu Press), pp. 293-300.

Bindoli, A., Rigobello, M.P. and Galzigna, L., 1989, Toxicity of aminochromes. Toxicology Letters, (in press).

Goodman, J., and Hochstein, P., 1977, Generation of free radicals and lipid peroxidation by redox cycling of adriamycin and daunomycin. Biochemical Biophysical Research Communication, 77, 797-803.

Gudbjarnason, S., Doell, B., Oskarsdottir, G. and Hallgrimsson, J., 1978, Modification of cardiac phospholipids and catecholamine stress tolerance. In: Tocopherol, Oxygen and Biomembranes, Amsterdam, edited by C. de Duve and O. Hayaishi, (Amsterdam: Elsevier North-Holland Biomedical Press), pp. 297-310.

Ursini, F., Hochstein, P. and Ernster, L., 1989, Microsomal lipid peroxidation: mechanism of initiation. The role of iron and chelators. Free Radical Biology and Medicine, 6, 31-36.

Wills, E.D., 1972, Effects of vitamin K and naphthoquinones on lipid peroxide formation and oxidative demethylation by liver microsomes. Biochemical Pharmacology, 21, 1879-1883.

EFFECTS OF COENZYME Q_1 ON ALKALINE PHOSPHATASE ACTIVITY

A. Mordente, G.E. Martorana, E. Meucci, G.A.D. Miggiano and G.P. Littarru*

Institute of Biological Chemistry, Faculty of Medicine, Catholic University, Roma, Italy.
*Institute of Biochemistry, Faculty of Medicine, University of Ancona, Italy.

INTRODUCTION

Coenzyme Q is an exceedingly versatile molecule which diplays antioxidant properties against lipid peroxidation of reconstituted membrane systems, submitochondrial particles and mitochondria (Beyer et al., 1987). On the other hand, Forman and Boveris (1982) have suggested that the semiquinone form of the coenzyme is the autooxidable component of the electron transport chain responsible of oxyradical production in mitochondria. Since scarce reports concern possible ubiquinone effects on enzymatic proteins, we have investigated the action of oxidized and reduced CoQ_1, a rather water soluble quinone, on alkaline phosphatase (AP) which is particularly susceptible to inactivation by oxyradicals, generated by several mixed-function oxidation (MFO) systems (Mordente et al., 1987).

Keywords: coenzyme Q_1; alkaline phosphatase; inactivation.

RESULTS AND DISCUSSION

When the reduced form of CoQ_1 is preincubated with AP, a biphasic (instantaneous and time-dependent) loss of enzyme activity is observed (Fig.1). Copper and iron significantly enhance the CoQ_1 effect indicating that metals are involved in AP inactivation by reduced CoQ_1.

Conversely, oxidized CoQ_1 does not show any significant effect on alkaline phosphatase activity.

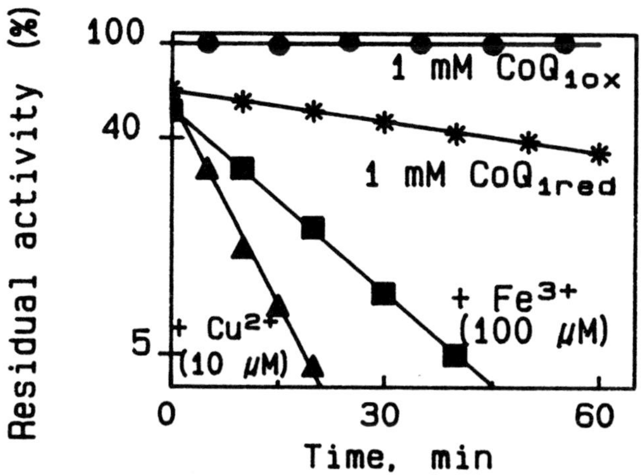

Fig.1. Alkaline phosphatase inactivation by reduced CoQ_1.

The oxidized CoQ_1 does not significantly shield (less than 3 %) the enzyme from known MFO systems such as xanthine/xanthine oxidase and NADH/NADH oxidase. Some protective effect can be observed only on the initial phase of AP inactivation by ascorbate system (Fig. 2).

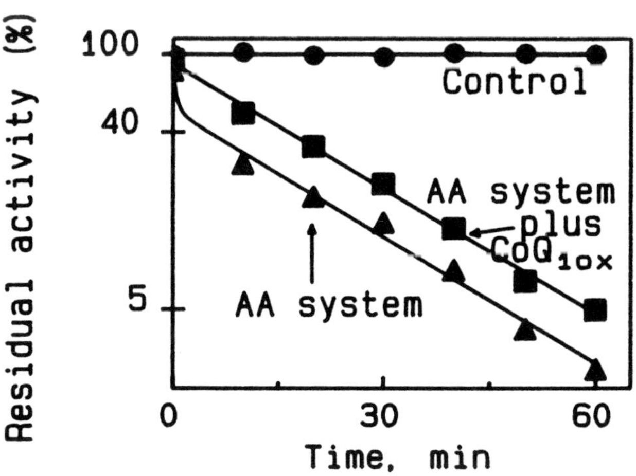

Fig.2. Effect of oxidized Coenzyme Q_1 on AP inactivation by ascorbate system (ascorbate/oxygen/Cu^{2+}).

CoQ_1 effect seems therefore linked to reduced coenzyme autooxidation. In oxygenated aqueous solutions, quinols have been shown to undergo spontaneous oxidation leading to semiquinone radical, superoxide anion and hydrogen peroxide (Cadenas et al., 1977).

Superoxide dismutase (SOD) enhances the instantaneous phase of inactivation, at the same time annulling the time-dependent one (Table 1). Catalase exerts a protective effect only on the first phase of AP inactivation. Desferrioxamine (DFO), a powerful ferric chelator, and dimethyl-sulfoxide (DMSO), an $OH^{\bullet}$ scavenger, do not significantly effect enzyme inactivation (Table 1).

Table 1. Effect of radical scavengers and antioxidants on AP inactivation by reduced coenzyme Q_1.

Incubation with			AP residual activity (%) 0 min	AP residual activity (%) 60 min
Red CoQ_1			68	35
	+	SOD	41	41
	+	Catalase	88	36
	+	DFO	67	42
	+	DMSO	69	43
	+	Ox CoQ_1	65	35

Final concentrations are: 1 mM reduced and oxidized CoQ_1, 1 mM DFO, 100 mM DMSO, 4.4 μM Catalase and SOD.

CONCLUSIONS

Metal-catalyzed autooxidation of reduced CoQ_1 provokes a strong inactivation of alkaline phosphatase with a biphasic pattern of decay. Catalase, albeit partially, diminishes only the first phase of inactivation. SOD accelerates AP activity decay so that the whole effect of the reduced coenzyme is actually accomplished almost instantaneously during the mixing time. Ubisemiquinone radical seems involved in the inactivation directly or, indirectly, by triggering a modified Haber-Weiss cycle with oxyradical production (Nohl and Jordan, 1987).

This in vitro behaviour resembles the "univalent leak" observed in respiratory chain blockade in mitochondria and the generation of free radicals during PMN phagocytosis. Coenzyme Q has been indeed implicated in superoxide anion production during the respiratory burst of activated human neurophils (Crawford and Schneider, 1984). At this regard, Oliver (1987) has already reported that stimulated human neutrophils inactivate at least four of the enzymes which are susceptible to oxidative modifications by MFO ystems.

REFERENCES

Beyer, R.E., Nordebrand, K. and Ernster, L., 1987, The Function of Coenzyme Q in Free Radical Production and as an Antioxidant: A Review. Chemica Scripta, 27, 145-153.

Cadenas, E., Boveris, A., Ragan, I. and Stoppani, A.O.M., 1977, Production of Superoxide Radicals and Hydrogen Peroxide by NADH-Ubiquinone Reductase and Ubiquinol-Cytochrome c Reductase from Beef-Heart Mitochondria. Archives of Biochemistry and Biophysics, 180, 248-257.

Crawford, D.R. and Schneider, D.L., 1982, Identification of Ubiquinone-50 in Human Neutrophils and Its Role in Microbicidal Events. Journal of Biological Chemistry, 257, 6662-6668.

Forman, H.J. and Boveris, A., 1982, Superoxide radical and Hydrogen Peroxide in Mitochondria. In Free radicals in biology, vol V, edited by W.A. Pryor (New York: Academic Press Inc), pp 65-90.

Mordente, A., Miggiano, G.A.D., Martorana, G.E., Meucci, E., Santini, S.A. and Castelli, A., 1987, Alkaline Phosphatase Inactivation by Mixed Function Oxidation Systems. Archives of Biochemistry and Biophysics, 258, 176-185.

Nohl, H. and Jordan, W., 1987, The Involvement of Biological Quinones in the Formation of Hydroxyl Radicals via the Haber-Weiss Reaction. Bioorganic Chemistry, 15, 374-382.

Oliver, C.N., 1987, Inactivation of Enzymes and Oxidative Modification of Proteins by Stimulated Neutrophils. Archives of Biochemistry and Biophysics, 253, 62-72.

CoQ10 PROTECTIVE EFFECT ON LINOLEIC ACID MICELLAE THERMAL AUTO-OXIDATION

G.P. Littarru, G.M. Wilkins, F. Tanfani, R. Staffolani, L. Mazzanti and B. Swoboda*.

Institute of Biochemistry, University of Ancona, Ancona, Italy and *Department of Chemistry, University of Warwick, Coventry, England.

INTRODUCTION

In addition to its central role in the respiratory chain Coenzyme Q10 (CoQ10) is proposed to function an antioxidant (1). The majority of evidence for antioxidant activity of CoQ10 comes from experiments with reconstituted membranes, mitochondria and submitochondrial particles, and lipid vesicles. The weight of evidence favours antioxidant activity of the reduced form of CoQ10 (semiquinone or quinol), either produced in vitro by electron transfer or added under controlled conditions to minimize auto-oxidation.

In the present study we have investigated the effect of CoQ10, added in the oxidised form, on the thermal auto-oxidation of linoleic acid micellae, which are considered to be simple model membranes (2).

KEY WORDS: Linoleic acid, Auto-oxidation, Antioxidant.

METHODS

Linoleic acid micellae were prepared, and auto-oxidised, as described by Hicks & Gebicki (2) to give a final concentration of 35mM. CoQ10 was added at a "standard" molar ratio of 1:100 with linoleic acid, unless otherwise stated. When present, 5-nitroxystearate (5NS) was added at 0.017 mol% relative to linoleic acid (6uM). Thiobarbituric acid reactants (TBAR) were measured according to Yagi (3); cojugated dienes were measured as the absorbance at 233nm of extracted fatty acids; ESR spectra were measured using a Varian E104 spectrometer; CoQ10 was measured by the HPLC method of Littarru et al. described in this volume.

RESULTS AND DISCUSSION

The data shown in table 1 demonstrate the effect of CoQ10 on the retention of 5NS ESR signal. In the presence and absence of CoQ10 the ESR signal declines as 5NS exerts a "chain breaking" antioxidant effect (2), but in the presence of CoQ10 the decline appears to be retarded slightly suggesting that CoQ10 may act upon the formation of the alkyl radicals which quench 5NS.

Table 1: Effect of CoQ10 (1:100) on the disappearance of 5NS ESR signal

	% 5NS ESR signal retained					
TIME (hours)	24	28	47	50	54	70
-CoQ10	68	63	45	37	36	25
+CoQ10 (1:100)	74	67	48	44	43	37

In figure 1 are the data for conjugated diene and TBAR formation in another experiment in the absence of 5NS. It can be seen that 1:100 CoQ10 completely eliminates the formation of both peroxidation products. It is also apparent that, as would be expected, conjugated dienes are formed rapidly after a short period of thermal treatment (20h) whereas the rapid formation of TBAR occurs after 48h.

It was observed that upon reducing the level of CoQ10 (to 1:200 and 1:400 relative to linoleic acid) the protective effects against TBAR formation is almost totally eliminated (figure 2). This figure also shows that the protective effect of 1:100 progressively decays to be lost after 100h incubation. CoQ10 is consumed with time, as shown in figure 3. It is of interest that CoQ10 appears not to be significantly consumed during the 46h when there is complete protection from all three levels of CoQ10.

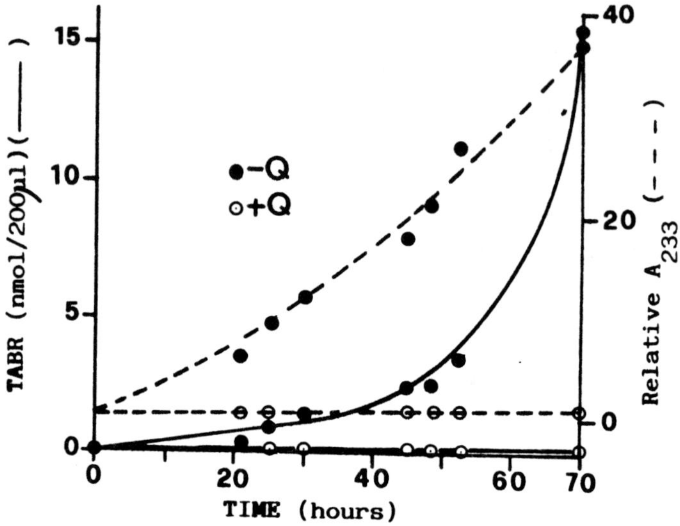

FIGURE 1. Effect of 1:100 CoQ10 on TBAR and conjugated diene (relative A_{233}) formation in linoleic acid micellae.

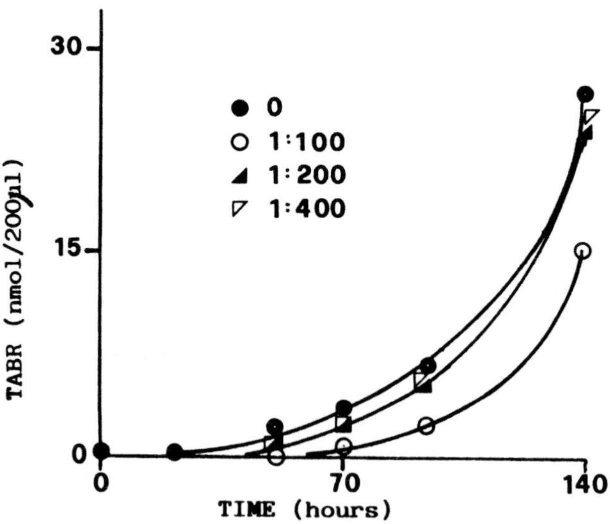

FIGURE 2. Effect of varying CoQ10 concentration on TBAR formation in linoleic acid micellae.

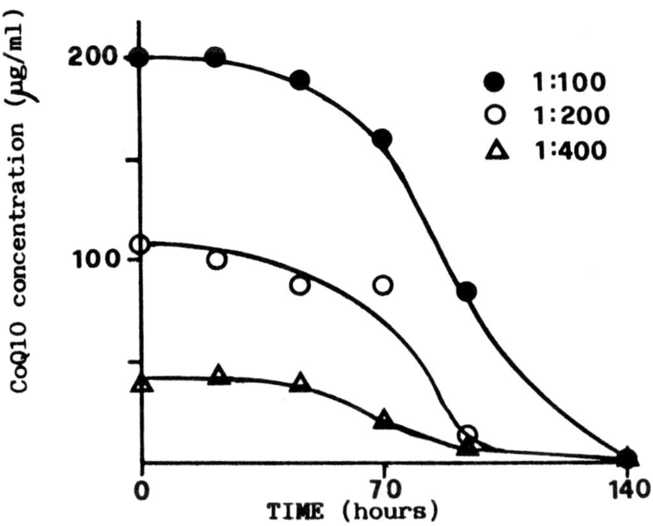

FIGURE 3. Reduction of CoQ10 concentration with time for the experiment shown in figure 2.

In conclusion these findings suggest that CoQ10 added to linoleic acid vesicles, in the oxidised state, in vitro acts as an antioxidant. At present the mechanism of this antioxidant activity of oxidised CoQ10 is unclear, but may be clarified by identification of a CoQ10 adduct.

REFERENCES

(1) Beyer,R.E., Nordenbrand,K. and Ernster,L., 1987, The Function of Coenzyme Q in Free Radical Production and as an Antioxidant: A Review. Chemica Scripta, 27, 145-153.

(2) Hicks,M. and Gebicki,J.M., 1981, Inhibition of Peroxidation in Linoleic Acid Membranes by Nitroxide Radicals, Butylated Hydroxytoluene, and alpha-Tocopherol. Archives of Biochjemistry and Biophysics, 210, 56-63.

(3) Yagi,K., 1984, Biological Damage Imposed by O_2. Assay for Blood Plasma or Serum. Methods in Enzymology, 105, 328-331.

Recent Contributions to the Role of Coenzyme Q in the Clinical Etiology and Treatment of Myopathies and Remarks on Its Nomenclature

Robert E. Beyer

Department of Biology, The University of Michigan, Ann Arbor, Michigan 48109-1048, U.S.A.

Since the early indications that coenzyme Q may counteract the damaging effects resulting from the administration of anthracycline cancer chemotherapeutic drugs such as adriamycin and daunorubicin (Kishi et al., 1977; Zbinden et al., 1977), thus widening the therapeutic window of such drugs, a number of disease states involving coenzyme Q have been reported in the clinical literature and in a number of symposia devoted to this interesting compound (for references see Ernster and Beyer, this volume). The involvement of free radicals formed as a result of the one-electron reduction of these drugs (Powis, 1989) and the antioxidant properties of coenzyme Q (Beyer, 1989) have focused attention on the underlying mechanisms of such effects and have provided a basis for the interpretation of data previously difficult to explain fully. For example, the decline accompanying ageing in cardiac function (Starnes et al., 1981; 1983a; Beyer and Starnes, 1985), protein synthesis (Starnes et al., 1983b), oxidative enzyme activities (Beyer et al., 1982), and coenzyme Q concentrations (Beyer et al., 1985; Beyer and Starnes, 1985), and the restoration of such defects by the administration of coenzyme Q to such aged animals (Beyer and Starnes, 1985) can now be interrelated and understood in terms of free radical damage due to loss of coenzyme Q and antioxidation upon administration of coenzyme Q (Beyer, 1988). The restoration of myocardial oxidative activities by coenzyme Q administration in aged rats is shown in Table 1.

Table 1. Effect of coenzyme Q administration on the rates of substrate oxidations catalyzed by myocardial homogenates from 28.5-30 month old rats. Rates as uatoms O_2/min/mg protein + SE. (From Beyer and Starnes, 1985.)

Substrate	n	Control	n	+ CoQ	%
Endogenous	6	8.8 ± 0.5	6	9.7 ± 1.9	+10
Succinate	6	111.8 ± 11.3	6	147.9 ± 13.0	+32
Glutamate	5	48.7 ± 7.9	4	50.5 ± 7.1	+ 4
Pyruvate-Malate	5	62.0 ± 10.8	6	93.2 ± 19.2	+50
Palmitoyl-carn.	5	27.0 ± 4.3	6	49.7 ± 10.5	+84

Lowered coenzyme Q concentrations, either due to inhibition of biosynthesis or stimulation of degradation, have been shown to be involved in several encephalomyopathies which respond to the administration of coenzyme Q (Ogasahara and Engel, 1988). For example, serum and muscle of patients with Kearns-Sayre syndrome and with ophthalmoplegia plus have been shown to be deficient in coenzyme Q (Zierz et al., 1989). Treatment of such patients with coenzyme Q_{10} results in functional improvement (Ogasahara et al., 1985; 1986). Muscle mitochondria from patients with familial mitochondrial encephalomyopathy catalyzed normal rates of electron transfer over individual complexes I, II, III, and IV while overall rates of NADH and succinate oxidation were severely reduced. Such mitochondria contained normal levels of all electron transfer chain components except for coenzyme Q which was 3.7% of normal (Ogasahara, 1989). Preliminary results indicate that the patients in this study respond positively to oral coenzyme Q therapy. These findings establish CoQ deficiency as a cause of familial mitochondrial cytopathy, suggest that the disease results from a tissue-specific defect of coenzyme Q biosynthesis and provide additional motivation for the Western medical profession to provide their patients with the benefits of coenzyme Q therapy when appropriate, thus avoiding severe side effects of many of the synthetic drugs manufactured by the pharmaceutical industry.

Since its discovery, coenzyme Q has been called various trivial names, two of which, coenzyme Q and ubiquinone, are used today. We prefer the trivial term coenzyme Q to ubiquinone as a more accurate descriptor of 2,3-dimethoxy-5-methyl-6-multiprenyl-1,4-benzoquinone for the following reasons: 1) the term ubiquinone is derived from the notion and implies that the compound is ubiquitous in life forms (Morton et al., 1957). It is not present in a number of organisms (Morton, 1961) including gram positive bacteria (Bishop, 1962), at least one species of photosynthetic bacterium (Hiraishi, 1989) and some fungi (Crane, 1965). It is also not present in methanogenic bacteria (Thauer, 1977); 2) A coenzymic function for this benzoquinone, originally suggested by Green (1959), has received considerable experimental support (King, 1982; Yu and Yu, 1980a,b) as a result of the isolation of mitochondrial coenzyme Q apoproteins or binding proteins; and 3) The objection to the use of the term coenzyme Q on the basis that it may be confused by students of biochemistry with a coenzyme of Q-enzyme (Morton, 1961) is no longer valid as this term for amylo (1,4-1,6)-transglycolase is no longer used in modern texts of biochemistry. A survey of recent textbooks of biochemistry indicates that the designation coenzyme Q is preferred.

ACKNOWLEDGEMENTS

The author received travel funds from The University of Michigan Office of the Vice President for Research, the LSA College Associate Dean for Research, and the Elsa U. Pardee Foundation.

REFERENCES

Beyer, R. E., 1988, Inhibition by coenzyme Q of ethanol- and carbon tetrachloride-stimulated lipid peroxidation in vivo and catalyzed by microsomal mitochondrial systems. Free Radical Biology & Medicine, 5, 297-303.

Beyer, R. E., 1989, The role of coenzyme Q in free radical formation and antioxidation. In CRC Handbook of Free Radicals and Antioxidants in Biomedicine, Vol. II, edited by J. Miquel, A. T. Quintanilha and H. Weber (Boca Raton: CRC Press), pp. 45-62.

Beyer, R. E., Burnett, B.-A., Cartwright, K. J., Edington, D. W., Falzon, M. J., Kreitman, K. R., Kuhn, T. W., Ramp, B. J., Rhee, S. Y. S., Rosenwasser, M. J., Stein, M. and An, L. C.-I., 1985, Tissue coenzyme Q (ubiquinone) and protein concentrations over the life span of the laboratory rat. Mechanisms of Ageing and Development, 32, 267-281.

Beyer, R. E., Edington, D. W., Starnes, J. W., Compton, R. T., Kwasman, M. A., Lipton, R. J., and Ponte, L. D., 1983, Exercise induced restoration of age associated decline of exergonic activities in skeletal muscle. The Gerontologist, 22, 167-168.

Beyer, R. E. and Starnes, J. W., 1985, Coenzyme Q and myocardial function in aging and exercise. In Pathobiology of Cardiovascular Injury, edited by H. L. Stone and W. B. Weglicki (Boston: Martinus Nijhoff Publishers), pp. 489-511.

Bishop, D. H. L., Pandya, K. P. and King, H. K., 1962, Ubiquinone and vitamin K in bacteria. Biochemical Journal, 83, 606-614.

Crane, F. L., 1965, Distribution of quinones. In Biochemistry of Quinones, edited by R. A. Morton (London: Academic Press), pp. 183-206.

Green, D. E., 1959, Structure-function interrelationships in mitochondrial electron transport and oxidative phosphorylation. Discussions of the Faraday Society, 27, 206-216.

Hiraishi, A., 1989, Occurrence of menaquinone as the sole isoprenoid quinone in the photosynthetic bacterium Heliobacterium chlorum. Archives of Microbiology, 151, 378.

King, T. E., 1982, Ubiquinone proteins in cardiac mitochondria. In Function of Quinones in Energy Conserving Systems, edited by B. L. Trumpower (New York: Academic Press), pp. 3-25.

Kishi, T., Kishi, H. and Folkers, K., 1977, Inhibition of cardiac CoQ_{10}-enzymes by clinically used drugs and possible prevention. In Biomedical and Clinical Aspects of Coenzyme Q, Vol. 1, edited by K. Folkers and Y. Yamamura (Amsterdam: Elsevier), pp. 47-61.

Morton, R. A., 1961, Isolation and characterization of ubiquinone (coenzyme Q) and ubichromenol. In Quinones in Electron Transport, edited by G. E. W. Westenholme and C. M. O'Connor (Boston: Little Brown), pp. 5-25.

Morton, R. A., Wilson, G. M., Lowe, J. S. and Leat, W. M. F., 1957, Ubiquinone. Chemical Industry, 1649.

Ogasahara, S. and Engel, A. G., 1988, Muscle coenzyme Q deficiency in a patient with familial mitochondrial encephalomyopathy. Neurology, 38 [Suppl. 1], 269.

Ogasahara, S., Engel, A., Frens, D. and Mack, D., 1989, Muscle coenzyme Q deficiency in familial mitochondrial encephalomyopathy. Proceedings of the National Academy of Sciences, U.S.A., 86, 2379-2382.

Ogasahara, S., Nichikawa, Y., Yorifuji, S., Soga, F., Nakamura, Y., Takahashi, M., Hashimoto, S., Kono, N. and Tarui, S., 1986, Treatment of Kearns-Sayre syndrome with coenzyme Q_{10}. Neurology, 36, 45-53.

Ogasahara, S., Yorifuji, S., Nishikawa, Y., Takahashi, M., Wada, K., Hazama, T., Nakamura, S., Kono, N. and Tarui, S., 1985, Improvement of abnormal pyruvate metabolism and cardiac conduction defect with coenzyme Q_{10} in Kearns-Sayre syndrome. Neurology, 35, 372-377.

Powis, G., 1989, Free radical formation by antitumor quinones. Free Radical Biology & Medicine, 6, 63-101.

Starnes, J. W., Beyer, R. E. and Edington, D. W., 1981, Effects of age and cardiac work in vitro on mitochondrial oxidative phosphorylation and (^{3}H)-leucine incorporation. Journal of Gerontology, 36, 130-135.

Starnes, J. W., Beyer, R. E. and Edington, D. W., 1983a, Myocardial adaptations to endurance exercise in aged rats. American Journal of Physiology, 245, H560-H566.

Starnes, J. W., Edington, D. W. and Beyer, R. E., 1983b, Myocardial protein synthesis during aging and endurance exercise in rats. Journal of Gerontology, 38, 660-665.

Thauer, R. K., Jungerman, K. and Decker, K., 1977, Energy conservation in chemotrophic anaerobic bacteria. Bacteriological Review, 41, 100-180.

Yu, C. A. and Yu, L., 1980a, Identification of ubiquinone binding proteins in ubiquinol-cytochrome c reductase by arylazido ubiquinone derivative. Biochemical and Biophysical Research Communications, 96, 286-292.

Yu, C. A., and Yu, L., 1980b, Interaction between succinate dehydrogenase and ubiquinone-binding protein from succinate-ubiquinone reductase. Biochimica et Biophysica Acta, 293, 24-38.

Zbinden, G., Bachman, E. and Bolliger, H., 1977, Study of coenzyme Q in toxicity of adriamycin. In Biomedical and Clinical Aspects of Coenzyme Q, Vol. 1, edited by K. Folkers and Y. Yamamura (Amsterdam: Elsevier), pp. 219-228.

Zierz, S., Jahns, G. and Jerusalem, F., 1989, Coenzyme Q in serum and muscle of 5 patients with Kearns-Sayre syndrome and 12 patients with ophthalmoplegia plus. Journal of Neurology, 236, 97-101.

ARE UBIQUINONES CHAIN-BREAKING ANTIOXIDANTS?

L. Landi, D. Fiorentini, L. Cabrini, C. Stefanelli, A.M. Sechi and G. Pedulli *

Dipartimento di Biochimica e *Dipartimento di Chimica Organica dell'Università di Bologna

INTRODUCTION

The detoxication of reactive oxygen species is one of the prerequisite of aerobic life; the antioxidant defense system comprises both preventive antioxidants such as catalase and glutathione peroxidase that destroy hydroperoxides without generating free radicals, and low molecular weight free radical scavengers such as the chain-breaking antioxidants.

These molecules owe their activity to the ability to trap peroxyl radicals. Vitamin E, known as a major lipophilic chain-breaking antioxidant in biological membranes, suppresses lipid peroxidation. Ubiquinones may also play this important role within the membrane where they are located.

We have been studying the peroxidation of egg lecithin vesicles and its inhibition by ubiquinone-3 (Q_3). This quinone was chosen since its side chain length is similar to that of α-tocopherol and, in any case, it has been previously shown that the antioxidant activity of ubiquinones does not depend on their side chain length (Landi _et al._ 1985). Two experimental systems were employed in order to induce lipid peroxidation: free radicals produced i) by ultrasonic irradiation and ii) by thermal decomposition of 2-2'-azobis-(2,4-dimethylvaleronitrile) (DMVN), a lipid-soluble radical initiator.

Key words: antioxidants, ubiquinone-3, liposomes, peroxidation, azo-initiators.

MATERIAL AND METHODS

Q_3 and α-tocopherol were kindly supplied by Eisai Co., Tokio, Japan. Egg yolk lecithin (EYL) was obtained from Lipid Products (Redhill, U.K.). DMVN was purchased from Polysciences, Inc. (Warrington, PA).

System i. Monolamellar vesicles were prepared by sonication under nitrogen in double distilled water. The final concentration was 4.2 mM for phospholipid and 60-70 μM for Q_3, when present.
System ii. Multilamellar vesicles were prepared by adding DMVN and, when present, Q_3 or α-tocopherol and EYL. After evaporation under nitrogen the samples were vortex stirred with an aliquot of 0.01 M phosphate buffer, pH 7.4, in order to obtain 4 mM DMVN, 0.072 mM Q_3 or 0.016 mM α-tocopherol and 15 mM EYL dispersions. The reaction vessels were immersed in a water bath at 37°C and kept in the air under dark.

Vesicle autoxidation was measured by determining spectrophotometrically conjugate diene formation. The increase in absorbance at 233 nm (Sessa et al., 1977) was taken as an indication of the appearance of conjugated dienes. After peroxidation unreacted Q_3 was determined by reversed-phase HPLC according to Takada et al. (1982). Fatty acid composition of phospholipid vesicles was determined by GLC, after methanolysis in acid-methanol.

RESULTS AND DISCUSSION

With the aim to evaluate the antioxidant efficiency of Q_3 we have investigated the peroxidation of lipid vesicles using two different systems.
System i. The principal reactive species formed during sonication are hydrated electrons, hydrogen atoms and hydroxyl free radicals. In the presence of oxygen, the hydrated electron and hydrogen atom react at diffusion controlled rates to form the superoxide anion and the perhydroxyl radical, both concurring by different pathways to the formation of H_2O_2 (Anbar and Pecht, 1964). Hydroxyl radicals, in particular, are known to be capable of abstracting hydrogen atoms from membrane lipids and of causing the formation of conjugated dienes. The oxidative free radical damage to the vesicles under experiment is reported in Fig. 1. As sonication proceeded, the increase in absorbance at 233 nm considerably increased in control vesicles, while it was noticeably lower in Q_3-containing vesicles. In these experimental conditions 15 nmol of quinone disappeared during the irradiation.

In order to calculate how many molecules of polyunsaturated fatty acid residues are effectively protected by 1 molecule of Q_3, the fatty acid composition of the vesicles before and after sonication was determined. In control vesicles 330 nmol of polyunsaturated fatty acids (PUFA) were lost. On a molar basis it can be calculated that 20-30 molecules of PUFA are effectively protected by 1 molecule of

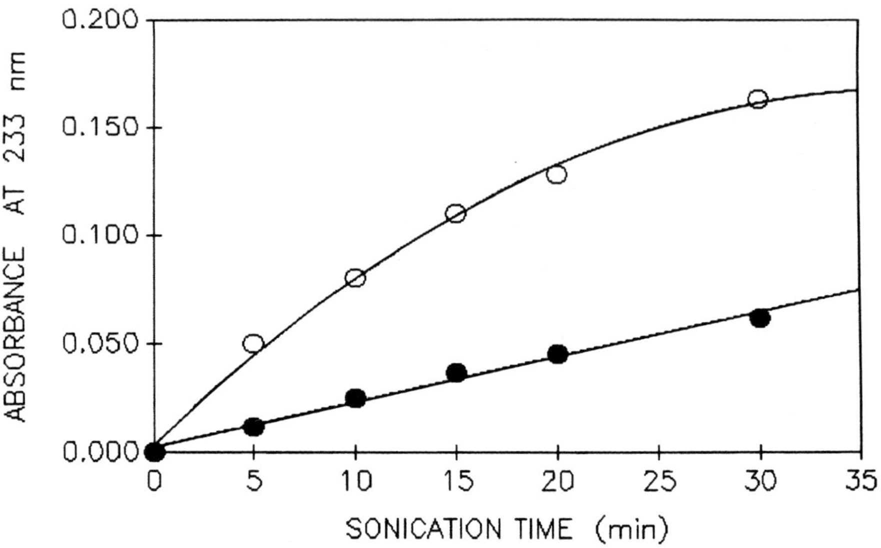

Fig. 1. Effect of Q_3 on the rate of lipid peroxidation caused by sonication of EYL vesicles. (o—o) control vesicles; (●—●) Q_3-containing vesicles.

oxidized ubiquinone. This ratio much higher than 1 indicates that the oxidation proceeds by a free radical chain mechanism and suggests that the antioxidant effect of Q_3 mostly resides in its ability of trapping lipid peroxyl radicals.

System ii. It was observed (Barclay and Ingold, 1981) that a uniform rate of the initiation step of lipid peroxidation is a prerequisite in order to obtain quantitative data on the activity of a chain-breaking antioxidant. The thermal decomposition of DMVN yields useful quantities of peroxyl radicals at a known and constant rate within the lipid bilayers. The rate of chain initiation was determined by the conventional inhibited method (Boozer *et al.*, 1955), i.e. by measuring the length of time the formation of conjugated diene is almost suppressed by a known concentration of α-tocopherol incorporated into liposomes.

From the equations reported by Barclay and Ingold (1981) the stoichiometric factor, *n*, i.e. the number of radicals trapped by each inhibitor and the rate constant of inhibition can be calculated. These values, reported in Table 1, show that the rate constant of inhibition for ubiquinone is smaller than that of α-tocopherol by two orders of magnitude. The mechanism to explain the chain-breaking activity of Q_3 might involve an initial electron transfer from

Table 1. Inhibition rate constant and stoichiometric factor for α-tocopherol and ubiquinone-3 at 37°C.

Experimental system	Antioxidant	$k_{inh}M^{-1}s^{-1}$	n
Vesicles + DMVN	0.016mM α-tocopherol	$3.2x10^4$	2.00
"	0.072mM Q_3	$2.3x10^2$	3.70

peroxyl radicals to give semiquinone radical. This radical may react with another peroxyl radical both by disproportionation and by radical combination. In the first case n equals infinity, in the second case n equals 2.

Such a mechanism is compatible with the stoichiometric factor larger than 2 and with the low reduction potential of quinones.

ACKNOWLEDGEMENTS

The work was supported by a grant from M.P.I., Rome, Italy.

REFERENCES

Anbar, M. and Pecht, I., 1964, On the sonochemical formation of hydrogen peroxide in water. Journal of Physical Chemistry, 68, 352-355.

Barclay, L.R.C. and Ingold, K.U., 1981, Autoxidation of biological molecules. 1. The antioxidation activity of vitamin E and related chain-breaking phenolic antioxidants in vitro. Journal of American Chemical Society, 103, 6472-6477.

Boozer, C.E., Hammond, G.S. Hamilton, C.E. and Sen, J.N., 1955, Air oxidation of hydrocarbons. II. The stoichiometry and fate of inhibitors in benzene and chlorobenzene. Journal of American Chemical Society, 77, 3233-3237.

Landi, L., Cabrini, L., Tadolini, B., Sechi, A.M. and Paquali, P., 1985, Incorporation of ubiquinones into lipid vesicles and inhibition of lipid peroxidation. It J. Biochem., 34, 356-363.

Sessa, D.J. Gardner, H.W., Kleiman, R. and Weisleder, D., 1977, Oxygenated fatty acid constituents of soybean phosphatidylcholines. Lipids, 12, 613-619.

Takada, M., Ikenoya, S., Yuzuriha, T. and Katayama, K., 1982, Studies on reduced and oxidized coenzyme Q (ubiquinones). Biochimica et Biophysica Acta, 679, 308-314.

DISTRIBUTION OF COENZYME Q_9 IN HEART MEMBRANES AFTER ISCHEMIA AND REPERFUSION STRESS IN AGED RATS.

C.Muscari, L.Biagetti, C.Stefanelli, C.Finelli, C.Guarnieri, C.M.Caldarera

Department of Biochemistry, University of Bologna, Italy.

INTRODUCTION

The protective effect of coenzyme Q_{10} (CoQ_{10}) against the ischemic myocardial injury has been described in several studies (Nagai et al., 1985; Okamoto et al., 1986), but few data are available on the possible modifications of CoQ levels in the cardiac tissue and membranes in response to ischemic stress (Katagiri et al., 1986). In fact, the change of CoQ content in the senescent heart under ischemia and reperfusion stress has not been reported. Moreover, Beyer et al. (1985) observed that myocardial CoQ level are lower in the aged rats. Since ubiquinone is a physiologically essential compound localized in the electron transfer system of several membranes such as the inner mitochondria or microsomes, it is conceivable that its deficiency might be in part responsible for the loss of myocardial adaptive response to various stress during ageing.

The aim of this study was to evaluate the age-dependent modifications of CoQ_9 concentrations in rat heart as a function of ischemic and reperfusion conditions.

Keywords: coenzyme Q, ageing, ischemia, reperfusion, mitochondria, microsome.

METHODS

Young (3 months) and old (24 months) male Wistar rats, fed "ad libitum", were used for this study.

The hearts were perfused in a retrograde non-recirculating circuit according to the Langendorff tecnique and paced at 340 b.p.m. The perfusion medium was equilibrated with O_2/CO_2 (95:5) at 37°C and contained (in mM): NaCl 115.0; $NaHCO_3$ 25.0, KCl 4.0; KH_2 PO_4 0.9; Mg SO_4 1.1; $CaCl_2$ 1.5;

glucose 11.0. After a 20 min equilibration period a group of hearts was perfused in aerobic conditions for 50 min at a constant coronary flow of 8 ml/min/g tissue; a second group of hearts was exposed to two repeated underperfusion-repersufion cycles in which the coronary flow was first reduced to 0.2 ml/min/g tissue for 10 min and then restored at the initial aerobic flow value for 15 min.

At the end of each perfusion the heart was rinsed in ice--cold saline and then mitochondrial and microsomal membranes were prepared according to the method described by Das et al. (1986). Tissue and membrane CoQ_9 contents were determined by HPLC. About 50-100 mg portion of myocardial tissue was homogenized in 1 ml of double distilled water in a glass homogeniter for 20 sec. A 0.5 ml aliquot of homogenate was extracted twice for 30 min with 7 vol of N-hexane/ethyl alcohol (5:2, v/v). The extraction of CoQ_9 from the membranes was carried out by adding 10 µl of 1 N HCl to 100 µl (25 mg protein/ml) of the membrane suspension followed by two additions of 2 ml of the above mentioned solvent mixture. The organic phases were evaporated under a flux of N_2. The residues were dissolved in 100 µl of isopropylalcohol and 25 µl were injected into an HPLC apparatus (Waters Associates). The separation was carried out with a 3 µm ultrasphere ODS column (75x4.6 mm I.D.) eluted with methanol/acetonitrile/isopropanol (45:30:25) at a flow rate of 1.8 ml/min. CoQ_9 was detected at 275 nm.

The proteins were estimated by the method of Bradford et al. (1976) using bovine serum albumin as a standard.

RESULTS AND CONCLUSIONS

Table 1 shows that there are not significative modifications in cardiac tissue and membrane CoQ_9 levels of old rats in comparison to young rats after 50 min of aerobic perfusion. On the contrary, when the hearts were isolated and perfused with two repeated cycles of ischemia and reperfusion, tissue and mitochondrial concentration of CoQ_9 were markedly reduced in both age-groups tested, with a more pronounced effect in the young rats. In contrast to mitochondria, we observed that the content of CoQ_9 in the microsomes did not change after the ischemia-reperfusion stress, irrespective to the age of the animals (Table 1).

These data indicate that in the isolated rat heart perfused in ischemia conditions there is a decrease in CoQ_9 content at the mitochondrial but not at the microsomal level. Moreover, this diminution is more pronounced in the young rats in comparison to the old rats.

Since no CoQ_9 was detected in coronary effluents of ischemic and reperfused rat hearts (data not shown), we support the notion that the tissue CoQ_9 deficiency observed in these conditions could be attributable to an alterated synthesis or degradation rate of CoQ_9. In fact, it has been recently found that the rat heart can synthetize ubiquinone from precursors of both the quinone ring and the polyisoprenoid side chain and that under anaerobic conditions, they are not completely generated (Yamamoto et al., 1989).

Table 1. CoQ_9 content (nmols/mg prot) of isolated rat heart perfused in aerobic or ischemic/reperfusion conditions.

Age (months)		Aerobic perfusion	Ischemia + Reperfusion
3	Tissue	1.85 ± 0.21	1.07 ± 0.09
	Mitochondria	3.26 ± 0.38	0.91 ± 0.05
	Microsomes	2.14 ± 0.26	2.88 ± 0.67
24	Tissue	1.77 ± 0.14	1.24 ± 0.16
	Mitochondria	3.11 ± 0.27	1.99 ± 0.5
	Microsomes	1.86 ± 0.13	2.15 ± 0.36

Values are expressed as mean ± S.E. of 5 separate experiments.

REFERENCES

Beyer, R.E., Burnett, B., Cartwright, K.S., Edington, W., Falzon, M.J., Kreitman, K.R., Kuhn, T.W., Ramp, B.J., Yung Shik Rhee, S., Rosenwasser, M.J., Stein, M. and Chin I An, L., 1985, Tissue coenzyme Q (ubiquinone) and protein concentrations over the life span of the laboratory rat. *Mechanisms of Aging and Development*, 32, 267-281.

Bradford, M.A., 1976, A rapid and sensitive method of the quantitation of microgram quantities of protein utilizing the principle of protein dye binding. *Analitical Biochemistry*, 72, 248-254.

Das, D.K., Engelman, R.M., Ronson, J.A., Breyer, R.H., Otani, H. and Lemeshow, S., 1986, Role of membrane phospholipids in myocardial injury induced by ischemia and reperfusion. *American Journal of Physiology*, 251, H71-H79.

Katagiri, T., Konno, N., Yanagishita, T., Tanno, F., Geshi, E., Kitsu, T., Akiyama, K. and Niitani, H., 1986, Protection of ischemic myocardial injury by coenzyme-Q_{10}-mechanism of action. In *Biomedical and Clinical Aspects of*

Coenzyme Q, vol.5, edited by K.Folkers and Y.Yamamura (Amsterdam: Elsevier Science Publishers), pp. 167-177.

Nagai, S., Miyazaki, Y., Ogawa, K., Satake, T., Sugiyama,S. and Ozawa, T., 1985, The effect of coenzyme Q_{10} on reperfusion injury in canine myocardium. *Journal of Molecular and Cellular Cardiology*, 17, 873-884.

Yamamoto, T., Shimizu, S., Sujawara, H., Momose, K. and Rudney, H., 1989, Identification of regulatory sites in the synthesis of ubiquinone in the perfused rat heart. *Archives of Biochemistry and Biophysics*, 269, 86-92.

Okamoto, F., Allen, B.S., Buckberg, G.D., Leaf, J. and Bugyi, H., 1986, Studies of controlled reperfusion after ischemia. X.Reperfusate composition: supplemental role of intravenous and intracoronary coenzyme Q_{10} in avoiding reperfusion damage. *Journal of Thoracic and Cardiovascular Surgery*, 92, 573-582.

DO NON-MITOCHONDRIAL ELECTRON ACCEPTORS INTERFERE IN MITOCHONDRIAL ELECTRON TRANSPORT AT THE LEVEL OF REDOX-CYCLING UBIQUINONE?

Nohl, H.

Institute of Pharmacology and Toxicology, Veterinary University of Vienna, Linke Bahngasse 11, A-1030 Vienna

INTRODUCTION

Dioxygen has long been suggested to operate as an electron acceptor of the respiratory chain at sites different from cytochrome oxidase.

This was concluded from the release of univalently reduced dioxygen ($O_2^{\cdot}$) and its dismutation product H_2O_2 under conditions where respiratory components between the antimycin A (AA) and rotenone block exhibited highest states of reduction. (Loschen et al, 1973; Nohl and Hegner, 1978; Nohl and Jordan, 1986).

One of the candidates most frequently reported to interact with dioxygen was redox-cycling ubiquinone (UQ), e.g. Trumpower and Simmons (1979) have suggested ubisemiquinone at center out ($SQ_o^{\cdot}$) to autoxidize in the presence of thenoyltrifluoroacetate (TTFA) and AA (SQ_o was not measured).

Boveris and Chance (1973) have concluded from extraction/reconstitution experiments that H_2O_2 formation depends on the presence of UQ. Turrens et al (1985) have reported that myxothiazole which prevents $SQ^{\cdot}$ formation at center out also inhibits generation of $O_2^{\cdot}$ radicals. From this observation they concluded on the involvement of this $SQ^{\cdot}$-species in the univalent reduction of molecular oxygen.

Since inhibition of $SQ_o^{\cdot}$ formation also prevents reduction of other possible candidates at the oxidant site of redox-cycling UQ it was unclear whether or not this e^- carrier may serve as an one-e^- donor for dioxygen. Thermodynamic considerations based on the one-e^- redox-potential of UQ and oxygen exclude the feasibility of such a reaction when assuming redox cycling UQ to operate in an aprotic environment.

Keywords: ubisemiquinone, superoxide, hydrogen peroxide, hydroxyl-radicals, cytochrome b_{566}, EPR.

RESULTS

We have proved this hypothesis experimentally by a chemically defined reaction system, where UQ-o-was dissolved in water-free acetonitrile together with an $O_2^{\overline{\cdot}}$-generator (KO_2 in crown ether) (see Fig 1).

Figure 1. Demonstration of the requirement of protons for ubisemiquinone to undergo autoxidation.

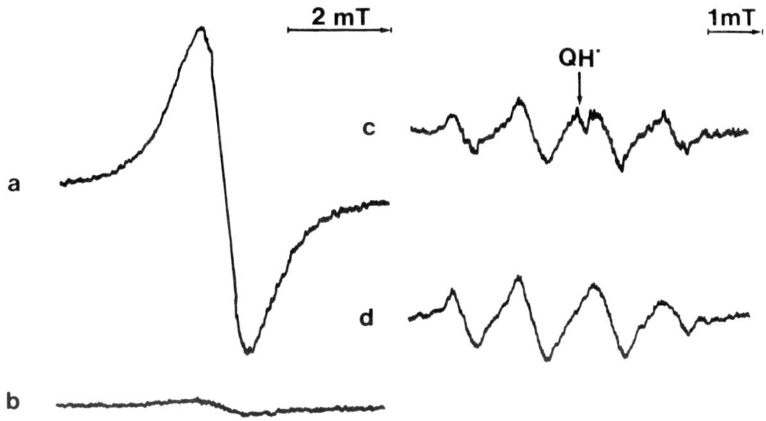

Final concentrations; UQ-o: 6 mM; 10 mM; crown ether: 20 mM. Spectra c and d are DMPO (5,5' dimethyl-1-pyrroline-N-oxide: 150 mM) spin adducts with $O_2^{\overline{\cdot}}$ from KO_2. EPR-measurements were carried out with a Bruker ER 414 s. Microwave power: 12 mW; modulation amplitude: 0.1 mT (a,b), 0.04 mT (c,d).

Despite the presence of air-oxygen the EPR-detectable SQ˙ signal remained stabile (Fig 1a) until the reaction systems was protonated by the addition of small amounts of water. In that case SQ˙-related EPR signals disappeared (Fig 1b) and simultaneously $O_2^{\overline{\cdot}}$ formation was seen (Fig 1c). The radicals were detected by spin-trapping with DMPO. The respective 4-line spectrum obtained was identified by a model reaction in which $O_2^{\bar{\circ}}$ radicals in the same reaction medium were formed from KO_2 in the presence of DMPO (Fig 1d).

From these experiments it may be concluded that redox-cycling UQ does not establish a redox couple with molecular oxygen as long as UQ operates as an e carrier in an aprotic environment of the inner membrane.

Figure 2. Correlation between steady state concentrations of mitochondrial ubisemiquinones and OH• formation in the presence of H_2O_2.

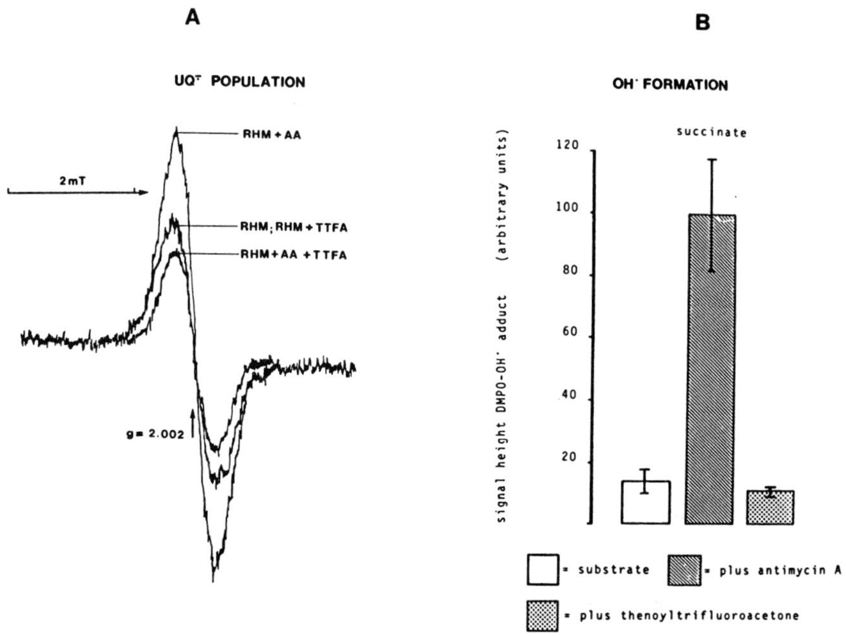

15 mg mitochondrial protein was suspended in 340 µl isolation buffer (125 mM KCl; 50 mM HEPES, pH 7.4) at room temperature and ubisemiquinone (UQ⁻)-related EPR spectra were recorded in the absence and presence of AA (15 µg); and TTFA (4.5 mM). Microwave power was 45 mW (A). B. OH• formation was inferred from EPR related DMPO-OH• spin adducts. (200 mM DMPO) in the presence of 0.15 mM H_2O_2 and 6 mM succinate as substrate for respiration. Inhibitor concentrations were as in A.

Fig 2 demonstrates the existence of a correlation between EPR-signal hights of e⁻ transferring SQ˙ and the amount of OH˙-formed when H_2O_2 was added.

H_2O_2 was therefore suggested to undergo a reductive homolytic cleavage after having accepted the unpaired e⁻ of redox cycling SQ˙ in this systems, according to the following equation:

$$SQ^{\bullet} + H_2O_2 \longrightarrow Q + OH^{\bullet} + OH^{-}$$

Figure 3. Influence of H_2O_2 on the redox state of cyt b_{566} in the presence of AA.

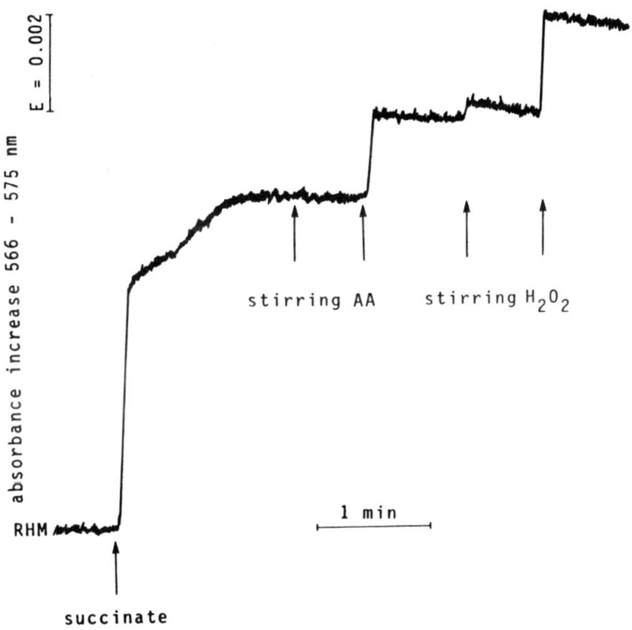

1.5 mg/ml rat heart-mitochondria were suspended in the reaction medium of Fig 2 at room temperature. Additions: 6 mM succinate; 0.15 mM H_2O_2. 1.5 µg AA.

A possible reaction of redox-cycling SQ˙ with H_2O_2 was further tested indirectly by following changes in the reduction state of cyt b_{566} (Fig 3).

This experimental concept was derived from the observation of Wikström and Berden (1972) that in AA-inhibited respiration, redox-changes of b-type cytochromes are governed by the ubiquinol/ubisemiquinone redox couple. Oxidants of SQ˙ may increase AA-sensitive oxidation of SQ˙ to UQ and consequently an extra-reduction of b-type cyt will occur under these conditions.

The extra-reduction of cyt b observed after the addition of H_2O_2 strongly supports the possibility of H_2O_2 to operate as an oxidant of redox-cycling SQ˙.

In contrast to dioxygen, data of the literature do not contradict the feasibility of such a reaction. Furthermore, there is indirect evidence that OH˙-formation from mitochondrial H_2O_2 may also occur in the living cell (Nohl et al 1982).

REFERENCES

Boveris, A. and Chance, B., 1973, The cellular production of hydrogen peroxide. Biochem. J., 128, 617-630.

Loschen, G., Azzi, A. and Flohe, L., 1973, Mitochondrial H_2O_2 formation at site II. Hoppe Seyler' Z. Physiol. Chem., 354, 791-794.

Nohl, H. and Hegner, D., 1978, Do mitochondria produce oxygen radicals *in vivo*? Eur. J. Biochem., 82, 563-567.

Nohl, H. and Jordan, W., 1986, The mitochondrial site of superoxide formation. Biochem. Biophys. Res. Com., 138, 533-539.

Nohl, H., Jordan, W., and Hegner, D., 1982, Mitochondrial formation of hydroxyl radicals by an ubisemiquinone-dependent reaction. An alternative pathway of the iron-catalysed Haber-Weiss cycle. Hoppe Seyler' Z. Physiol. Chem., 363, 599-607.

Trumpower, B.L., and Simmons, Z., Diminished inhibition of mitochondrial electron transfer from succinate to cytochrome c by thenoyltrifluoroacetone induced by antimycin. J. Biol. Chem., 254. 4608-4616.

Turrens, J.F., Alexandre, A., and Lehninger, A.L. 1985, Ubisemiquinone is the electron donor for superoxide formation by complex III of heart mitochondria. Arch. Biochem. Biophys. 237, 408-414.

Wikström, M.K.F., and Berden, J.A., 1972, Oxidoreduction of cytochrome b in the presence of antimycin, Biochim. Biophys. Acta, 283, 403-420.

CONCLUSIONS

Our investigation suggest that H_2O_2, which is a byproduct of respiration, can serve as an e^--acceptor of redox-cycling UQ.

Whether or not molecular oxygen will accept e^- from UQ seems to depend on the availability of protons at domains, where $SQ^{\bullet}$ undergo one-e^- redox-cycling.

CARDIOTOXICITY CAUSED BY ANTRACYCLINES AND/OR RADIOTHERAPY IN PATIENTS AFFECTED BY SMALL CELL LUNG CANCER.

RAVENNA F., MELONCELLI U., CASTELLETTI M.L., TURBA E.*, ABBATI A.*, CIACCIA A.
Cattedra di Mal. dell'App. Resp. Università di Ferrara - Divisione di Pneumologia, Ospedale Enzo Boeri USL34, P.le Forlanini, 44039 Tresigallo - Ferrara . Direttore: Prof. A. Ciaccia.
* Ser. di Medicina Nucleare, Ospedale Maggiore, Bologna USL27. Primario: Prof. A. Abbati.

Key words: Angiocardioscintigraphy, antracyclines cardiotoxicity, Ubiquinone, L- carnitine.

INTRODUCTION

The possible cardiotoxic effects of some antiproliferative drugs are known and among these in the last few years Doxorubicin and Epirubicin have attracted more attention. In the same way the myocardiotoxic effects of radiotherapy, when the cardiac muscle is included in the irradiation field, are well known. The treatment of small cell lung cancer includes both polychemotherapy (PCT) and radiotherapy; thus it is necessary to select rigorously the patients, to oversee with adequate diagnostic assessments and, if it is possible, to direct therapies that prevent or reduce the incidence of collateral effects.

The aim of this study was monitoring the left cardiac function with angiocardioscintigraphy in patients treated with PCT to whom supportive care was administered with L-carnitine 2-3 g/day and Ubiquinone 60 - 100 mg/day.

METHOD

This study was performed on 15 patients affected by small cell lung cancer (SCLC) in both limited (12 patients) or extensive stages (3 patients). All potential subjects had to have an adequate performance status (score > 50 as Karnofsky and Burchenal), a normal bone marrow, renal and hepatic function. Patients with active cardiac diseases, symptomatic valvular heart diseses, congestive heart failure, myocardial infarction occuring within 24 months and active angina pectoris were exluded.

After initial staging (Standard Chest x-ray, Computer Tomography, Bronchofibrescopy, total body Scintigraphy), the cardiac functionality of all patients admitted to the therapeutic protocol was studied with accurate visits, electrocardiogram and angiocardioscintigraphy with Tc99m labeled albumin.

The patients were treated with chemiotherapy with Doxorubicin, Vincristine and Cyclophosphamide or with Epirubicin, Vincristine and Cyclophosphamide. Eigth patients were also treated with pulmonary and mediastinic radiotherapy.

The prevention of both acute and chronic cardiotoxicity was made with: -- ECG before and after the chemiotherapy; -- angiocardioscintigraphy with or without baseline isometric effort test, after 3 cycles of therapy, after 6 cycles and after 12 months; -- PCT interruption if Ejection Fraction (EF) decrease is more than 10% as regards to baseline values; -- supportive care all time long exposition to cardiotoxic drugs or to radiotherapy with ubiquinone (60 - 100 mg/day) and L-carnitine (2-3 g/day); -- limitation of total dose for Doxorubicin to 500 mg per square metre and for Epirubicin to 900 mg per square metre.

RESULTS

Fifteen male patients with a mean age of 64 years (range 55 - 74 years; standard deviation 6 years) were admitted to the study. The average follow up was of 11 months (range 6-8 months; standard deviation 3.4 months).

In fig. 1 average values of the baseline EF after 3 and 6 cycles of therapy are reported. It is possible to observe that the mean values with standard deviation are always higher than the lowest normal value and that there is a very light trend to the increase of EF after PCT (paradox effect connected with the smaller number of observations than the baseline data).

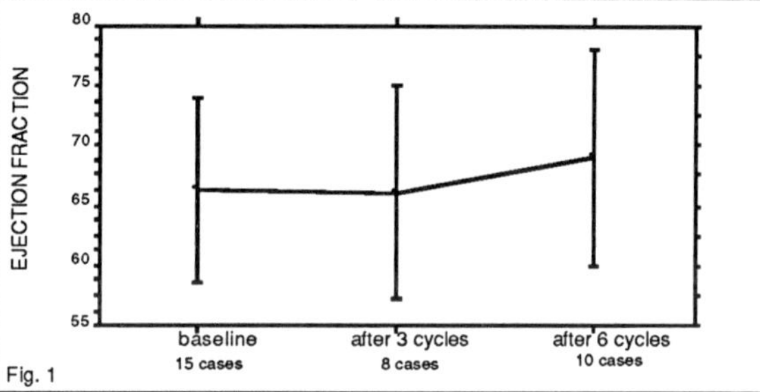

Fig. 1

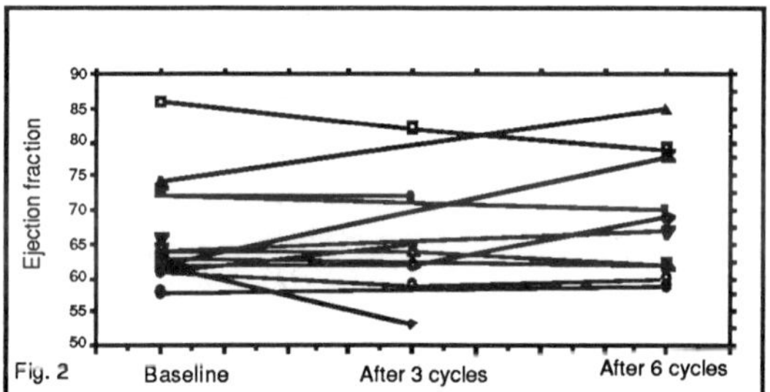

Fig. 2

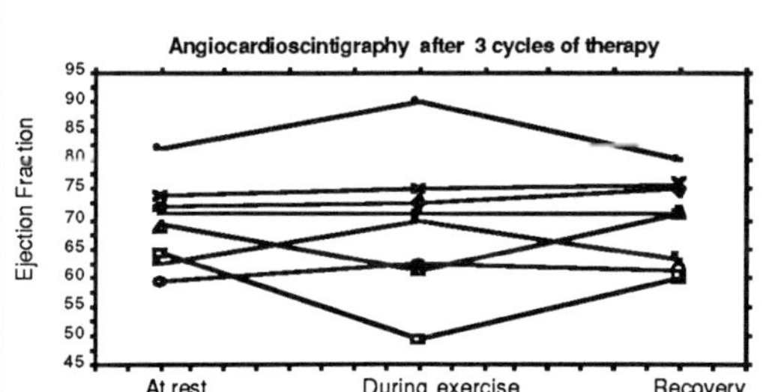

Fig. 3

In fig. 2 the graphic schematization is reported with line chart of each single patient EF relative to the first three angiocardiographic tests of protocol (baseline, after 3 cycles and after 6 cycles of therapy). As it is possible to note, in one case only, there is a significant decrase of EF, after 3 cycles of therapy with a final value lower than 55%. Moreover the parametric image showed a decrease in anterior and septal proximal areas. Finally the Thallium 201 myocardial perfusion scintigraphy showed a decreased uptake of inferior and anteroseptal proximal regions. We believed useful to change the polychemiotherapeutic protocol to this patient, with a precautional intent also without clinic manifestation.

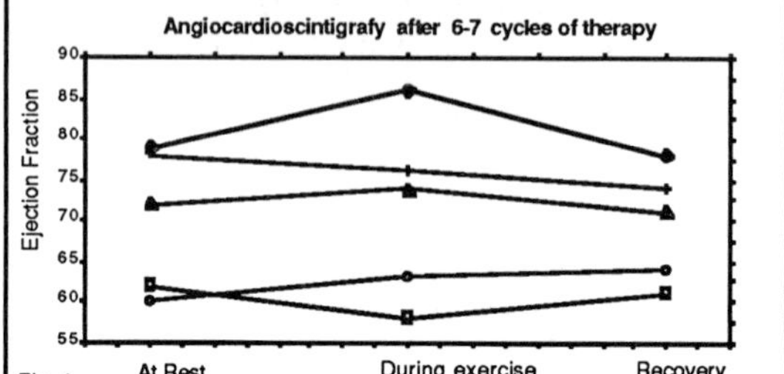

Fig. 4

In 8 patients the EF was evaluated with an isometric exercise after 3 cycles of PCT (Fig. 3) and in 5 patients after 6-7 cycles of PCT (Fig. 4). In two cases a decrese of EF during the isometric effort was proved. The protocol was broken; at the following control (after other three cycles of PCT) the EF did not change significantly at rest; in only one patient the effort test was repeated with an unchanged result.

Finally in 4 patients baseline data have been compared with those obtained after radiotherapy (48 Gy). The EF values are reported in fig. 5.

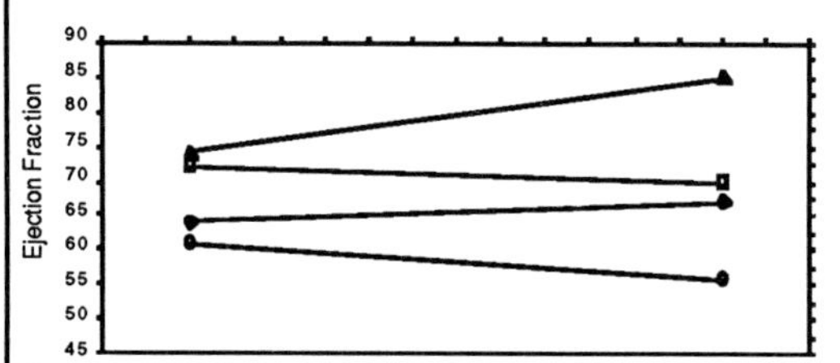

Fig. 5

Significant changes of EF and thus of cardiac function have not been demonstrated.

Only three noteworthy, but readily controlled with medical theraphy, episodes of acute toxicity caused by antracyclines occured: two attacks of ventricular tachycardia resolved with verapamil and one episode of ventricular extrasystoles treated and resolved in another hospital.

CONCLUSIONS

The low incidence of cardiotoxic effects after PCT including Antracyclines and/or radiotherapy, which are well known cardiotoxic therapies, shows the utility of the angiocardioscintigraphy above all in selecting the patients before the beginning of PCT. The angiocardioscintigraphy may be less useful in the follow up because the treatment has been stopped for the reduction of ejection fraction in one case only.

The angioscintigraphy which is a bloodless and easily practicable test, is surely suitable in selecting patients affected by SCLC, who must be treated with PCT including Antracyclines and/or radiotherapy.

Moreover supportive care with ubiquinone and L-carnitine can have played an important role.

Nevertheless the most important prevention of chronic cardiotoxic effects is represented by not overcoming the allowable highest dose of antracyclines and moreover by administration of these drugs only to selected patients. At this point in order to increase the safety, it is possible to prescribe a supportive care, to organize and program a monitoring protocol of the function of the target organs and in this specific case of the cardiac function.

In fact the absence of a control group of patients cannot allow us to demonstrate that the maintenance therapy with Ubiquinone and L-carnitene has influenced the incidence and seriousness of cardiotoxic effects of antracyclines. Nevertheless we can affirm that in the studied cases the incidence of undesidered cardiac effects is lower than those reported by literature.

REFERENCES

Azuma J.,Spereiakis N., Hasegawa H., 1981, Doxorubicin cardiotoxicity: possible pathogenic mechanisms. Journal of molecular and cellular cardiology 13,381.

Bachur N.R.,Gordon S.L., Gee M.V., 1978, A general mechanism for microsomial activation of quinone anticancer agents to free radicals, Cancer Research 38, 1745.

Bonadonna G., 1984, Advances in Anthracycline chemotherapy: Epirubicin. Masson Italia Editori, Milano.

Cortes E.P.,Lutman G., 1975, Doxorubicin cardiotoxicity: a clinical pathologic correlation. Cancer Chemotherapy Report 6,215.

Ghione M., 1978, Cardiotoxic effects of antitumors agents, Cancer Chemoterapy Pharmacology,1, 25.

Isner J.M. , Ferrans V.J., Cohen S.R., Witkind B.G., Virmani R., Gottdiener J.S., Beck G.R., Roberts W.C., 1983, Clinical and morphologic cardiac findings after Anthracycline chemoterapy. American Journal of Cardiology, 51,1167.

Littarru G.P., 1988, Metabolismo energetico del cuore. Edizioni Minerva Medica Torino.

McKillop J.H., Bristow M.R., Goris M.L., Billingham M.E., 1983, Sensitivity and specificity of radionuclide ejection fractions in doxorubicin cardiotoxicity. American Heart Journal, 106, 1048.

Morgan G.W., McIlveen B.M., Freedman A., 1981, Radionuclide ejection fraction in Doxorubicin cardiotoxicity. Cancer Treatment Reports, 65, 629.

Neri B., Cini-Neri G., Bandinelli M., Pacini P., 1988, Doxorubicin and epirubicin cardiotoxicity: experimental and clinical aspects. In: Progress in cancer chemotherapy. Edizioni Medico-Scientifiche, Pavia, edit by Robustelli della Cuna G. and Nagel G.A., p.9.

Judy W.V., Hall J.M., Dugan W., Toth P.D., Folkers K., 1984, Coenzimo Q10 reduction of Adriamycin cardiotoxicity. In: Biomedical and clinical aspects of Coenzime Q10. Amsterdam, edit by Folkers K. and Yamamura Y., Elsevier p.231.

CLINICAL USE OF COENZYME Q_{10} IN ESSENTIAL ARTERIAL HYPERTENSION

V. Digiesi, F. Cantini*, B. Brodbeck

Third Institute of Clinical Medicine and Medical Therapy, Chair of Systematic Medical Therapy, Faculty of Medicine, University of Firenze (Italy).
*Second Division of Internal Medicine, General Hospital of Prato (Italy).

Keywords: Coenzyme Q_{10}, essential arterial hypertension, diastolic blood pressure, systolic blood pressure

Several experimental studies on animals with induced arterial hypertension (Igarashi et al., 1972; Yamagami et al., 1974) and on patients with essential arterial hypertension (Yamagami et al., 1975, 1976, 1986; Hata et al., 1977; Folkers et al., 1981; Richardson et al., 1980) report a beneficial therapeutic effect of Coenzyme Q_{10} (CoQ) upon systolic blood pressure (SBP) and diastolic blood pressure (DBP).

It should be noted that a reduction of enzymatic activity of CoQ has been observed in several hypertensive patients, as compared to control patients (Iwamoto et al., 1974), and that the antihypertensive effects of the drug are correlated to the augmentation of serumal levels of CoQ (Yamagami et al., 1986).

The mechanism of the antihypertensive effect of CoQ is fewly known; perhaps it is related to an antialdosteronic effect or to the reduction of the peripheral resistance, as observed by Richardson et al.,1980, and Kumagai et al.,1970

MATERIAL AND METHODS

A cross-over, placebo controlled, randomized study was designed to evaluate the antihypertensive efficacy of CoQ in patients with essential arterial hypertension in stages I and II as defined by the World Health Organization, lacking other significant diseases. Furthermore patients with age above 70 years, or with renal failure, or with body weight exceeding 90 kg have been excluded from the trial.

Eighteen subjects were selected, among those with essential arterial hypertension observed, four females and fourteen males, averaging 55.9 of age (ranging from 42 to 66 years).

At the moment of entry into the trial all treatment with drugs was suspended. After a washout period of two weeks the eighteen patients were randomly assigned to either the group A, treated with CoQ (100 mg daily, orally), or to the group B, treated with placebo. The clinical observation lasted for ten weeks; at the end of this first period the treatment was suspended for two weeks, and then each patient of group A crossed over to placebo treatment and each patient of group B to CoQ treatment (at the same dosage) for an additional ten weeks.

At the beginning of the trial basic haematologic and urinary examinations regarding blood crasis, renal and liver function, blood lipid composition and serum electrolytes were carried out. The control examinations effected at the end of the trial did not demonstrate statistically significant variations.

The arterial pressure readings were carried out with a Riva-Rocci sphygmomanometer at the same time in the morning, by the same examiner and on the same arm, after the subject had passed ten minutes reclining in a position of complete rest. The definitive pressure reading was obtained by calculating the mean of five successive readings at one minute intervals. For the statistical evaluation of data the Student's t test for dependent samples was utilized. $P < 0.05$ has been accepted as statistically significant value.

RESULTS

The initial values of blood pressure, expressed as means $\pm$ standard error, were 166.6 $\pm$ 2.6 mmHg for SBP and 102.9 $\pm$ 1.2 mmHg for DBP. After treatment with CoQ the blood pressure, either systolic than diastolic, decreased significantly in 17 of 18 patients. At the end of the trial the values of blood pressure, expressed as means $\pm$ standard error, resulted to be 156 $\pm$ 2.25 mmHg for SBP ($P < 0.001$) and 95.2 $\pm$ 1.04 mmHg for DBP ($P < 0.001$). Therefore the decrease is statistically significant compared to the initial values. Placebo treatment did not offer significant pressure modification in any patient. The mean values of SBP resulted to be 166.3 $\pm$ 2.4 mmHg (P: n.s.) and those of DBP 103.3 $\pm$ 1.07 mmHg (P: n.s.) without any statistically significant variation compared to the initial values.

Al last the results of CoQ treatment have been compared with those of placebo treatment: the differences of pressure values of SBP (156 ± 2.25 mmHg) and of DBP (95.2 ± 1.04 mmHg) after CoQ treatment and those after placebo treatment (SBP 166.3 ± 2.4 mmHg; DBP 103.3 ± 1.07 mmHg) resulted to be statistically significant ($P < 0.001$ for SBP and for DBP).

COENZYME Q_{10} IN ESSENTIAL ARTERIAL HYPERTENSION: EFFECTS ON SYSTOLIC AND DIASTOLIC BLOOD PRESSURE

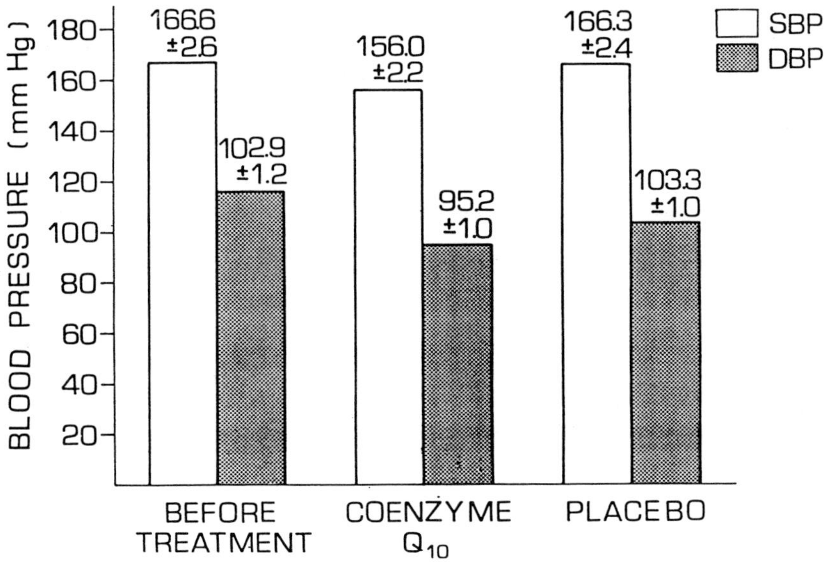

It should be noted that the antihypertensive effect of CoQ was observed during the third and fourth week of treatment, remained constant for the entire duration of the trial and stopped 7-10 days after suspending the drug.

Treatment was optimally tolerated and none of the patients referred significant disturbances to require suspension of the drug.

CONCLUSIONS

The results of this controlled clinical study confirm the therapeutic utility of CoQ in patients with essential arterial hypertension: the drug induced a significant decrease of both systolic and diastolic blood pressure. The small number of patients treated during the trial does not allow definitive indications and we suggest multicenter trial.

REFERENCES

Folkers K., Drzewoski J., Richardson P., 1981, Bioenergetics in clinical medicine.XVI. Reduction of hypertension in patients by therapy with coenzyme Q_{10}. Research Communications in Chemical Pathology and Pharmacology, 13 (1), 129-140.

Hata T., Kunida H., Oyama Y., 1977, Antihypertensive effects of coenzyme Q_{10} in essential hypertension. Clinical Endocrinology, 25, 1019-1023.

Igarashi T., Tanake Y., Nakajima Y., et al., 1972, Effect of coenzyme Q_{10} on experimental hypertension in the desoxycorticosterone acetate-saline loaded rats. Folia Pharmacologica Japonica, 68, 460-472.

Iwamoto Y., Yamaguchi T., Folkers K. et al., 1974, Deficiency of coenzyme Q_{10} in hypertensive rats and reduction of deficiency by treatment with coenzyme Q_{10}. Biochemical and Biophysical Research Communications, 58, 734-748.

Kumagai A., Mishino K., Kinuu T. et al., 1970, Inhibitory effect of ubiquinone on biosynthesis of aldosterone in rat adrenal in vitro. Endocrinologia Japonica 17, 143.

Richardson P., Baker L., Folkers K. et al., 1980, Clinical studies on coenzyme Q_{10} for treatment of cardiovascular disease.In: Yamamura Y., Folkers K., Ito Y.et al.,(eds.), 1980, Biomedical and clinical aspects of coenzyme Q, vol.2, Amsterdam Elsevier/North Holland Biomedical Press, pp. 301-311.

Yamagami T., Iwamoto Y., Folkers K.,et al.,1974, Reduction by coenzyme Q_{10} of hypertension induced by deoxycortocosterone and saline in rats. International Journal for Vitamin and Nutrition Research. 4, 487-496.

Yamagami T., Shibata N., Folkers K., 1975, Bioenergetics in clinical medicine. Studies on coenzyme Q_{10} and essential hypertension. Research Communications in Chemical Pathology and Pharmacology, 11, 273-288.

Yamagami T., Shibata N., Folkers K., 1976, Bioenergetics in clinical medicine. VIII. Administration of coenzyme Q_{10} to patients with essential hypertension. Research Communications in Chemical Pathology and Pharmacology, 14 (4), 721-728.

Yamagami T., Takagi M., Akagami H.et al., 1986, Effect of coenzyme Q_{10} on essential hypertension: a double blind controlled study. In: Folkers K., Yamamura Y. (eds.), Biomedical and clinical aspects of coenzyme Q, vol.5, Amsterdam, Elsevier Science Publications BV, pp. 337-343.

DOES UBIQUINONE HAVE ANY INFLUENCE ON STRUCTURAL/FUNCTIONAL MITOCHONDRIAL MEMBRANE PROPERTIES DURING γ-IRRADIATION?

E. Ferri*, M. Battino°, E. Gattavecchia*, V. Barzanti** and G. Lenaz§.

*Istituto di Scienze Chimiche, Via S. Donato, 15 - Bologna
°Istituto di Biochimica, Via Monte D'Ago - Ancona
**Dipartimento di Biochimica, Via Irnerio, 48 - Bologna
§Dip. Biologia Ev. Sper., Via Irnerio, 42 - Bologna
ITALY

The involvement of quinones in energy transducing processes in the inner mitochondrial membrane, chloroplasts and bacterial membranes is only the best known activity of these extremely hydrophobic molecules. In eukaryotic cells ubiquinone is also present in the membranes of the Golgi apparatus, like a component of non-phosphorylating redox chains, and in the plasma membrane; its wide distribution associated with the peculiar redox properties suggested that ubiquinone can be involved, in different ways, in other pathways, like free radicals producing or scavenging processes.

At the beginning its capability to produce free radicals was considered, owing to the possibility, for the quinone ring, to undergo a reversible monoelectronic reduction to a radical molecule, i.e. $CoQ^{\cdot -}$ or $CoQH^{\cdot}$. These semiquinone forms, and also the quinol one, were postulated as the major responsible for the formation of the superoxide radical during oxidative phosphorylation or in some metabolic conditions (reoxygenation after ischemic attacks, activation of other NADH oxidase systems in human neutrophils (Beyer, 1989)). However the lack of direct evidences that could allow to support this hypothesis, the acquirements of data suggesting an important role of other components like cytochromes (Nohl, 1986) or flavoproteins in this oxy-radicals production and, on the other hand, the demonstrations that ubiquinone acts as an effective quencher of singlet oxygen, led to change the opinion on the effect of ubiquinone on free radicals equilibrium and to seriously evaluate its antioxidant properties.

Evidences supporting the effectiveness of ubiquinone to prevent the lipid peroxidation, by reaction with oxygen or lipid free radicals, are now large and convincing.

In vitro experiments have been performed with model systems, mitochondria and submitochondrial particles where lipid peroxidation was induced in different manners (ultrasonic irradiation, NAD(P)H ADP-Fe+++ chelate addition, etc.) and in all cases it was possible to demonstrate a clear antioxidant effect of ubiquinone, particularly in the reduced form (Cabrini et al., 1986; Minakami et al., 1982; Narabayashi et al., 1982).

In general the treatment of a biological system with ionizing radiations provokes damages due to direct ionization of macromolecules or, more frequently, to successive effects of free radicals produced, for example, by water radiolysis; in the case of membranes the result is principally a lipid peroxidation.

Our goal was to study the possible preventing effect of ubiquinone on the lipid peroxidation induced by γ-irradiation and to elucidate the effect of this treatment on the catalytic properties of the mitochondrial respiratory chain enzymes.

Keywords: lipid peroxidation, γ-irradiation, gas-chromatography, mitochondria, HPLC

METHODS

Beef heart mitochondria were prepared according to Blair 1967. The irradiation was performed in a γ-cell (^{60}Co) (Canadian atomic energy, Ltd.) to reach an irradiation dose of 8 or 15 Gray. The enzymatic activities were performed by standard spectrophotometric methods, under quasi-saturating substrate concentrations in a dual-wavelength SIGMA ZWS spectrophotometer; the mitochondria fatty acid compositions were accomplished by the gas-chromatographic analysis of the fatty acid methyl esters using a Varian 3700 with double column and a flame ionization detector.
Ubiquinone contents were evaluated by HPLC measurements

after extraction according to Tsai et al., 1985 and Kröger, 1978. The CoQ-depleted mitochondria were prepared according to Szarkowska, 1966 .

RESULTS AND DISCUSSION

The enzymatic activities considered in order to evaluate possible damages to respiratory complexes were: NADH-cyt. c reductase, ubiquinol 2-cyt. c reductase and cytochrome oxidase. The mitochondria were irradiated either without any treatment, or depleted of the endogenous CoQ_{10} by pentane extraction, or reintegrated with exogenous CoQ_6. In all cases the values of maximum velocity obtained for each couple of treated and untreated samples were quite the same, even at the higher dose of irradiation (15 Gray).

The gas-chromatographic analysis of the fatty acid composition revealed that there were no significant differences between controls and treated samples with the only exception of extracted samples where the irradiation induced a dramatic decrease of polyunsaturated fatty acids, whilst a partial protection has been found after CoQ_6 reintegration (see Table I).

Table I. Polyunsaturated fatty acid composition expressed as percentage of the total amount of fatty acids.

Fatty Acid	C	C*	Extr.	Extr*	Extr+Q_6^*
18:2	26.0	26.0	25.9	4.5	15.8
20:4	9.9	9.0	9.3	1.1	6.0
20:5	1.7	1.2	1.2	trace	0.5
22:5	0.48	0.2	0.28	trace	trace

C: control sample; C*: irradiated sample; Extr: Q-depleted sample; Extr*: irradiated Q-depleted sample; Extr+Q_6^*: extracted sample reintegrated with CoQ_6 and Irradiated.

The amounts of ubiquinone detected by HPLC measurements in the samples after irradiation were exactly the same as those contained in the respective controls.

These results might confirm that CoQ is essential for protection of membrane polyunsaturated fatty acids from peroxidative damage caused by free radicals, in this case generated externally during irradiation. In spite of the peroxidative damage the enzymatic activities were generally unaffected or rather they were higher in the treated samples. Finally the unchanged levels of ubiquinone after irradiation could indicate that, while carrying out its antioxidant activity, it does not suffer alterations in the molecular structure, as suggested in previous works (Narabayashi et al., 1982).

REFERENCES

Beyer R.E., 1989, The role of coenzyme Q in free radical formation and antioxidation. In CRC Handbook of free radicals and antioxidants in biomedicine, Vol. 2, edited by J. Miquel et al. (Boca Raton: CRC Press), pp. 45-62.

Cabrini L., Pasquali P., Tadolini B., Sechi A.M. and Landi L., 1986, Antioxidant behaviour of ubiquinone and ß-carotene incorporated in model membranes. Free Radical Research Communications, 2, 85-92.

Kröger A., 1978, Determination of contents and redox states of ubiquinone and menaquinone. Methods in Enzymology, 53, 579-591.

Minakami S., Takeshige K., Takayanagi R. and Narabayashi H., 1982, Reduced coenzyme-linked lipid peroxidation in SMP. In Oxigenases and oxigen metabolism, edited by M. Nozaki et al. (New York: Academic Press), pp. 593-603.

Narabayashi H., Takeshige K. and Minakami S., 1982, Alteration of inner membrane components and damage to electron transfer activities of bovine heart SMP induced by NADPH-dependent lipid peroxidation. Biochemical Journal, 202, 97-105.

Nohl H. and Jordan W., 1986, The mitochondrial site of superoxide formation. Biochemical and Biophysical Research Communications, 138, 533-539.

Szarkowska L., 1966, The restoration of DPNH oxidase activity by coenzyme Q (ubiquinone). Archives of Biochemistry and biophysics, 113, 519-525.

Tsai A.L., Kauten R. and Palmer G., 1985, Redox changes in coenzyme Q.... Analytical Biochemistry, 151, 131-136.

SKELETAL MUSCLE AND BLOOD CoQ_{10} IN HEALTH AND DISEASE

Jan Karlsson[*)], Bertil Diamant[#)], Karl Folkers[@)], P-O Edlund[$)], Bernt Lund[$] and Henning Theorell[+)]

[*)]Department of Clinical Physiology, Karolinska Hospital, Stockholm, [#)]Department of Pharmacology, Copenhagen University, Copenhagen, [@)]Institute of Biomedical Research, University of Texas at Austin, Austin, [$)]Kabi Pharma, Stockholm and [+)]GP, Stockholm.

INTRODUCTION

During recent years an exponentially increasing interest has been shown to the biology of ubiquinone, coenzyme Q_{10} or CoQ_{10} in man. The reason for that is the intrinsic role of CoQ_{10} in mitochondrial electron transport but also its role as an unspecific antioxidant (for ref see Karlsson 1987). Its application as both a food supplement and as a drug has caused some debate as its pharmacological properties has been disputed upon related to the compounds extreme lipophilic character (for ref see Greenberg and Frishman 1988). Recent controlled studies of CoQ_{10} treatment programs in patients with cardiomyopathy (Mortensen et al 1986) seem to reveal though improvements in patients. Simultaneously heart muscle CoQ_{10} content increased and a statistical relationship could be proven. An other recent study on a rat model with experimental myocardial infarction revealed equally good effects with orally administered CoQ_{10} as with an angiotensin converting enzyme (ACE) inhibitor (Captopril[R]) (Diamant et al 1989).

Comparative data on skeletal muscle and blood contents in healthy man and in patients are scarce in the literature. The present report is a presentation of data gathered during the 80-ties as a collaborative effort between Austin, Copenhagen and Stockholm. Some of them have already been published but new data have been added as well as preliminary information concerning CoQ_{10} treatment in healthy volunteers.

METHODS

Blood samples for whole blood or plasma CoQ_{10} determinations have been drawn from deep vein catheters or from a cubital vein. Muscle specimens have been taken from the thigh muscle (m vastus lateralis) with the needle biopsy technique (Bergstrom 1962, Karlsson 1979 and 1986). CoQ_{10} analyses have been done according to Vadhanavikit et al (1984) and Edlund (1988). No systematic differences have been seen between the two methods (Karlsson to be published).

RESULTS AND COMMENTS

Skeletal muscle CoQ_{10} contents varied between 0.1-0.3 ug x mg^{-1} dry muscle expressed as means for patients with signs of cardiotoxic effects of Adriamycin[R] treatment, candidates for coronary by pass grafting, sedentary but healthy subjects, and endurance trained athletes. The corresponding blood values ranged 0.4-0.9 ug x ml^{-1} (Fig 1). The high muscle but low blood CoQ_{10} contents in endurance trained have to be stressed in particular.

Centrifugation and the plasma saving procedures did not reveal any cellrelated CoQ_{10} compartmentalization or sequestration as a linear relationship was present over a wide range of blood CoQ_{10} contents (Fig. 2).

Two studies on thigh muscle biochemistry and histology in patients with severe effort angina and candidates for coronary bypass surgery demonstrated significant differences in respect to muscle and blood CoQ_{10} contents (Fig. 3). These differences covaried with a higher proportion of the muscle fiber subgroup FTc in the patients with the most detoriated CoQ_{10} depots. A causal relationship has earlier been suggested (Karlsson 1988) and is not contradicted by the present results.

The appearance of the muscle fiber subgroup FTc is indicative of a fiber transition. The subgroup has also been linked to fiber trauma, cell death and decomposition (for ref see Karlsson 1979 and 1986).

A double blind cross over study in healthy volunteers including run in and wash out periods showed both skeletal muscle and blood CoQ_{10} increases. The treatment lasted for 6 weeks and consisted of 100 mg CoQ_{10} in an oil preparation a day. On an individual basis the increases varied considerably, but were approximately the same in relative terms (Fig. 4). Thus, the same dosage of CoQ_{10} did not respond in one and the same blood CoQ_{10} increment, but a response which was in accordance to the volunteers' run in and base line value.

In patients with either congestive heart failure or with muscle dystrophy observations have been made also indicating an individualized treatment response. They were administered the same CoQ_{10} formulation as above. Two extremes in respect to responses are presented in Fig. 5. Pat I with a relatively higher base line value responded on the open treatment and was after 1.5 months released from the hospital and referred to her GP. Pat II with a low run in value did not respond to the initial treatment and only a tendency to dose-response relationship was seen when the daily dosage was tripled. No major clinical effect of the oral CoQ_{10} treatment was seen in Pat II.

DISCUSSION

The present summery confirms the finding that ubiquinone, coenzyme Q_{10} or CoQ_{10} is depleted in many patient categories. It has been frequently stated that CoQ_{10} has a "vitamin" character and is available as other vitamins as a food ingredient (Crane 1986). In a recent report it was shown that CoQ_{10} in rats was present and synthesized in many organelles and/or subcellular systems (Elmberger 1989).

The controlled CoQ_{10} treatment study in healthy volunteers indicated that the pretreatment values ("base line levels") determined the result of the treatment. Thus, those with an initial low level showed in absolute terms a rather meager response. Earlier it has been suggested that the low blood values seen in endurance trained athletes could be explained on the basis of increased turnover as a result the higher average metabolic rate with training (a caloric intake of 8 - 10 000 kcal a day as compared to 2 - 3 000 kcal in sedentaries) (Karlsson 1988). It is wellknown that endurance training will result in a lowered plasma cholesterol level (e g Nikkila et al 1982) and subsequently decreased blood triglyceride contents. This means by definition a reduced storing and transporting capacity in blood for CoQ_{10}.

REFERENCES

Bergstrom, J., 1962, Muscle electrolytes in man. Scandinavan Journal of Laboratory and Clinical Investigations. Suppl 68, 1-107.

Crane, F.L., 1986, Physiological coenzyme Q function and pharmacological reactions. In: Biomedical and Clinical Aspects of Coenzyme Q, vol. 5, edited by Folkers, K. and Yamamura, Y., (Amsterdam: Elsevier Science Publishers), pp. 3-14.

Edlund, P.O., 1988, Determination of coenzyme Q_{10}, alpha-tocopherol and cholesterol in biological samples by coupled-column liquid chromatography with coulometric and ultraviolet detection. Journal of Chromatography. 425, 87-97.

Elmberger, P.G., 1989, Quantitation and metabolism of polyisoprenoid lipids. Thesis, Karolinska Institute, Stockholm, pp. 1-64.

Greenberg, S.M., and Frishman, W.H., 1988, Coenzyme Q_{10}: a new drug for myocardial ischemia. Medical Clinics of North America, 72, 243-258.

Karlsson, J., 1979, Localized muscular fatigue: role of muscle metabolism and substrate depletion. Exercise Sports Sciences Review, 7, 1-42.

Karlsson, J., 1986, Muscle fiber composition, metabolic potentials, oxygen transport and exercise performance in man. In: Biochemical Aspects of Physical Exercise., edited by Benzi, G., Packer, L. and Siliprandi N. (Amsterdam:Elsevier), pp. 3-13.

Karlsson, J., 1987, Heart and muscle ubiquinone or CoQ_{10} as a protective agent against radical formation in man. Advances in Myochemistry, I, 305-318.

Karlsson, J., 1987, Exercise capacity and muscle fibre types in effort angina. European Heart Journal, 8 (suppl. G), 51-57.

Karlsson, J., 1988, Onset of blood lactate accumulation, exercise capacity, skeletal muscle fibers and metabolism before and after coronary artery bypass grafting. American Journal of Cardiology, 62, 108E-114E.

Mortensen, S.A., Bouchelouche, P., Muratso, K., Folkers, K., 1986, Clinical decline and relapse of cardiac patients on coenzyme Q_{10} withdrawal. In: Biomedical and Clinical Aspects of Coenzyme Q, vol. 5, edited by Folkers, K. and Yamamura, Y., (Amsterdam: Elsevier Science Publishers), pp. 281-290.

Nikkila, E.A., Kuusi, T. and Taskinen M.-R., 1982, Role of lipoproteinlipase and hepatic endothelial lipase in the metabolism of high density lipoproteins: a novel concept on cholesterol transport in HDL cycle. In Metabolic Risk

Factors in Ischemic Cardiovascular Disease, New York, edited by Carlsson, L.A. and Pernow, B., (New York: Raven Press), pp. 205-215.

Vadhanavikit, S., Morishita, M., Duff, G.A., and Folkers, K., 1984, Micro-analysis for coenzyme Q_{10} in endomyocardial biopsies of cardiac patients and data on bovine and canine hearts. *Biochemica Biophysica Research Communication*, 123, 1165-1169.

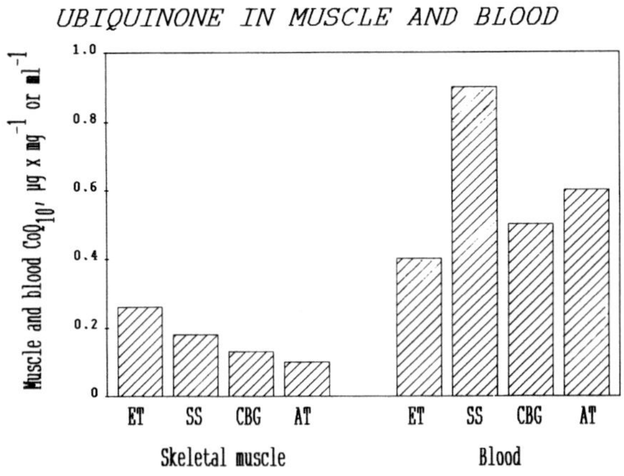

Fig.1 Skeletal muscle (m vastus lateralis) and venous blood contents of ubiquinone, coenzyme Q_{10} or CoQ_{10} in endurance athletes (ET), sedentary subjects (SS), patients with ischemic heart disease before coronary by-pass grafting (CBG) and patients with signs of cardiotoxicity related to Adriamycin^R treatment (AT).

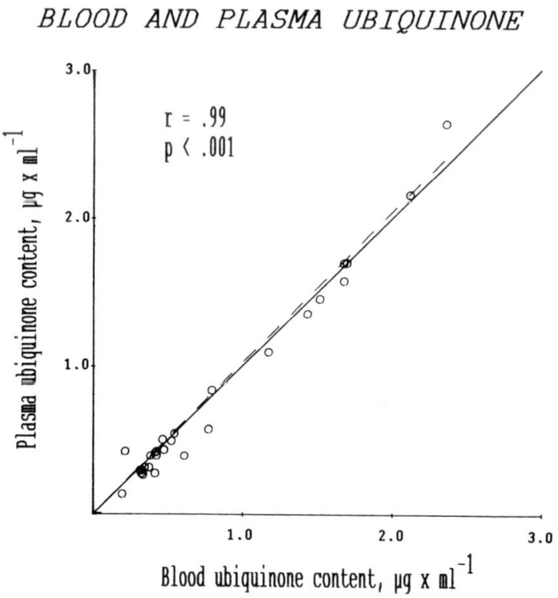

Fig. 2. Comparison of blood and plasma CoQ_{10} contents in healthy man as well as patients. The solid line depicts "line of identity" and the dashed line the regression line for the relationship plasma to blood CoQ_{10} contents.

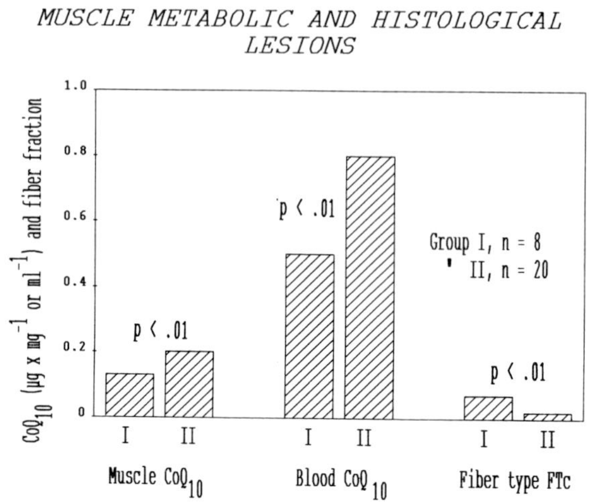

Fig. 3. Muscle and blood CoQ_{10} contents and muscle fiber FTc frequency in two patient groups, who both were candidates for CBG.

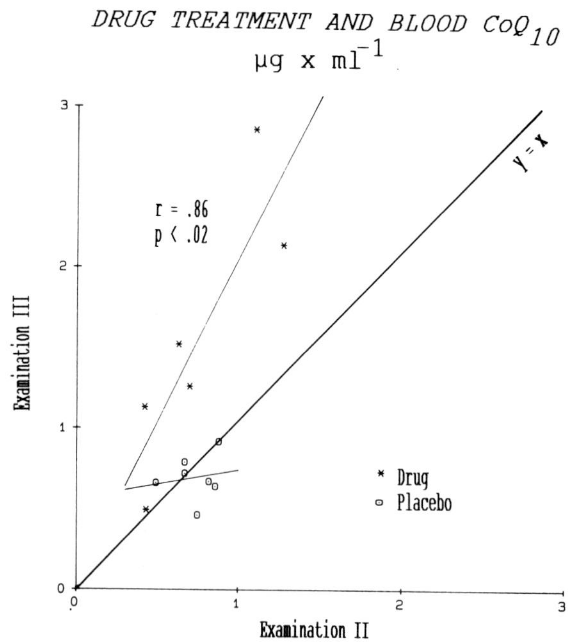

Fig. 4. The individual relationship in healthy man between blood CoQ_{10} contents before and after an oral CoQ_{10} treatment with 100 mg a day in an oil formulation for 6 weeks according to a double blind, cross over design.

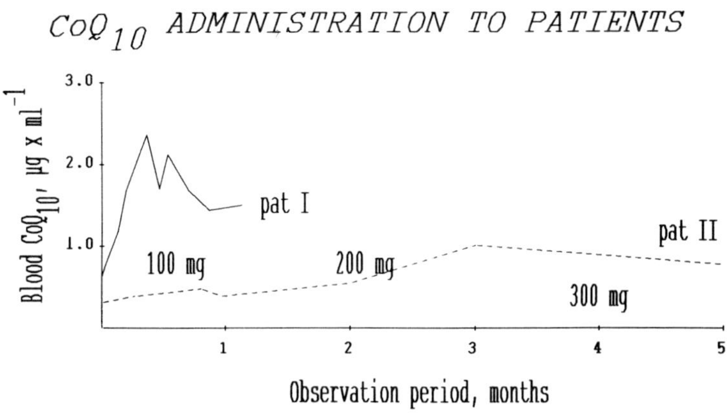

Fig. 5. The effect of CoQ_{10} treatment in two patients with the same formulation as in Fig. 4. The nonresponder did not show any clinical improvements either in contrast to pat I.

PLASMA UBIQUINONE AND CHOLESTEROL CONTENTS WITH AND WITHOUT UBIQUINONE TREATMENT

Jan Karlsson*), Bertil Diamant#), Karl Folkers@), P-O Edlund$), Bernt Lund$ and Henning Theorell+)

*)Department of Clinical Physiology, Karolinska Hospital, Stockholm, #)Department of Pharmacology, Copenhagen University, Copenhagen, @)Institute of Biomedical Research, University of Texas at Austin, Austin, $)Kabi Pharma, Stockholm and +)GP, Stockholm.

INTRODUCTION

The compound ubiquinone, coenzyme Q_{10} or CoQ_{10} has extreme lipophilic properties (Crane 1986). To be transported in the cell or from one cell to an other it has either to be attached to a "transporter" usually a protein or dissolved in a lipid phase. Such an opportunity is available in blood in the form of circulating triglycerides (TG). Bloodborne TG are either in the form of chylomicrons or linked to lipoproteins in the form of VLDL or LDL. Physiologically only the latter two fractions are of significance (for ref see Carlsson 1982). As they also contain cholesterol, quantitative blood cholesterol determinations could give an acceptable estimate of the storage and transporting capacity of CoQ_{10} in blood.

Recently, an other rational has been introduced by the Gustav Dallner group at the Huddinge Hospital, Stockholm, to clinically combine CoQ_{10} and cholesterol determinations. They are both endproducts of one and the same main biosynthetic pathway of mevalonate products with acetyl-CoA as the precursor (for ref see Elmberger 1989). Thus in the rat there is an almost linear relationship between tagged cholesterol and CoQ_{10} with infusion of tracer amounts of (^{3}H)mevalonate (Elmberger et al 1987). Moreover, inhibition of the mevalonate pathway exerts a reduced formation of both products.

The physiological implications of these findings are at this stage speculative but it seems reasonable to be openminded for a modification of the CoQ_{10} metabolism concept covering intake, synthesis, cellular incorporation and breakdown in mammalian tissues and organs.

METHODS

Blood samples for plasma CoQ_{10} and cholesterol determinations have been drawn from deep vein catheters or from cubital veins. All samples have after preparation etc been stored at -80°C for further analyses .
CoQ_{10} analyses have been done according to Vadhanavikit et al (1984) and Edlund (1988). No systematic differences are present between to two applied methods (Karlsson to be published). Cholesterol was determined as described by Edlund (1988).

RESULTS AND COMMENTS

Blood CoQ_{10} contents can vary considerably in man irrespective disease or not (Fig. 1). Thus in extreme endurance athletes equally low blood or plasma values can be seen as in patients after Adriamycin^R treatment and severe adverse effects as cardiotoxicity, skeletal muscle weakness, etc. (Karlsson et al 1986). CoQ_{10} treatment of patients and healthy man as well may or may not elevate blood values as reported in the same volume. One significant feature seems to be base line or run in values.

Plasma cholesterol analyses showed that the means mimicked the means for plasma CoQ_{10} in the different groups examined with the exception of those treated with CoQ_{10} (100 mg a day in an oil preparation) (Fig. 2).

Plasma CoQ_{10} content was linearly related to cholesterol levels when the individuals were allocated to the groups: CoQ_{10} treated patients and healthy controls, untreated patients and healthy controls,respectively (Fig. 3). The three groups appeared to represent three different populations in respect to the relationship. Thus, for the same plasma cholesterol level, treated patients/volunteers had the highest CoQ_{10} contents, whereas the controls (endurance athletes and sedentaries combined) had the lowest. It is apparent both from Figures 2 and 3 that many patients/patientgroups had low plasma cholesterol levels. Whether that was related to the pharmacological treatment or the nutritional status can not be evaluated at this stage.

DISCUSSION

The results demonstrated and confirm Edlund (1988) original observation that a statistical relationship exists between plasma CoQ_{10} and cholesterol levels providing a relevant allocation of patients and healthy controls. The nature of this relationship can, however, only be speculated upon.

CoQ_{10} has a "vitamin" character and is available as other vitamins as a food ingredient (Crane 1986). The mechanisms of its uptake from the visceral tract is virtually unknown, but it has been suggested to be related to the lipid uptake and possibly with the chylomicron transport (Karlsson 1987). In addition the endogenous synthesis has to be recognized although the magnitude has been disputed upon (Beyer et al 1986, Crane 1986).

Blood and plasma CoQ_{10} analysis have been undertaken in many clinical settings as it has been shown to be relevant in relation to cardiomyopathy and reflects heart muscle content and also is suggested to be related to the clinical status of the disease (Mortensen et al 1986). The lack of validity in that respect has, however, been documented in endurance athletes where low blood CoQ_{10} content is present simultaneously as skeletal muscle content is high (Karlsson 1987). The opposite pattern is present in e g sedentaries or prediabetes, where high blood values coincide with low muscle values (Karlsson 1987).

It is obvious from these observations that the lipid content in the blood could be explanatory for the different blood CoQ_{10} contents (Nikkila 1953). Interfaced with that an increased turn over rate in e g the athletes as well as in patients has to be included (Karlsson 1987).

As pointed out above the Dallner group has shown in rats that plasma CoQ_{10} content covaried with cholesterol and a joint pathway for biosynthesis of CoQ_{10} have been documented. The investigators also showed that a reduced pathway flux influenced and depressed both cholesterol and CoQ_{10} levels.

It is wellknown that endurance training will result in a lowered plasma cholesterol level (e g Nikkila et al 1982) and subsequently increased fractions of HDL. This means by definition a reduced storing and transporting capacity of bloodborne CoQ_{10}. In addition it is temptating although speculative as pointed out above to consider a reduced synthesis of polyisoprenoid moieties as an alternative feature. It is well documented that skeletal muscle in endurance athletes has a high combustion rate of lipids as the muscle is rich in type I or slow twitch (ST) muscle fibers. Moreover, the ST fiber is surrounded by more capillaries and to this is also confined a higher lipoprotein-lipase (LPL) activity (for ref see Karlsson 1979, Karlsson 1984, Karlsson 1987). Taken together this may imply a lower availability of acetyl-CoA as a precursor for the mevalonate pathway and a subsequently reduced CoQ_{10} and cholesterol synthesis. It is selfexplainable that drugs reducing cholesterol formation will exert a similar effect even on CoQ_{10} formation resulting in low blood CoQ_{10} and cholesterol contents.

To summarize, ubiquinone, coenzyme Q_{10} or CoQ_{10} is depleted in blood in many patient categories. This can depend on an increased break down, reduced synthesis or combinations. The low CoQ_{10} coincided with low plasma cholesterol levels, which offers an alternative explanation as cholesterol levels reflect bloodborne triglycerides and consequently a storing entity. A third, but speculative explanation, is that the low plasma CoQ_{10} and cholesterol contents might reflect a lower biosynthesis activity of the joint mevalonate pathway.

REFERENCES

Beyer, R.E., Nordenbrand, K., and Ernster, L., 1986, The role of coenzyme Q as a mitochondrial antioxidant: a review. In: Biomedical and Clinical Aspects of Coenzyme Q, vol. 5, edited by Folkers, K. and Yamamura, Y., (Amsterdam: Elsevier Science Publishers), pp. 3-14.

Crane, F.L., 1986, Physiological coenzyme Q function and pharmacological reactions. In: Biomedical and Clinical Aspects of Coenzyme Q, vol. 5, edited by Folkers, K. and Yamamura, Y., (Amsterdam: Elsevier Science Publishers), pp. 3-14.

Carlsson, L.A., 1982, Serum lipids and atherosclerotic diseases. In: Metabolic Risk Factors in Ischemic Cardiovascular Disease, New York, edited by Carlsson, L.A. and Pernow, B., (New York: Raven Press), pp. 205-215.

Edlund, P.O., 1988, Determination of coenzyme Q_{10}, alpha-tocopherol and cholesterol in biological samples by coupled-column liquid chromatography with coulometric and ultraviolet detection. Journal of Chromatography. 425, 87-97.

Elmberger, P.G., 1989, Quantitation and metabolism of polyisoprenoid lipids. Thesis, Karolinska Institute, Stockholm, pp. 1-64.

Elmberger, P.G., Kale'n, A., Appelkvist, E.-L. and Dallner, G., 1987, In vitro and in vivo synthesis of dolichol and other main mevalonate products in various organs of the rat. European Journal of Biochemistry, 168, 1-11.

Karlsson, J., 1979, Localized muscular fatigue: role of muscle metabolism and substrate depletion. Exercise Sports Sciences Review, 7, 1-42.

Karlsson, J., 1987, Heart and muscle ubiquinone or CoQ_{10} as a protective agent against radical formation in man. Advances in Myochemistry, I, 305-318.

Karlsson, J., 1987, Exercise capacity and muscle fibre type in effort angina. *European Heart Journal*, 8 (suppl. G), 51-57.

Karlsson, J., Folkers, K., Astrom, H., Jansson, E., Pernow, B., Holmgren, A., Mellstedt, A. and Diamant, B., 1986, Effect of Adriamycine[R] on heart and skeletal muscle coenzyme Q in man. In: *Biomedical and Clinical Aspects of Coenzyme Q*, vol. 5, edited by Folkers, K. and Yamamura, Y., (Amsterdam: Elsevier Science Publishers), pp. 3-14.

Mortensen, S.A., Bouchelouche, P., Muratso, K., and Folkers, K., 1986, Clinical decline and relapse of cardiac patients on coenzyme Q_{10} withdrawal. In: *Biomedical and Clinical Aspects of Coenzyme Q*, vol. 5, edited by Folkers, K. and Yamamura, Y., (Amsterdam: Elsevier Science Publishers), pp. 281-290.

Nikkila, E.A., 1953, Studies on the lipid-protein relationships in normal and pathological sera and the effect of heparin on serum lipoproteins. *Scandinavian Journal of Laboratory and Clinical Investigations*. Suppl. 8, 5, 1-101.

Nikkila, E.A., Kuusi, T. and Taskinen M.-R., 1982, Role of lipoproteinlipase and hepatic endothelial lipase in the metabolism of high density lipoproteins: a novel concept on cholesterol transport in HDL cycle. In *Metabolic Risk Factors in Ischemic Cardiovascular Disease*, New York, edited by Carlsson, L.A. and Pernow, B., (New York: Raven Press), pp. 205-215.

Vadhanavikit, S., Morishita, M., Duff, G.A., and Folkers, K., 1984, Micro-analysis for coenzyme Q_{10} in endomyocardial biopsies of cardiac patients and data on bovine and canine hearts. *Biochemica Biophysica Research Communication*, 123, 1165-1169.

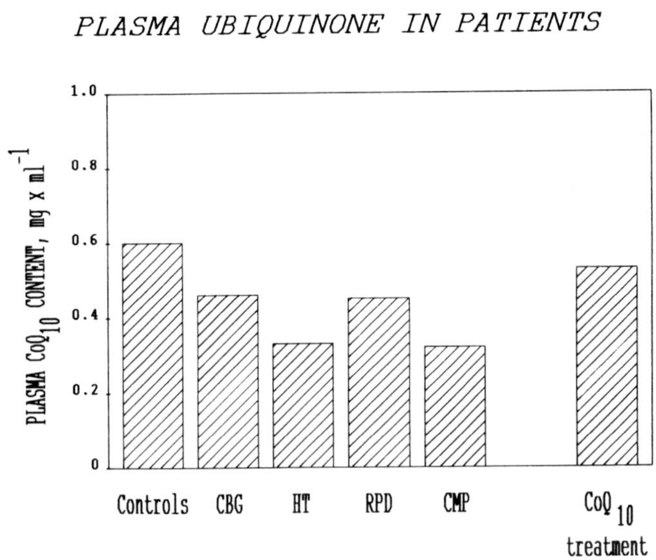

Fig. 1. Plasma CoQ_{10} contents in healthy male volunteers representing different physical conditioning levels (Controls), patients the day before coronary by-pass grafting (CBG), heart transplants (HT), patients treated with radical producing drugs such as Adriamycin^{R} (RPD), patients with cardiomyopathy (CMP), and patients and healthy volunteers, who have been treated in well controlled studies with an oral intake of 100 mg CoQ_{10} in an oil formulation.

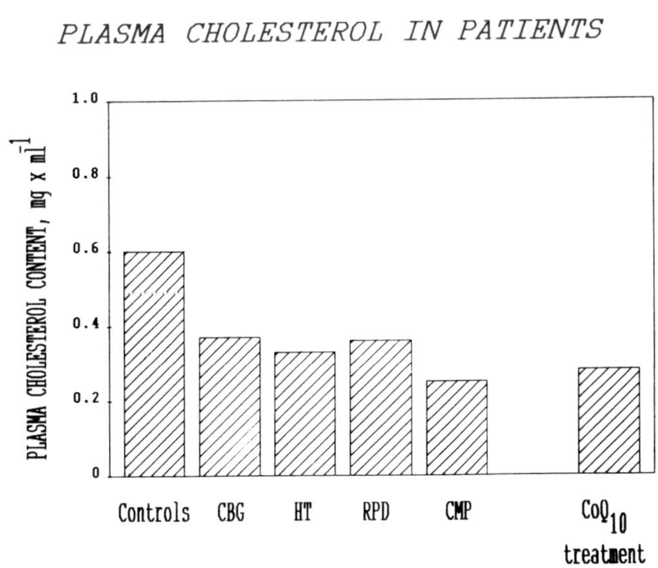

Fig. 2. Plasma cholesterol levels in the same patients/volunteers as in Fig. 1.

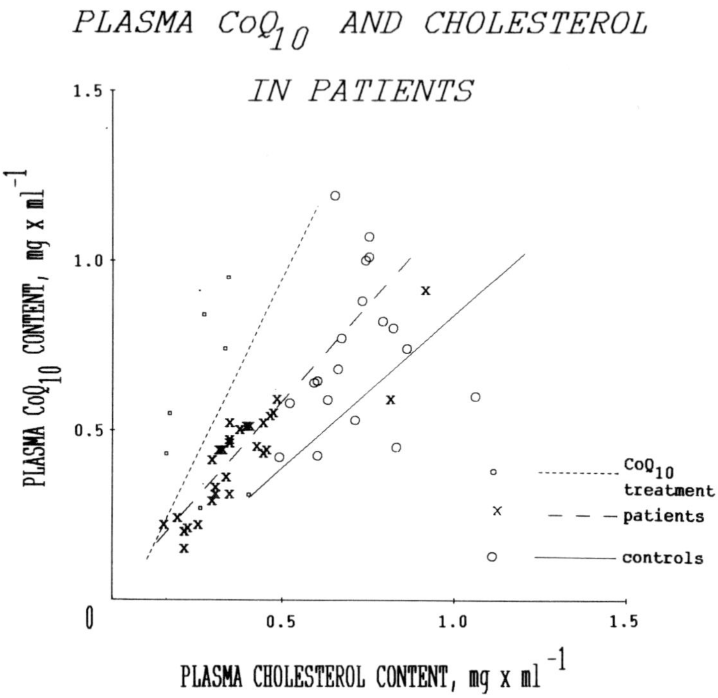

Fig. 3. Individual values for plasma CoQ_{10} versus plasma cholesterol contents allocated according to whether they are treated with CoQ_{10} (patients and volunteers), nontreated patients and healthy controls with different conditioning levels, respectively.

REMARKS ON PROLONGED UBIQUINONE ADMINISTRATION IN PHYSICAL EXERCISE

V. Wyss, T. Lubich*, G.P. Ganzit, D. Cesaretti*, P.L. Fiorella^, C. Dei Rocini*, A.M. Bargossi^, R. Battistoni^, A. Lippi^, G. Grossi^, G. Sprovieri^ and M. Battino°.

Chair of Sport Medicine, University of Torino.
*Chair of Sport Medicine, University of Bologna.
^Laboratorio Centralizzato, USL 28 Bologna Nord, Bologna.
°Institute of Biochemistry, University of Ancona.
ITALY

INTRODUCTION

The authors according to the biochemical rationale for Ubiquinone (UQ) treatment (Lenaz, 1985) and to more recent results on athletes (Guerra et al., 1987) have undertaken this study to stress effects of a cycle of UQ per os administration on myocardial and muscolar fitness in healthy young men engaged in strenuous exercise test with increasing work load.

Keywords: VO_2max, VCO_2max, respiratory quotient, systolic and diastolic blood pressure.

MATERIALS AND METHODS

A group of 18 men matched with regard to anthropometric

Table I. Anthropometric and functional parameters at the beginning of the study.

age	years	24.8	± 3.6
height	cm	181.4	± 7.4
weight	Kg	75.3	± 6.8
HR	b/m	62	± 8
SOBP	mmHg	118	± 10
DOBP	mmHg	70	± 5
VO_2max	l/min	3.76	± 0.5

HR= Heart Rate, SOBP and DOBP = Systolic and Diastolic Orthostatic Blood Pressure, VO_2max = maximal Oxigen consumption per minute.

features and functional parameters were tested (Table I).

Patients were also matched with regard to leisure time physical activity being accustomed to endurance running. Immediately before beginning the treatment each athlete has been visited and undergone ECG and determination of blood glucose, lactate, NEFA, glycerol.

All analytes were determined by routine commercial kits.

At the same time VO_2max of each participant to the study was measured in two successive days.

They were given 100mg Ubiquinone or placebo (PL) per os using a double blind crossover study for 30 days.

During the 5-7 following days after either the treatment with the drug or with the placebo again twice in successive days were performed tests to measure VO_2max of each athlete.

RESULTS

At the beginning of the study plasmatic UQ levels of our volountaries were lower than plasmatic levels of a matched control group whose members only differ for not being trained. At the end of the treatment period plasmatic UQ levels were significantly higher (unpublished data). During the period of aerobic exercise all the patients did not show any significant difference between UQ and PL groups with regard to: polmonary flow, VO_2max, VCO_2max, respiratory quotient (RQ), heart rate (HR), systolic and diastolic blood pressure (BP). Neither blood parameters did significant changes. Again no differences arose for the aerobic-anaerobic thresold (cf. Table III). On the contrary during the maximal exercise patients with the UQ treatment had a 33% longer work, with 5.3% higher watt, 7.4% higher VO_2max, 7.7% higher HR and better performance (cf. Table V). In Table IV it is also shown values of total O_2uptake, O_2 burnt and lactic acid O_2 equivalent. During the recovery period only blood lactate level was 21% higher in UQ administered patients. Table VI shows the trend of main considered plasmatic parameters during recovery.

CONCLUSIONS

Bioavailability of UQ in people with important leisure time physical activity is matter of a growing interest and speculation. In fact it is controversial why a decrease of

Table II. Basal average values ± s.d. before the test.

	H.R.	SOBP	DOBP	R.Q.
UQ	72 ± 8	113 ± 8	71 ± 8	0.81 ± 0.06
PL	73 ± 10	117 ± 10	73 ± 10	0.81 ± 0.04

Table III. Average values ± s.d. at the anaerobic thresold.

	Watt	H.R.	SOBP	DOBP
UQ	216 ± 26	173 ± 8	175 ± 12	78 ± 6
PL	207 ± 23	167 ± 11	174 ± 16	74 ± 4

Table IV. Average values ± s.d. at maximal work.

	UQ	PL
VO_2	24.6 ± 7.3*	18.8 ± 6.8
LA O_2 equivalent	2.86 ± 0.8*	1.89 ± 0.9
Total O_2	27.6 ± 7.7*	20.6 ± 7.1

* $p < 0.01$ vs PL

all values are in liters; VO_2= Total O_2 consumption; LA O_2 equivalent = O_2 amount need to oxide lactate; Total O_2 = Ventilated O_2.

Table V. Average values ± s.d. at maximal work.

		UQ	PL
maximal work load	(W)	258.00 ± 41*	247.00 ± 38
Total work	(Kj)	243.00 ± 61*	217.00 ± 56
VO_2 max	(l/min)	4.04 ± 0.5*	3.79 ± 0.6
R.Q.		0.99 ± 0.05	0.98 ± 0.07

*$p < 0.05$ vs PL.

plasmatic values of UQ has been observed in athletes at the end of a training cycle; it may occur by consumption in ergogenics cycle and may lead to an impairement with new growth muscle masses.

Our results seem allow to suggest that the administration of UQ increases plasmatic levels so that might have effect both by increasing the myocardial and muscular efficiency and, by increasing tolerance for higher blood lactate levels, on the fatigue.

Table VI. Average values ± s.d. of main analytes during recovery.

	5th min		20th min	
	UQ	PL	UQ	PL
Glucose	6.9±0.9*	5.9±0.8	5.1±0.8	5.2±0.7
NEFA	386±142	429±179	245±111	326±173
Glycerol	171±43^	147±56	138±44	142±39
Lactate	10.9±1.8^	9.1±2.1	5.3±1.6	5.5±2.1

Glucose, Lactate = mmol/l
NEFA, Glycerol = micromol/l
* = $p < 0.01$ vs PL
^ = $p < 0.05$ vs PL

REFERENCES

Guerra G.P., Ballardini E., Lippa S., Oradei A. and Littarru G.P., 1987, Effetto della somministrazione di Ubidecarenone nel consumo massimo di ossigeno e sulla performance fisica in un gruppo di giovani ciclisti. Medicina dello Sport, 40, 359-364.

Lenaz G., 1985, Abiochemical rationale for the therapeutics effects of Coenzyme Q. In "Coenzyme Q" (edited by G. Lenaz), John Wiley & Sons, New York, pp. 485-491.

Closing Lecture delivered by K. Folkers

PROGRESS IN BIOCHEMICAL APPROACHES TO CLINICAL THERAPY WITH COENZYME Q_{10}

Karl Folkers*

Institute for Biomedical Research, The University of Texas at Austin, Welch 4.304, Austin, TX 78712

INTRODUCTION

I believe that a Closing Lecture for this International Symposium could best be a combination of perspectives, prospectives with a modicum of data. I have been advised by the late Professor Lee Irwin Smith of the University of Minnesota to make Lectures interesting over and above data, and surely a Closing Lecture should be interesting.

Professor Giorgio Lenaz deserves a tribute of the appreciation of all of us for having inspired this International Symposium on the Biochemistry, Bioenergetics and Clinical Applications of Ubiquinone. Research on all aspects of ubiquinone will surely continue at an accelerated level for decades to come, and because of this superabundance of information in the future, Giorgio´s Symposium could be repeated, but alternatively with the existing Symposia on the Biomedical and Clinical Aspects of Coenzyme Q. Giorgio´s Symposium could continue to emphasize the biochemistry and bioenergetics and the existing Symposia could continue to emphasize the biomedical and clinical aspects. The financial needs of both Symposia would likely be the only limitation.

DISEASES OF BIOENERGETICS

It is common in therapy with drugs that one drug is for treatment of one disease. In the past, if one drug were promoted for treatment of several diseases and particularly when the several diseases were quite different, the impression was gained by patients and physicians alike that the drug was a fake. Sometimes, such misrepresentations have been a---"kiss of death"--- for the

CoQ. 401

drug.

Since bioenergetics supports very diverse life functions of the human body, and since coenzyme Q_{10} (vitamin Q_{10}) is an indispensable component of bioenergetics, it follows that coenzyme Q_{10} can truly be medically useful to treat several similar and even several dissimilar diseases.

For example, coenzyme Q_{10} has been used to treat cardiomyopathy, muscular dystrophy, periodontal disease, hypertension, etc. It is scientifically and medically acceptable that coenzyme Q_{10} may indeed be useful to treat diverse diseases. Bioenergetics supports the functions of the beating of the heart, the health of the gingival tissue, and the physical performance of muscles which can be impaired by genetic defects.

The status and promise of coenzyme Q_{10} to treat such diverse diseases is critiqued in this presentation and begins with cardiomyopathy.

CARDIOMYOPATHY

Background of CoQ_{10} and the Heart and Cardiac Disease.- Crystalline CoQ_{10} was isolated by Crane *et al.* (1) from the beef heart in 1957 at Wisconsin and by my group from the human heart in 1959 at Merck (2). At Liverpool, the early ubiquinone preparations of Morton *et al.* were artifacts of the monoethoxy and diethoxy homologues of CoQ_{10}, because ethanol had been used in a hydrolysis step. The organic chemistry at Merck, 1959, revealed the methoxy groups of the quinone are labile in contrast to the methoxy groups of the hydroquinone which are more stable. Morton *et al.* in 1958 at Liverpool isolated the dimethoxy-CoQ_{10}. I have questioned whether one of the methoxy groups of CoQ_{10} has a biochemical function. We have heard in this Symposium about the possibility that CoQ_{10} may have interesting biochemical functions in addition to its bioenergetic and antioxidant activities.

The localization and biochemistry of CoQ_{10} in the mitochondria of beef heart (3) forecasted the essentiality of CoQ_{10} to human myocardial bioenergetics. Looking back, it was to be expected that the essentiality of CoQ_{10} to the bioenergetics of the human heart when considered in conjunction with the magnitude and fatality of human cardiac disease would lead to clinical trials with CoQ_{10} for cardiac disease, and it did.

The Vitamin Aspects of Q_{10} and Deficiencies.- The vitamin and nutritional aspects of Q_{10} were extensively critiqued in 1969 (4). The widespread occurrence of CoQ_{10} in foodstuffs, the biochemical activity of Q_{10} as a coenzyme, the biosynthetic origin of CoQ_{10} from the amino acid, tyrosine, by enzymic mechanisms involving vitamins B_6, B_{12}, folic acid, etc. served as rationale for seeking deficiencies of CoQ_{10} in the human body, and such deficiencies were found.

Human Urinary Deficiency of CoQ_{10}.- Analysis (5) of 160 urine specimens representing 63 males and females revealed an average excretion of ca. 55 ug/24 hours for males and 22 ug/24 hours for females. 35% of the 63 individuals excreted less than 10 ug/24 hours on one or more occasions, which may have been the earliest indication in 1960 of human deficiencies of CoQ_{10}.

Human Cardiac Deficiencies of CoQ_{10} in Biopsies from Surgery.- It was found that about 75% of 132 patients undergoing cardiac surgery showed deficiencies in cardiac biopsies in the activity of the succinate dehydrogenase-CoQ_{10} reductase. There were 13 categories of these 132 patients and 12/13 categories showed deficiencies of CoQ_{10} ranging from 20-44%. Cardiopathies associated with diabetes showed 42% deficiencies (6,7,8).

Human Blood Deficiency of CoQ_{10}.- Blood samples from 406 cardiac patients were analyzed (9) for deficiencies of CoQ_{10}. The assay was based upon a differential determination of the specific activities (S.A.) of the succinate dehydrogenase-CoQ_{10} reductase (SDQR). Approximately 20% or 87/406 showed a mean lower basal S.A. and a mean higher percent deficiency than the corresponding values of a control group. Both differences were significant, $p<0.001$. The remaining 80% showed a lower mean basal S.A., $p<0.001$. The desirable state for such cardiac patients is a normal level of CoQ_{10}-enzymes for bioenergetics, and no CoQ_{10} deficiency.

Deficiency of CoQ_{10} in Surgical Biopsies of the Human Heart.- A CoQ_{10}-enzymic analysis of biopsy specimens from surgery of 66 diseased human hearts showed substantial increases in enzymic activity on addition of CoQ_3, but such increases were not evidenced for control biopsies from the normal heart (6).

Littarru in 1972 (7,8) found that about 75% of 123 cardiac patients showed varying degrees of deficiency of CoQ_{10} in their heart tissue.

Deficiency of CoQ_{10} in Human Endomyocardial biopsies.- Endomyocardial biopsies and blood from 43 patients with cardiomyopathy were obtained for assays of CoQ_{10} levels. The biopsies were obtained during heart catherizations. Blood levels for patients of Class IV were lower ($p<0.01$)

than those of Class I. Blood levels for Classes III and IV were lower ($p<0.001$) than those of Classes I and II. The CoQ_{10} levels of biopsies after treatment of patients with CoQ_{10} for 2-8 months were higher ($p<0.02$) than before treatment. In summary, the myocardial deficiency of CoQ_{10} was higher with increasing severity of disease and was reduced by therapy. Such data of Mortensen *et al.* (10) correlated with the effective treatment of cardiomyopathy with CoQ_{10}.

Commercial Production of CoQ_{10} in Japan.- Apparently, the Nisshin Flour and Milling Co. in Japan was first to commercially synthesize CoQ_{10} and which supported the clinical investigations of Eisai Co., Ltd. of Tokyo. On a laboratory basis, Page *et al.* in 1960 (11) isolated CoQ_{10} from a bacterial fermentation which forecasted the ultimate technology for commercial fermentation which produced CoQ_{10} by Kanegafuchi Chemical Industry Co. Inc. in Osaka.

Therefore, CoQ_{10} from partial synthesis and from fermentation in Japan made possible diverse clinical studies to treat different diseases with CoQ_{10}.

The First Ten Years of Clinical Treatment with CoQ_{10} of Cardiac Disease.- During 1967-76, about 25 studies appeared in the Japanese medical literature which stemmed from the efforts of 110 physicians in 41 medical institutions. Regardless of the diversified protocols and results over this decade, the consensus was that the administration of CoQ_{10} to patients with congestive heart failure was beneficial; Folkers *et al.* (12).

Clinical Studies on CoQ_{10} and Cardiomyopathy, 1976-1987.- The availability to me at the University of Texas of kilogram quantities of pure CoQ_{10} from the generosity of Mr. Tsuyoshi Sakata of the Kanegafuchi Chemical Industry Co. Inc. in Osaka made possible many and diverse clinical studies with capsules of CoQ_{10} through many cooperations in the United States and Europe, particularly for cardiomyopathy. The consensus of this second decade of clinical research was that CoQ_{10} is a beneficial treatment of cardiomyopathy over and above all conventional therapy.

This Symposium of Professor Lenaz updated such clinical studies by Dr. Mortensen of Denmark, Dr. Oda of Japan, and Dr. Judy of the U.S.

Long-Term Efficacy and Safety of CoQ_{10} to Treat Cardiomyopathy.- In addition to the clinical studies described in this Symposium, I am very pleased to add the newest results from my cooperation with Dr. Per Langsjoen of Texas on the long-term efficacy and safety of CoQ_{10} to treat cardiomyopathy. In summary, this study covered 6 years and 143 patients for a total of 402-patient years of therapy. There were no significant side reactions. The low control blood levels of CoQ_{10} increased to the normal range on 100 mg daily within one month and then remained stable. Myocardial contractility improved in 85% of all patients and was sustained for up to 6 years and correlatedwell with improvement in the quality of life. Clinical improvement of 1 or 2 NYHA Classes were recorded for 85% of the cases. Survival figures are highly encouraging. The concept that deficiencies of intracellular bioenergetics may be involved in the enigma of myocardial failure is clearly supported by these 6-year data, and present a very important medical opportunity to improve the serious cardiac health problem by a new and sound biochemical approach.

New Clinical Challenges on CoQ_{10} for Cardiac Disease.- Cardiomyopathy is a minor aspect of cardiac disease. Other continuing clinical challenges for CoQ_{10} in the field of cardiac disease include the following.

1) Additional clinical studies on CoQ_{10} are needed for treatment of ischemic and rheumatic disease. The past observations of Per Langsjoen and other cardiologists strongly indicate positive therapeutic results for the treatment of ischemic and rheumatic disease with CoQ_{10}.

2) It is estimated that about 40% of cases of myocardial failure are due to diastolic dysfunction which are difficult to diagnose and are not effectively treated by conventional medications according to Per Langsjoen. Encouraging therapeutic results are available for diastolic dysfunction and which merit expanded efforts to improve diagnostic accuracy, quantitation of impairment and conclusively establish the value of CoQ_{10} to treat diastolic dysfunction with high priority, again according to the wisdom of Per Langsjoen.

3) The proper dose of CoQ_{10} to treat cardiac disease is not really established. Per Langsjoen has observed that 200 mg can be more effective than 100 mg and both of those doses are empirical and higher levels should be tried.

4) Because of the established indispensable status of CoQ_{10} in myocardial function, all conventional drugs which are presently described for human cardiac disease should be biochemically tested for possible effects on depression of CoQ_{10} biosynthesis and coenzyme function.

The Importance of Electrical Impedance to Monitor Cardiac Function.- Conventional invasive techniques to measure cardiac function such as that of FICK and indicator dilution (thermal, dye, and radioisotope) have significant limitation. Cardiac criteria are highly variable and can be measured only infrequently by conventional techniques.

The electrical impedance technique is simple, painless, harmless, and noninvasive, and has been correlated by several investigators with measurements from conventional techniques (13). Therefore, the electral impedance technique is ideal for multiple measurements which can be statistically analyzed. As unbelievable as it may be, the electrical impedance technique can entirely replace some conventional invasive techniques. Professor Lee Baker was one of the pioneers of the impedance technique for cardiology and Dr. William Judy, both of whom are with us here, is a master of this impedance technique.

The solidarity of the evidence that CoQ_{10} greatly improves the declining cardiac function is well-based upon the impedance technique. Were it not for the solidarity of the impedance technique and the associated statistical data which are superior to the limited and less reliable conventional measurements of cardiac function from clinical investigations, we might still be arguing about whether CoQ_{10} benefited cardiac disease or not. The electrical impedance technique is not yet well known and the instrumentation is not widely available, but its superiority will be recognized in due time.

MUSCULAR DYSTROPHIES AND NEUROGENIC ATROPHIES

Background.- At least 8 studies during 1963-1985 documented the association of cardiac disease with muscle disease (14,15,16,17,18,19,20,21). It is difficult to measure accurately and reproducibly the physical performance of typical patients with any of the muscle dystrophies. In contrast, it was evident that we could accurately and reproducibly monitor the cardiac function of such patients with muscular dystrophies by the electrical impedance technique, and we have done so. In summary, we have completed a second double blind trial on the oral administration of CoQ_{10} to patients with progressive muscular dystrophies and neurogenic atrophies. These diseases included the Duchenne, Becker and limb-girdle dystrophies, myotonic dystrophy, Charcot-Marie Tooth disease, fascioscapulohumeral muscular disease, hypotonia congenitale and Welander's disease. Solely by significant change or no change in stroke volume and cardiac output, all patients were correctly assigned during the blind and crossover periods (22). The cardiac function was significantly increased for all patients on CoQ_{10} and very gratifyingly to a normal range for some patients. Also of importance, there was evidence of improved physical performance and the quality of life for about one-half of the patients. Since such diverse dystrophies undoubtedly includes both reversible and irreversible stages of the disease, it was indeed a success to find that physical performance and the quality of life was improved for one-half of the patients.

Deficiency of CoQ_{10} and Mitochondrial Encephalomyopathy.- The genetic aspects of the muscular dystrophies and the physical handicaps of such patients, whether teens or adults, overwhelms with difficulties the clinical assessment of physical performance of the administration of CoQ_{10} to such patients. In contrast, the newer investigations which emphasize the mitochondrial biochemistry of cases with myopathy and encephalomyopathy have the advantage of basic biochemistry to elucidate the possible cause of such diseases and guide rational treatment. In 1986, Fischer *et al.* (23) and in 1989, Ogasahara and Engel *et al.* (24) made important contributions as follows. Fischer *et al.* found that a mentally retarded boy with epileptic seizures and myoclonic contractions revealed the expected morphologic signs of a myopathy in a quadriceps muscle biopsy. Further, they observed a defect in the proximal part of the respiratory chain involving the region of CoQ_{10}. Ogasahara and Engel *et al.* investigated the electron transport system of muscle mitochondria of two sisters, 14 and 12 years, who had suffered from progressive muscle weakness for years. Their data established a CoQ_{10} deficiency as a cause of a familial mitochondrial cytopathy and which indicated to them that the disease resulted from a tissue-specific defect of CoQ_{10} biosynthesis.

These two studies support the projection for combining a biochemical diagnosis of the electron transport system of muscle mitochondia from patients with the classic muscle dystrophies, Duchenne, Becker, limb-girdle, etc. with a repetition of the clinical studies that my associates and I have conducted in two clinical trials, one of which was described above. Our clinical success does seem to forecast that a companion analysis of the electron transport system of muscle mitochondria from such patients will in the future show that the clinical improvement was based upon the improved bioenergetics from CoQ_{10} therapy.

Since all the muscle dystrophies have never had any truly beneficial therapy, the treatment of such muscle dystrophies with CoQ_{10} is an immediate, practical and safe life-time therapy for the countless patients who are handicapped by such genetic dystrophies. The therapy of muscle dystrophies with CoQ_{10} can begin as soon as the pharmaceutical industry does so — even today.

Impaired Immunity in Vitamin Deficiencies.-

Nutritional and vitamin deficiencies are known to be the most common cause of immunodeficiencies. Individuals having vitamin and nutritional deficiencies have impaired cell-mediated immunity and the microbicidal activity of phagocytes or they are much more susceptible to infections, etc. The severity of the impaired immunity correlates with the degree of nutritional deficiencies. Consequently, it was anticipated that a deficiency of CoQ_{10} could significantly impair immunological mechanisms.

Immunological Activity of Coenzyme Q_{10} in Animal Models. - Six studies during 1970-82 documented the immunological activity of CoQ_{10}. I cite four of these studies which Bliznakov and I respectively conducted. Of stuctural importance, CoQ_{10} increased the phagocytic rate in rats, but $H_{20}CoQ_{10}$, the 6-chromenols of CoQ, and vitamin E were ineffective (25). CoQ_{10} reduced the incidence of tumors from dibenzpyrene in mice and increased survivors. CoQ_{10} increased survival of mice infected with the Friend leukemia virus (26). Surprisingly, the Friend leukemia virus actually caused a deficiency of a CoQ_{10}-enzyme in mice (27). A deficiency of a CoQ_{10}-enzyme escalated in the thymus of mice with increases in age, and the gland itself actually decrease (28).

In summary, the antioxidants did not show immunological activity but CoQ_{10} did stimulate the host defense system which was interpreted on the basis of bioenergetics separate from antioxidant mechanisms.

Immunological Activity of CoQ_{10} in Humans. - It was recorded in 1982 that CoQ_{10} increased the levels of IgG in patients with cancer, cardiovascular disease, and diabetes; Folkers *et al.* (29).

Deficiencies of CoQ_{10} in Patients with AIDS, ARC and HIV-infection. - Dr. Peter Langsjoen of the U.T. Health Center in Tyler, Texas, and I (30) conducted an exploratory study on patients with AIDS or ARC. All but one of 12 patients had extremely low blood levels of CoQ_{10}. These low levels and the known immunological function of CoQ_{10} served as rationale for the pilot study on treating 3 patients with AIDS and 4 with an AIDS-related complex with CoQ_{10}. All of these patients had subjective improvement in symptoms and objective evidence of improvement in painful adenopathy, diarrhea, fever, and night sweats. The AIDS patients become noncompliant and stopped taking CoQ_{10} and ultimately expired. One patient with ARC with pulmonary tuberculosis had excellant clinical improvement. A second ARC patient became essentially asymptomatic and 2 more ARC patients ultimately had no evidence of opportunistic infection and were leading normal active lifestyles after

16 and 18 months, respectively; new data from Peter Langsjoen (31).

Obviously, this pilot study on treating AIDS and ARC patients with CoQ_{10} is extraordinarily promising and has high priority for continuance, but access to patients and their noncompliance are difficulties. Dr. Langsjoen considered some of the clinical responses to these AIDS and ARC patients to be even striking. It could be in the future that the clinical use of CoQ_{10} to treat AIDS could become a breakthrough (30). The treatment would be safe for long-term therapy and based on the intrinsic biochemistry of immunology.

Deficiency of CoQ_{10} in Cancer Patients. - On the basis of the assumption that a deficiency of CoQ_{10} could be included in the biochemical abnormalities of cancer patients, we sought blood samples from newly diagnosised cancer patients before they were treated with chemotherapy or radiation.

To date, blood samples from 22 newly diagnosed and untreated cancer patients have become available, and the data on the patients and their diverse diagnoses are in Table I. The cancer pathology included the thyroid, esophagus, colon, lung, breast, skin and neck, and uterus.

The statistical analyses of the CoQ_{10} levels are in Tables II, III, and IV. The statistical analyses were based upon comparing CoQ_{10} blood levels of groups of ordinary subjects and the cancer patients. We also compared ordinary subjects and cancer patients who were above 32 years of age and again we compared those who were above 39 years of age.

In summary, the statistical anaylses on these three bases showed the higher incidence, $p<0.005$-<0.001-<0.05, of CoQ_{10} deficiency in cancer patients than in ordinary subjects. Consequently, it seems to me that CoQ_{10} has a very urgent and a very high priority for immunotheraputic trials on cancer patients. I have been advised that these trials will be difficult. Doubtless, these trials will be difficult but the potentiality for cancer therapy is more important than the difficulty; Folkers *et al.* (32).

Potential Therapy with CoQ_{10} in General Immunology. - The existing animal data also point to the potentiality of chemotherapy with CoQ_{10} to stimulate the host defense system for resistance to bacterial, viral, and protozoal infections.

SENESCENCE

If we consider that countless old people in the ages of senescence have a complex of unknown nutritional deficiencies and other disease states, then it can be believed that CoQ_{10} has potentiality for slowing down senescence, particularly through improvement of the immune mechanisms, and even extending survival. Background for this projection was the study which Bliznokov and I conducted (33) on monitoring the specific activities of the succinate dehydrogenase-CoQ-reductase in aged mice. We found a significant and steep escalation of the deficiency of coenzyme activity in the thymus of three groups of aged mice. Also, the ratio of thymus weight to body weight decreased significantly in the three groups of aged mice on the basis of comparision of mice of about two years in age to those of about 10 weeks.

PERIODONTAL DISEASE IN DENTISTRY

Many of my colleagues from Japan in Dentistry and I, in cooperation with Dr. Edward Wilkinson, have studied CoQ_{10} and periodontal disease for many years and with positive therapeutic results. It has been said that about 60% of young adults and 80% of middle-aged Americans have periodontal disease. Once, I was told in Italy that practically every Italian has some degree of periodontal disease. Even if these magnitudes of periodontal disease are excessive, the magnitude is still very high. Periodontal disease never seems to attract the same biomedical and clinical attention as cardiac disease, and probably because cardiac disease has high fatality and periodontal disease only results in the extraction of teeth.

It seems plausible that the therapy of periodontal disease with CoQ_{10} can become widely accepted and practiced when, and perhaps only when, there is a new technique to monitor accurately and reliably gingival health by a new technique which can be as novel for periodontal health as the electrical impedence method has been novel for cardiac disease. My colleague, Dr. Shizukuishi, and his associates of the Faculty of Dentistry of Osaka University have made remarkable progress on pursuing gingival tissue reflectance spectrophotometry as a superior instrument of choice to evaluate the oxygen utilization in the gingiva during CoQ_{10} therapy (34).

DIABETES MELLITUS

Even in 1976, it became known by Drs. Takeo and Hiroe Kishi, Drs. Watanabe, and Bowers, and myself that the specific activity of the succinate dehydrogenase-CoQ_{10} reductase from as many as 120 diabetic patients was significantly lower ($p<0.001$) and the percent deficiency was significantly higher ($p<0.001$) than that of the controls (35). Open and double blind trials on CoQ_{10} as adjunctive treatment of diabetes appears justified, and with some urgency.

ESSENTIAL HYPERTENSION

Even in 1975, the medical studies of Drs. Yamagami and Shibata and myself (36) found that the mean specific activity of the succinate dehydrogenase-CoQ_{10} reductase of a group of 65 Japanese adults with essential hypertension was significantly lower ($p<0.001$) and the mean percent deficiency of enzyme activity was significantly higher ($p<0.001$) than the values for the control group of 59. These data on hypertensives in Japan were in agreement with comparable data on Americans in Dallas.

PERSPECTIVES ON DIABETES AND HYPERTENSION

There is such a plethora of drugs from the pharmaceutical industries to treat diabetes and hypertension that the motivation to conduct clinical trials with CoQ_{10} for these diseases is less than that for diseases where fatality is high and where there is no therapy. However, this plethora of existing drugs for diabetes and hypertension is really no basis for downgrading clinical trials for diabetes and hypertension, because of the possiblity that CoQ_{10} may benefit the cause of these diseases in contrast to the possiblity -- even probability -- that many of these existing drugs are only treating symptoms.

INHIBITION OF COQ_{10}-ENZYMES BY CONVENTIONAL DRUGS

The International Pharmaceutical Industries market many drugs for diabetes, hypertension and other disease states with little or no consideration of existing data on the inhibition by some of these drugs of mitochondrial CoQ_{10}-enzymes.

For example, we know (35) that dymelor, glyburide, phenformin, and tolazamide inhibit the CoQ_{10}-enzyme, NADH-oxidase. Also, we know that some of the beta-blockers, particularly propranolol, which are so commonly used to treat hypertension, do indeed inhibit CoQ_{10}-enzymes. Also, we know that the anti-tumor drug, adriamycin, inhibits CoQ_{10}-enzymes.

In the future, it does seem that the International Pharmaceutical Industry should test their new drugs under development for inhibition, as side effects, of the life-essential CoQ_{10}-enzymes.

BIOCHEMICAL RESEARCH ON CoQ_{10} VERSUS CLINICAL RESEARCH ON CoQ_{10}

Some investigators are intellectually attracted to that research which is considered theoretical and some investigators are attracted to that research which is considered practical. Basic biochemistry on CoQ_{10} may be considered theoretical, and clinical research to save lives and to prolong survival may be considered practical.

It is not really meaningful to discuss or to argue about whether it is more important to be theoretical than to be practical or whether it is more important to be practical than to be theoretical. Progress requires both the theoretical and the practical aspects, particularly in chemistry and medicine and particularly for the biochemistry of CoQ_{10} and the clinical uses of CoQ_{10}.

APPRECIATION

In closing, may I express the appreciation from all of us to Professor Barnabei, Professor Lenaz, and to Professor Rabbi for their having organized this Symposium.

THANK YOU

ACKNOWLEDGEMENT

Appreciation is expressed to the Robert A. Welch Foundation of Houston, Texas for their support of this research.

REFERENCES

1. Lester, R.L., Crane, F.L., Hatefi, Y. (1958) J. Am. Chem. Soc. 80, 7451.
2. Linn, B.O., Page, Jr., A.C., Wong, E.L., Gale, P.H., Shunk, C.H., Folkers, K. (1959) J. Am. Chem. Soc. 81, 4007.
3. Crane, F.L., Hatefi, Y., Lester, R.L., Widmer, C. (1957) Biochim. et Biophys. Acta 25, 220.
4. Folkers, K. (1969) Intern. J. Vitamin Res. 39 (3), 334.
5. Koniusky, F.R., Gale, P.H., Page, Jr., A.C., Folkers, K. (1960) Archives of Biochem. and Biophys. 87 (2), 298.
6. Folkers, K., Littarru, G.P., Ho, L., Runge, T.M., Havanonda, S., Cooley, D. (1970) Int. J. Vitam. Nutr. Res. 40, 380.
7. Littarru, G.P., Ho., L., Folkers, K. (1972) Int. J. Vitam. Nutr. Res. 42, 291.
8. Littarru, G.P., Ho., L., Folkers, K. (1972) Int. J. Vitam. Nutr. Res. 42, 413.

9. Goto, Y., Wada, Y., Katsushima, I. (1969) Rinsho To Kenkyu 46, 1192-1197.
10. Mortensen, S.A., Vadhanavikit, S., Folkers, K. (1984) Drugs Exptl. Clin. Res. 10, (7), 497-502.
11. Page, A.C., Gale, P., Wallick, H., Walton, R.B., McDaniel, L.E., Woodruff, H.B., Folkers, K. (1960) Arch. Biochem. Biophys. 89, 318.
12. Folkers, K., Watanabe, T., Kaji, M. (1977) J. Molecular Med. 2, 431-460.
13. Baker, L.E., Mistry, G.D. (1981) In Biomedical and Clinical Aspects of Coenzyme Q, Vol. 3, Elsevier/ North-Holland Biomedical Press, K. Folkers and Y. Yamamura, eds., 193-206.
14. Welsh, J.D., Lynn, T.N., Haase, G.F. (1963) Arch. Intern. Med. 112, 199-206.
15. Venco, A., Grandi, A., Barzizza, F., Malamani, G., Ghisoni, A., Basano, D. (1984) Neuropsychiatr. Eta. Evol. 4, Suppl. 1, 61-66.
16. Kuhn, E., Fiehn, W., Schroder, J.M., Assmus, H., Wagner, A. (1979) Neurology. 29, 1144-1149.
17. Carstens, V., Behrenbeck, D.S. (1985) Herz Kreisl. 17, 147-153.
18. Hopkins, L.C., Jackson, J.A., Elsas, L.J. (1981) Ann. Neurol. 10, 230-237.
19. Goldberg, S.J., Stern, L.Z., Feldman, L., Allen, H.D., Sahn, D.J., Valdes-Cruz, L.M. (1982) Neurology. 32, 1101-1105.
20. Hawley, R.J., Gottdiener, J.S., Gay, J.A., Engel, W.K. (1983) Arch. Intern. Med. 143, 2134-2136.
21. Saviotti, M. Casazza, F., Fiorista, F., Murpurgo, M. (1984) Neuropsychiatr. Eta. Evol. 4, Suppl. 1, 67-69.
22. Folkers, K., Komorowski, J., Komorowska, A., Lyson T., Simonsen, R. (1988) submitted.
23. Fischer, J.C., Ruitenbeek, W., Gabreels, F.J.M., Janssen, A.J.M., Renier, W.O., Sengers, R.C.A., Stadhouders, A.M., ter Laak, J.H., Trijbels, J.M.F., Veerkamp, J.H. (1986) Eur. J. Pediatr. 144, 441-444.
24. Ogasahara, S., Engel, A., Frens, D., Mack, D. (1989) Proc. Natl. Acad. Sci. USA 86, 2379-2382.
25. Casey, A.C., Bliznakov, E.G. (1972) Chem.-Biol. Interactions 5, 1-12.
26. Bliznakov, E.G. (1973) Proc. Nat. Acad. Sci. USA 70 (2), 390-394.
27. Bliznakov, E.G., Casey, A.C., Kishi, T., Kishi, H., Folkers, K. (1975) Internat. J. Vitam. Nutr. Res. 45 (4), 388-395.
28. Bliznakov, E.G., Watanabe, T., Saji, S., Folkers, K. (1978) J. Med. 9 (4), 337-346.

29. Folkers, K., Shizukuishi, S., Takemura, K., Drzewoski, J., Richardson, P., Ellis, J., Kuzell, W.C. (1982) Res. Commun. Chem. Path. Pharm. 38 (2), 335-338.
30. Folkers, K., Langsjoen, P., Nara, Y., Muratsu, K., Komorowski, J., Richardson, P.C., Smith, T.H. (1988) Biochem. Biophys. Res. Commun. 153 (2), 888-896.
31. New and unpublished correspondance from Peter Langsjoen to Karl Folkers (1989).
32. Folkers, K., Ellis, J., Yang, O., Tamagawa, H., Nara, Y., Nara, K., Ye, C.-Q., Shen, Z. (1989) submitted to Contemporary Issues in Clinical Nutrition.
33. Bliznakov, E.G., Watanabe, T., Saji, S., Folkers, K. (1978) J. Med. 9 (4), 337-346.
34. Shizukuishi, S., Hanioka, T., Tsunemitsu, A, Fukunaga, Y., Kishi, T., Sato, N. (1986) Biomedical and Clinical Aspects of Coenzyme Q, Vol. 5, Elsevier/North-Holland Biomedical Press, K. Folkers and Y. Yamamura, eds., 359-368.
35. Kishi, T., Kishi, H., Watanabe, T., Folkers, K., Bowers, C.Y. (1976) J. Med. 7 (3,4), 307-321.
36. Yamagami, T., Shibata, N., Folkers, K. (1975) Res. Commun. Chem. Path. Pharamacol. 11 (2), 273-288.

INDEX

A

B

C

F

G

H

K

L

M

N

O

P

U

V

W

X